HOMOGENEOUS

CATALYSIS

Second Edition

HOMOGENEOUS CATALYSIS

THE APPLICATIONS AND CHEMISTRY OF CATALYSIS BY SOLUBLE TRANSITION METAL COMPLEXES

Second Edition

GEORGE W. PARSHALL
STEVEN D. ITTEL

Central Research and Development
E. I. du Pont de Nemours and Company
Experimental Station
Wilmington, Delaware

A Wiley-Interscience Publication
JOHN WILEY & SONS, INC.
New York • Chichester • Brisbane • Toronto • Singapore

In recognition of the importance of preserving what has been
written, it is a policy of John Wiley & Sons, Inc. to have books of
enduring value published in the United States printed on acid-free
paper, and we exert our best efforts to that end.

Library of Congress Cataloging-in-Publication Data

Parshall, George William, 1929-
    Homogeneous catalysis: the applications and chemistry of
catalysis by soluble transition metal complexes/George W.
Parshall, Steven D. Ittel. -- 2nd ed.
      p.   cm.
"A Wiley-Interscience publication."
Includes index.
ISBN 0-471-53829-9 (cloth)
1. Catalysis. 2. Transition metal complexes. 3. Organic
compounds-- Synthesis. I. Ittel, Steven D. II. Title.

QD505.P37   1992   92-5108
660'.2995--dc20   CIP

PREFACE

In the 12 years since the preparation of the first edition of *Homogeneous Catalysis*, the chemical community has become much better informed about the use of soluble transition metal catalysts. The number of industrial applications has more than doubled as a result of sophisticated new processes for making fine chemicals such as pharmaceuticals, flavors, and fragrances. Simultaneously, university chemistry faculties have increased the teaching of organometallic chemistry and, to a lesser extent, homogeneous catalysis. Because of the greater familiarity of chemists with these topics, the second edition of this book does not discuss the general principles of coordination or organometallic chemistry. Descriptions of mechanism are confined to the specific processes being discussed. The reader is referred to some excellent recent texts for the general principles underlying catalytic reaction mechanisms.

The increased breadth and sophistication of homogeneous catalysis has also meant that no one individual is likely to have the expertise necessary to describe all aspects of the field. For this reason, as well as timeliness in production, the second edition is coauthored. The objective, however, remains the same as for the first edition. We hope to present a balanced description of homogeneous catalytic reactions that are used in industry or that have broad application in the organic synthesis laboratory. We have tried to include every documented example of a homogeneous catalytic reaction used in a current commercial process. The compilation includes some reactions such as ethylene polymerization, which operate by organometallic mechanisms even though the catalyst is evidently insoluble in the reaction medium. We have surely missed some relevant examples; for these omissions, we beg the reader's indulgence and offer an invitation to the reader to bring them to our attention for future editions.

GEORGE W. PARSHALL
STEVEN D. ITTEL

Wilmington, Delaware
May 1992

CONTENTS

HOMOGENEOUS
CATALYSIS
Second Edition

1 | TRENDS IN HOMOGENEOUS CATALYSIS IN INDUSTRY

University research in organometallic chemistry has grown spectacularly in recent decades and has provided much of the fundamental understanding for homogeneous catalysis as we know it. Industrial scientists charged with applying this knowledge in terms of practical process technology are encountering several challenging trends. Some of the trends listed below, which became prominent in the 1980s, promise to be dominant in the 1990s [1].

Economics - Economically competitive process technology has always been critical for success in the chemical industry. This factor has become even more decisive as the industry has become global and competition has crossed national boundaries. Innovative processes developed in Japan [2] are likely to have significant impact elsewhere.

Environment - Potential impact on the environment has become as large a factor as economics in the design of new chemical products and processes. New plastics must be not only safe and cheap, they must also be recyclable or readily degradable. The chlorofluorocarbons (CFCs) used in refrigerants and "blowing agents" for foamed polymers are being replaced with new materials that are innocuous to the ozone layer and have minimal "greenhouse effect" (Chapter 12). New processes are evaluated for energy efficiency to conserve fossil fuels and feedstocks and to reduce global warming by CO_2 emissions.

Enantioselectivity - Many new products designed for specific biological activity must be made with a specified chirality. New science and technology to achieve enantioselective synthesis is and will continue to be one of the most exciting areas in homogeneous catalysis.

Enzymes - Enzymatic catalysis is outside the scope of this book, but it has become an important new tool for the industrial chemist, especially when enantioselectivity is required. For products accessible by fermentation or enzymatic catalysis, the process designer must weigh the economic and environmental merits of biocatalysis vs. conventional chemical catalysis.

The relative impact of these factors will differ between conventional large-scale chemicals and the rapidly growing fine chemicals segment of the chemical industry. These industry segments are discussed separately below.

1.1   LARGE-SCALE CHEMICAL INTERMEDIATES

Rapid growth has continued in the use of homogeneous catalysis to make large volume chemical intermediates and polymers (Table 1.1). This growth has occurred despite the fact that relatively few new large-scale processes have been brought into commercial production. Notable new processes are Tennessee Eastman's acetic anhydride from coal-based synthesis gas, Ruhrchemie's hydroformylations with a water-soluble catalyst commercialized by Rhone-Poulenc, and Kuraray's hydroformylation technology for making 1,4-butanediol from allyl alcohol. The last process is a significant example of innovative technology developed in Japan, which has now been commercialized in the United States by Arco Chemical.

One common characteristic of these three large new processes is that all use CO as a reactant. Carbon monoxide is relatively inexpensive and can be made from a wide variety of feedstocks - coal, oil, or natural gas. The feedstock versatility should ensure its availability at a reasonable price, even if petrochemicals vary wildly in price owing to crises in petroleum supply. The combination of price and availability make it a reagent of choice. It is highly unlikely that anyone will build additional acetic acid plants based on Wacker ethylene-based technology because the Monsanto CO-based process is much more economical. The advantages of CO as a reactant may increase the chance that some small-scale CO-based products such dimethyl carbonate and dimethyl oxalate (Chapter 5) will grow to become large-volume products.

The dearth of totally new processes does not mean that innovation has ceased. On the contrary, several new processes now in development or the early stages of commercialization promise to have commercial impact in the 1990s [2,3]. For example, the CFC replacements described in Chapter 12 are expected to become major products.

In addition to the new processes mentioned above, superior new catalysts have been developed for many existing processes. This situation is especially relevant to the catalysts used for polymerization of ethylene and propylene. These industrial processes now employ "third generation" catalysts that are a far cry from the mixtures of alkylaluminum and chlorotitanium compounds discovered by Ziegler and Natta. The catalysts, which are now coordination complexes supported on inorganic solids, are extremely efficient. The quantities used are so small that they need not be removed from the product, a major cost saving. Although the catalysts are not soluble in the reaction media, the chemistry parallels that of homogeneous polymerization catalysts and is discussed in detail in Chapter 4.

Table 1.1   Homogeneous and Organometallic Catalysis in the US Chemical Industry

Chapter	Major Reactions and Products	Approximate Production[a] (thousands of metric tons)	
		1980	1990
3	**Olefin additions**		
	Adiponitrile	200	420[b]
4	**Olefin polymerizations and oligomerization**		
	α-Olefins	105	550[b]
	Propylene dimers	136	1364[b]
	Polybutadiene (coordination catalysis)	239	310[c]
	Cyclododecatriene	10	8[b]
	Polyethylene (HDPE and LLDPE)	2339	5743[bc]
	Polypropylene	1658	3781[c]
	Ethylene/propylene copolymers	144	256[c]
5	**Carbonylations**		
	Oxo alcohols (hydroformylation)	1297	1818[b]
	Acetic acid (from methanol)	773	1164[b]
	Acetic anhydride (from methanol)	0	527[b]
6	**Olefin oxidation**		
	Acetaldehyde	409	273[b]
	Propylene oxide (Oxirane process)	362	815[b]
10	**Alkane and arene oxidation**		
	Terephthalic acid and esters	2752	3496[c]
	Adipic acid	668	746[c]
	Acetic acid	511	159[b]
	Benzoic, isophthalic acids	81	121[b]
11	**Condensation polymerization**		
	Polyester fiber	1813	1452
12	**Halocarbons**		
	Dichloroethane	5049	6045[c]
	Chloroprene	151	116[b]
	Chlorofluoromethanes	309	296[c]

[a] From *Synthetic Organic Chemicals-19*, US International Trade Commission, unless shown otherwise.
[b] From *Chemical Economics Handbook*, SRI International.
[c] From *Chemical and Engineering News*, June 24, 1991.

1.2 FINE CHEMICALS

The overall growth in the number of commercial homogeneous catalytic processes reported in this monograph owes much to the increased use of soluble catalysts in the fine chemicals industry [4]. During the 1980s many chemical companies put increased emphasis on small volume, high value products as a means to maintain profitability in the face of increased competition in commodity chemicals. These fine or specialty chemicals have included specialized polymers for electronic applications, intermediates for high performance structural materials, and many biologically active compounds. The latter class, which includes pharmaceuticals, crop protection chemicals, flavors, fragrances, and food additives, has provided an especially strong driving force for the use of homogeneous catalysis.

The bioactive compounds usually involve complex organic structures, sometimes containing one or more chiral centers. The complexity of the structures and the need for product purity has put a premium on the chemical selectivity associated with homogeneous catalysts, the same property that commends it for use in making highly pure polymer intermediates. Much of the selectivity observed with soluble catalysts arises from the process control that is attainable in the liquid phase. Not only are temperature and mixing better controlled than in heterogeneous systems, but also the nature of the active catalytic species is regulated more effectively. Control of catalyst and ligand concentration is better than that attainable on the surface of a solid. As a result, a soluble catalyst is apt to be homogeneous in the sense that only a few catalytic species are present, as well as in the traditional sense that it operates in a single phase.

Enantioselective Catalysis

The need for chiral specificity (enantioselectivity) in bioactive products reflects the fact that most enzymes have an inherent chirality ("handedness"). As a consequence, attempts to manipulate biological systems through therapeutic drugs or aroma enhancers often involve the use of chemicals containing chiral centers. The desired biological activity is usually associated with only one of the two or more stereoisomers of a chiral compound. In the extreme case, exemplified by thalidomide, one optical isomer is therapeutic and the other has serious undesired biological consequences (teratogenicity in the thalidomide example).

To improve product safety, the pharmaceutical industry is producing an increasing number of products in enantiomerically pure form. Naproxen®, a widely used anti-inflammatory, is currently sold as a single optical isomer because the "wrong" isomer is a liver toxin. The desired (S) isomer is now produced by a cumbersome optical resolution of a racemic mixture, but, by 1993, it is likely to be made by an enantioselective hydrogenation process (Section 3.4).

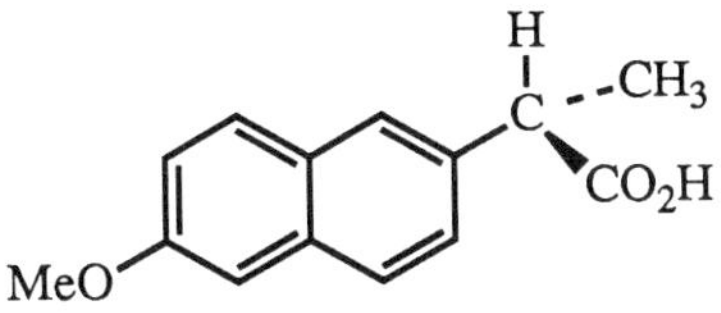

S-Naproxen

The opportunity to make chiral products specifically as the desired optical isomers has been the driving force for a burst of effort and creativity in enantioselective catalysis. Traditionally, these products have been made either by chemical or physical resolution of a racemate or directly by a fermentation process (biocatalysis). One of the major developments in homogeneous catalysis in the 1980s has been the development and practical application of enantioselective catalysts that display enzyme-like selectivity. Some current and likely commercial applications are listed in Table 1.2.

Table 1.2   Industrial Applications of Enantioselective Catalysis

Chapter	Reaction	Product	Use
2	Isomerization of allylic amine	L-Menthol	Aroma and flavor chemical
3	Hydrogenation of enamides	Levodopa L-Phenylalanine	Pharmaceutical Food additive
	Hydrogenation of substituted acrylic acid	S-Naproxen[a]	Pharmaceutical
6	Oxidation of allylic alcohol	Disparlure Glycidol	Insect attractant Intermediate
9	Cyclopropanation	Cilastatin	Pharmaceutical

[a] Commercial production likely.

The first industrial example of an enantioselective process using a soluble catalyst was Monsanto's production of Levodopa, which was commercialized in 1971. Their use of chiral ligands to create an asymmetric catalytic site on a transition metal ion stimulated extensive academic research on new ligands and new catalyst systems.

This research has made possible a wide range of enantioselective hydrogenation, isomerization, oxidation, and C-C bond-forming reactions [5-7]. As shown in Table 1.2, many of these reactions are being applied in industry. Many more commercial applications can be expected in the 1990s. As noted earlier, however, biocatalytic processes will provide stiff competition in many applications. A fermentation process for making phenylalanine may displace the catalytic process [8].

Other Areas of Rapid Change

Apart from the exciting topic of enantioselective catalysis, many other applications of chemically selective homogeneous catalysis are being reported [4]. Numerous examples are found in transformations of the terpenoid compounds widely used in perfumes and food additives [9]. New crop protection chemicals such as the synthetic pyrethroid insecticides and the sulfonylurea herbicides are based on complex chemical structures necessary to achieve previously unattainable degrees of activity and safety.

One of the most dynamic areas has been catalysis by metal complexes containing alkylidene or carbenoid ligands (Chapter 9). The mechanism of olefin metathesis is now understood thoroughly and new catalyst systems have been designed on the basis of this understanding. Several new industrial processes for specialty polymers and for pharmaceuticals are based on soluble catalysts for olefin metathesis and for production of cyclopropanes by addition of carbenoid fragments to C=C bonds.

The pattern of intellectual development in olefin metathesis and cyclopropanation is much like that in other areas of homogeneous catalysis. Mechanistic studies have followed commercial application, but the information from these studies has been valuable in the optimization of reaction conditions and in the development of new catalysts. Studies of the mechanisms have advanced rapidly because most of the techniques of physical organic chemistry can be used with little modification. Many of the mechanistic principles developed with soluble olefin metathesis catalysts also apply to industrial heterogeneous catalysts for which much less mechanistic information is available.

GENERAL REFERENCES

I. Wender and P. Pino, *Organic Synthesis with Metal Carbonyls,* Wiley, New York, 1st ed., 1968; 2nd ed., 1977, provide expert coverage of a broad variety of topics in homogeneous catalysis, not just metal carbonyl reactions.

R. Ugo, *Aspects of Homogeneous Catalysis*, D. Reidel, Dordrecht (1971-1984), covers a continuing series of volumes that contain chapters on many aspects of catalysis by workers in the field.

Homogeneous Catalysis. Advances in Chemistry Series **70** and **132**, American Chemical Society, Washington, D.C., 1968, 1974, are collections of symposia papers with a strong industrial flavor.

L. H. Pignolet, Ed., *Homogeneous Catalysis with Metal Phosphine Complexes*, Plenum Press, New York, 1983; R. L. Augustine, Ed., *Catalysis of Organic Reactions*, Marcel Dekker, New York, 1985, includes chapters by experts on several aspects of homogeneous catalysis.

M. Graziani and M. Giongo, Eds., *Fundamental Research in Homogeneous Catalysis*, Vol. 4, Plenum Press, New York, 1984, contains papers from prominent researchers on many currently interesting aspects of homogeneous catalysis.

C. Masters, *Homogeneous Transition-Metal Catalysis*, Chapman and Hall, London, 1981, offers good descriptions of several homogeneous catalytic processes currently used in industry as well as projections about future process technology.

A. Mortreux and F. Petit, Eds., *Industrial Applications of Homogeneous Catalysis*, D. Reidel, Dordrecht, 1988, offers up-to-date chapters on both commercial and fundamental topics in selected areas of homogeneous catalysis.

Several serial publications contain chapters on homogeneous catalysis. Some prominent examples are:
Advances in Catalysis, Academic Press.
Advances in Organometallic Chemistry, Academic Press (particularly Vol. 17, 1979).
Organometallic Reactions, Wiley-Interscience.
Catalysis-Specialist Periodical Reports, Chemical Society.
Catalysis Reviews, Marcel Dekker.

Books on Special Topics

J. Falbe, *Carbon Monoxide in Organic Synthesis*, Springer-Verlag, Berlin, 1970.

J. C. W. Chien, *Coordination Polymerization,* Academic Press, New York, 975.

C. E. Schildknecht and I. Skeist, *Polymerization Processes*, Wiley-Interscience, New York, 1977.

G. Henrici-Olive and S. Olive, *The Chemistry of the Catalyzed Hydrogenation of Carbon Monoxide*, Springer-Verlag, Berlin, 1984.

C. L. Hill, *Activation and Functionalization of Alkanes*, Wiley-Interscience, New York, 1989.

In addition to these reviews on homogeneous catalysis, readers may wish to study the general principles of organometallic chemistry that underlie the use of soluble catalysts. We recommend the following texts as particularly appropriate:

J. P. Collman, L. S. Hegedus, J. R. Norton, and R. G. Finke, *Principles and Applications of Organotransition Metal Chemistry*, 2nd ed., University Science Books, Mill Valley, CA, 1987.

A. Nakamura and M. Tsutsui, *Principles and Applications of Homogeneous Catalysis*, Wiley-Interscience, New York, 1980.

A. Yamamoto, *Organotransition Metal Chemistry*, Wiley, New York, 1986.

G. Wilkinson, R. Gillard, and J. A. McCleverty, *Comprehensive Coordination Chemistry* (particularly Vol. 6 "Applications"), Pergamon Press, New York, 1987.

F. R. Hartley, Ed., *The Chemistry of the Metal-Carbon Bond*, Vol. 5, Wiley, New York, 1989.

SPECIFIC REFERENCES

1. A. T. Bell, Ed., *Catalysis Looks to the Future,* National Academy Press, 1992.
2. M. Misono and N. Nojiri, *Appl. Catal.*, **64**, 1-30 (1990).
3. J. N. Armor, *Appl. Catal.*, **78**, 141-73 (1991).
4. G. W. Parshall and W. A. Nugent, *Chemtech*, **18**, 184-90, 314-20, 376-84 (1988).
5. R. Noyori, *Science*, **248**, 1194-9 (1990); *Chem. Soc. Rev.*, **18**, 187-208 (1989).
6. E. Eliel and S. Otsuka, Eds., *Asymmetric Reactions and Processes in Chemistry*, American Chemical Society Advances in Chemistry Series, Washington, DC, 1982 (particularly Chapter 10).
7. J. D. Morrison, Ed., *Asymmetric Synthesis*, Vol. 5, Academic Press, New York, 1985.
8. R. Sheldon, *Chem. Ind.*, 212-9 (1990).
9. H. Pommer and A. Nurrenbach, *Pure Appl. Chem.*, **43**, 527-51 (1975).

2 | ISOMERIZATION OF OLEFINS

Double-bond migration in olefins is one of the simplest and most thoroughly studied catalytic reactions [1,2]. Soluble catalysts are used industrially to isomerize olefins that are involved as intermediates in other homogeneous catalytic processes. For example, Du Pont's synthesis of adiponitrile from butadiene and HCN (Section 3.6) includes two olefin isomerization steps:

$$H_2C=CHCH\begin{smallmatrix}CH_3\\|\\|\\CN\end{smallmatrix} \rightleftharpoons CH_3CH=CHCH_2CN$$

$$CH_3CH=CHCH_2CN \rightleftharpoons H_2C=CHCH_2CH_2CN$$

The first step, conversion of a branched chain to a linear chain, involves cleavage of a C-C bond, a relatively uncommon mode of isomerization. The second step is a more common type in which an internal olefin equilibrates with a terminal olefin by hydrogen migration without disruption of the carbon skeleton of the olefin. This example is notable in that the C-C isomerization occurs in the presence of another potentially reactive, functional group. In recent years, several isomerizations of functionally substituted olefins have become important in the manufacture of specialty chemicals [3], as discussed throughout this chapter.

Isomerization of unconjugated to conjugated polyenes appears to be a key step in the selective hydrogenation of compounds such as 1,5-cyclooctadiene to monoenes (e.g., cyclooctene). This process is discussed in Section 3.3. The isomerization mechanisms are believed to be analogous to those of simple olefins.

9

2.1 ISOMERIZATION OF SIMPLE OLEFINS

The double-bond migration in simple olefins may be exemplified by the isomerization of 1-octene to a mixture of predominantly internal isomers [4]:

$$C_6H_{13}CH=CH_2 \longrightarrow \begin{array}{l} 2\% \ \text{1-octene} \\ 36\% \ \text{2-octene} \\ 36\% \ \text{3-octene} \\ 26\% \ \text{4-octene} \end{array}$$

This conversion of a terminal olefin to a near-equilibrium mixture of internal olefins (both cis and trans isomers) is carried out on a massive scale as one step in the SHOP process described in Section 4.3. In industrial practice, a heterogeneous catalyst (potassium supported on alumina) is used [4], but a wide variety of soluble catalysts are effective for isomerization of simple olefins.

The SHOP process [5], which produces both linear α-olefins and linear detergent alcohols, also involves equilibration of internal and terminal olefins by a soluble catalyst. Linear aldehydes can be prepared from internal olefins such as 2-decene by using a catalyst that is active for both isomerization and hydroformylation of olefins. If the catalyst generates a terminal olefin rapidly and hydroformylates the terminal olefin preferentially, respectable yields of linear aldehyde form by the sequence

$$C_7H_{15}CH=CHCH_3 \rightleftharpoons C_8H_{17}CH=CH_2 \xrightarrow[CO]{H_2} C_8H_{17}CH_2CH_2CHO$$

The linear aldehyde undecanal is of interest both as a perfume intermediate [6] and as a precursor to the C_{11} fatty alcohol, which is an intermediate in detergent manufacture.

A standard hydroformylation catalyst, $HCo(CO)_4$, is moderately effective in the sequential isomerization and hydroformylation of olefins [2], as described in Section 5.4. This catalyst can also isomerize olefins without hydroformylation when the partial pressures of CO and H_2 are reduced to the minimum amount necessary to stabilize the complex. In the SHOP process practiced by Shell, it appears that a modified catalyst, $HCo(CO)_3(PBu_3)$, is used to effect hydrogenation of the product aldehyde as well as isomerization and hydroformylation [7].

The industrial synthesis of ethylidenenorbornene 2, a widely used comonomer for ethylene/propylene elastomers (Chapter 4), also involves olefin isomerization. The monomer synthesis comprises two major steps [8]:

1   2

Vinylnorbornene 1 is produced by the uncatalyzed Diels-Alder reaction of butadiene with cyclopentadiene, which comes from thermolysis of dicyclopentadiene, a refinery byproduct. The isomerization of 1 to 2 is catalyzed either with a strong base (Na on

Al$_2$O$_3$ or potassium-*tert*-butoxide) [9] or with a Ti-based Ziegler catalyst [10]. The desired product **2** is formed to the extent of 98% or more at equilibrium. It is likely that the basic catalysts abstract an allylic proton from **1** to form a delocalized anion that is reprotonated to give **2**.

The Ziegler-type catalysts are prepared by reducing TiCl$_4$ or Ti(OR)$_4$ with LiAlH$_4$ or R$_2$AlH to form a titanium hydride species [10]. The isomerization is presumed to occur by an M-H addition/elimination reaction like those discussed in the next section. Addition of a Ti-H bond to **1** forms an alkyltitanium compound **3**. β-Hydrogen elimination from **3** either regenerates **1** or yields the desired product **2**.

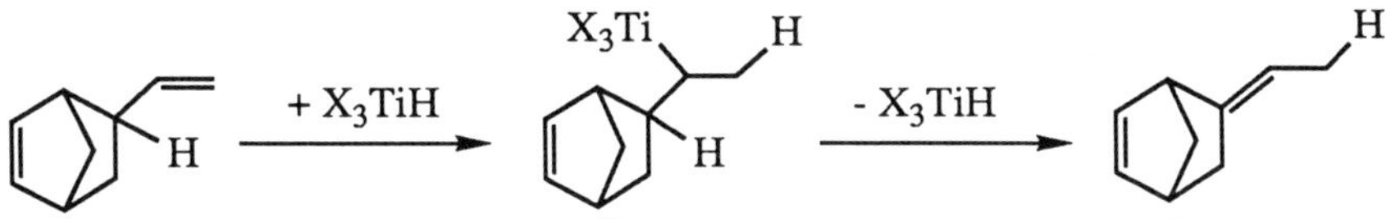

Double-bond migration catalyzed by soluble metal complexes is also useful in laboratory scale organic synthesis. Some difficultly accessible unsaturated steroids have been prepared by RhCl$_3$-catalyzed isomerization reactions [11]. This salt forms a soluble hydrate, RhCl$_3$•3H$_2$O, which is probably the most convenient olefin isomerization catalyst for laboratory use. It is air-stable, commercially available, and easy to use. Typically, the olefin is heated with an ethanol solution of RhCl$_3$•3H$_2$O. After several hours, the mixture is cooled and diluted with water and the olefin is isolated by conventional means. Another commercially available isomerization catalyst is Wilkinson's compound, RhCl(PPh$_3$)$_3$. It has been used in several syntheses of natural products [12] and is faster than the simple RhCl$_3$ system [13]. A closely related rhodium complex bearing a chiral ligand is used commercially in an enantioselective isomerization of an allylic amine (Section 2.3).

Other commercially available materials such as Fe(CO)$_5$, Fe$_3$(CO)$_{12}$, and PdCl$_2$, are also useful isomerization catalysts [14] for the synthesis laboratory. These compounds catalyze double-bond migration by a different mechanism than do the rhodium or nickel catalysts and, hence, different products may be isolated in kinetically controlled experiments. Iron pentacarbonyl is the catalyst in a potentially attractive synthesis of anethole, a licorice fragrance chemical [15]. The isomerization of estragole **4** (*p*-allylanisole) to the arene-conjugated olefin anethole **5** is accomplished by simply heating **4** with a catalytic amount of Fe(CO)$_5$ at 140°C. A conversion of 96-99% is attained in eight hours. The desired trans isomer of anethole predominates over cis by approximately 87:13.

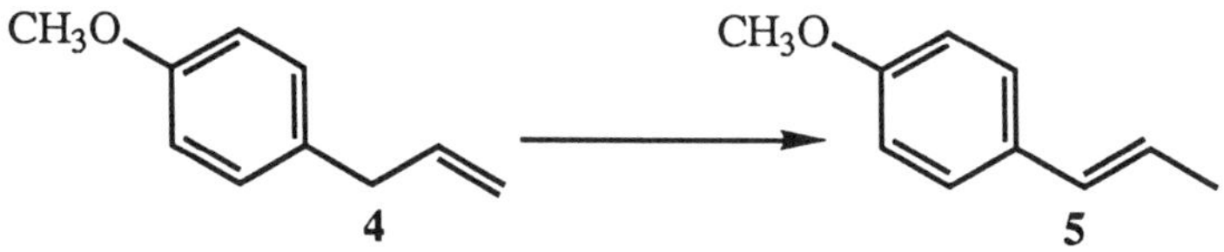

2.2 OLEFIN ISOMERIZATION MECHANISMS

Extensive studies of olefin isomerization catalysis in the 1960s and 1970s identified two major families of catalysts which function by different reaction mechanisms. The largest family is that of transition metal hydrides, which may be either preformed catalysts or catalytic species generated in situ. These species catalyze C=C bond migration by addition and elimination of an M–H bond, as described for the Ti-based Ziegler catalyst. A less common, but well documented, alternative mechanism involves a metal-mediated 1,3-hydrogen shift, which may involve allylic intermediates.

In addition to isomerizing olefins by moving hydrogens, metal ions can also catalyze olefin isomerization by moving a substituent such as Cl, OH, OAc, or CN. Some commercial applications of this kind of catalysis are discussed in Section 2.4.

It should be emphasized that, whatever the mechanism, catalysis of olefin isomerization is a kinetic phenomenon. If reactions are allowed to proceed to completion, equilibrium mixtures of olefins form. For instance, the ultimate product of 1-butene isomerization is an equilibrium mixture of 69% *trans*-2-butene, 25% *cis*-2-butene, and 6% 1-butene [16]. With many catalysts, however, the *cis*-2-butene is formed more rapidly than the trans isomer and can be isolated as a major product early in the reaction. Some of the more active catalysts facilitate equilibration of cis and trans isomers of olefins such as stilbene for which double-bond migration is impossible.

Hydride Addition-Elimination

The most common mechanism for moving a C=C bond involves addition of an M–H bond to give an alkylmetal complex that then undergoes nondegenerate β-hydrogen elimination to form a new C=C bond as illustrated in Figure 2.1. The function may be a preformed metal hydride complex such as $HCo(CO)_4$, mentioned earlier, or it may be formed in the reaction mixture. Some of the more important catalytic species are tabulated in Table 2.1.

All the compounds in the table have potential utility in organic synthesis and, as mentioned previously, the cobalt and titanium species are involved in industrially important processes. Cationic nickel species formed by protonation of nickel(0) complexes are also commercially significant. Du Pont's hydrocyanation technology, discussed in Section 3.6, is based on isomerization and HCN addition reactions catalyzed by $HNiL_3^+$ and $HNiCNL_3$ species in which L is a triaryl phosphite ligand. Studies [16] of the analogous triethyl phosphite complex shown in the table have shed much light on the mechanism of olefin isomerization.

A catalytic reaction of potential industrial importance is the selective reduction of a molecule containing multiple C=C bonds to monoolefins (e.g., 1,5,9-cyclo-dodecatriene to cyclododecene or linolenic acid esters to oleate esters) as discussed in Section 3.3. An initial step in the reaction is believed to be isomerization of the unconjugated polyene to a conjugated polyene, which undergoes hydrogenation in preference to an isolated C=C bond. The ruthenium and platinum hydrides shown in

Table 2.1 Isomerization Catalysts Based on Metal Hydrides

Catalyst Type	Examples	References
Preformed M-H bond formed by reduction with a hydride reagent	$HCo(CO)_4$ $HRuCl(PPh_3)_3$ $HPt(SnCl_3)(PPh_3)_2$	2 17 17
M-H formed by reduction with a hydride reagent	$HTiX_3$ species	10
M-H formed by H^+ addition	$HNi[P(OEt)_3]_3^+$	16
M-H formed by reduction and protonation in situ	$HRh^{III}Cl_x$ (ex $RhCl_3 \cdot 3H_2O$)	11,18

the table are good catalysts for this kind of reaction and, as expected, are good isomerization catalysts.

The metal hydride addition-elimination mechanism is nicely illustrated by $RuHCl(PPh_3)_3$, a hydrogenation catalyst that catalyzes isomerization of simple olefins as well as polyenes. The mechanism of isomerization of 1-pentene by this catalyst [17] is shown in Figure 2.1. A coordinatively unsaturated complex labeled H-Ru in the figure reacts with 1-pentene to form the olefin complex **6** shown at the 3 o'clock position in the cycle. Migratory insertion can occur in either of two ways. Addition of

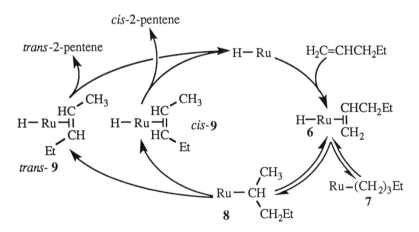

Figure 2.1 Catalytic cycle for isomerization of 1-pentene to *cis*- and *trans*-2-pentene by $RuHCl(PPh_3)_3$. The phosphine and chloride ligands are omitted for clarity.

Ru-H to the olefin to form the 1-pentyl derivative **7** is a nonproductive side reaction. Addition in the opposite sense to give the 2-pentyl derivative **8** opens pathways to isomerization (although β-hydrogen elimination from the methyl group leads back to 1-pentene). β-Hydrogen elimination from C-3 can occur in two ways to give either *cis*- or *trans*-2-pentene as in the complexes **9**. Dissociation of the 2-pentene ligand completes the catalytic cycle. Readditions of Ru-H to 2-pentene can occur to give a 3-pentyl complex. This step is unproductive for pentenes because the 3-pentyl group can only yield 2-pentene but, for long-chain olefins, repeated addition and elimination steps can move the double bond along the chain at random.

The β-hydrogen elimination to give *cis*-2-pentene is faster than that which gives the trans isomer. Early in the reaction (50°C in benzene), a 60:40 ratio of *cis*- and *trans*-2-pentene is observed. The factor that determines which isomer is formed in a single catalytic cycle is almost certainly the conformation of the 2-pentyl group at the time that β-hydrogen elimination occurs. The view along the C-2/C-3 bond axis may be represented as (C-3 in front):

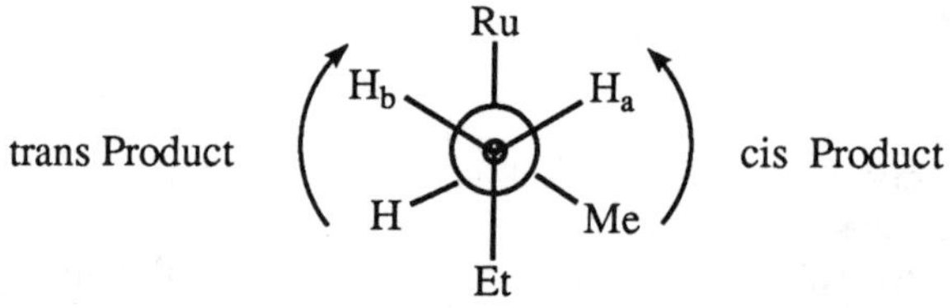

If the C-3 substituents in front rotate counterclockwise to place the Ru on C-2 and H_a on C-3 adjacent in eclipsed positions, cis elimination of Ru-H should yield *cis*-2-pentene. Rotation in the opposite sense would give the trans isomer.

These mechanistic principles seem to apply to metal hydride-based catalysts generally, but the details will vary with the catalyst species. In particular, the kinetic distribution of cis and trans isomers will depend on the steric and electronic factors that control the β-hydride elimination process.

1,3-Hydrogen Shift

A second major mechanism for olefin isomerization is a metal-assisted shift of an allylic hydrogen from the 3-position of an olefin to the 1-position, as illustrated for 3,3-dideuterio-1-pentene:

$$EtCD_2CH=CH_2 \rightleftharpoons EtCD=CHCH_2D$$

Experimental criteria for diagnosing the occurrence of this mechanism are:

- High (ca. 4) cis:trans ratios of isomerized olefin early in the reaction [19].
- 1,3-Deuterium shift in deuterated olefins with little or no deuterium on C-2 early in the reaction.

This kind of mechanism has been suggested for isomerizations catalyzed by Fe(CO)$_5$ [2], Fe$_3$(CO)$_{12}$ [14], PdCl$_2$(NCPh)$_2$ [14], and RhCl(PPh$_3$)$_3$ [19].

The palladium(2+) catalysts have been studied extensively because their stability to air and moisture make them attractive for kinetic investigation and for practical application. This kind of catalyst appears to be used in the industrially significant isomerization of 6-methyl-6-hepten-2-one **10** to the internal olefin **11** [20].

$$H_2C\!=\!\underset{\underset{CH_3}{|}}{C}\text{-}(CH_2)_3\text{-}\underset{\underset{O}{\|}}{C}\text{-}CH_3 \rightleftharpoons (CH_3)_2C\!=\!\underset{\underset{H}{|}}{C}\text{-}(CH_2)_2\text{-}\underset{\underset{O}{\|}}{C}\text{-}CH_3$$

10   **11**

The terminal olefin **10** is prepared on a large scale by BASF via condensation of isobutylene, acetone, and formaldehyde [20]. It and **11** are versatile intermediates for a wide range of flavors and fragrances as well as Vitamin A as discussed in Section 2.4.

Several detailed mechanisms have been proposed for the 1,3-hydrogen shift mechanism. The simplest is a 1,3-suprafacial shift in which a hydrogen ion migrates from C-3 to C-1 in a coordinated terminal olefin without any direct metal-hydrogen interaction. Another conceptually simple proposal, metal-assisted proton migration, is shown in Figure 2.2. This mechanism is proposed to operate in Pd^{2+}-catalyzed isomerization in nonpolar media such as benzene, the solvent chosen for mechanistic study [14]. Coordination of the chosen olefin (1-pentene) to the metal brings the allylic C-3 hydrogens close to the metal atom in an initial complex **12**. Transfer of hydrogen to the metal gives a π-allyl palladium hydride **13**. The metal-bound hydrogen may return to C-3 to reform 1-pentene, or it may migrate to C-1 to form 2-pentene. With this catalyst, the trans isomer **14** shown in Figure 2.2 is favored kinetically as well as thermodynamically.

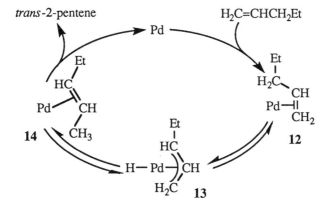

Figure 2.2   Isomerization of 1-pentene by PdCl$_2$ complexes. The chloride ligands are omitted for clarity.

One objection to this π-allylic mechanism is that it involves oxidative addition of a C-H bond to a Pd(II) complex to form a species **13** that is formally Pd(IV), an uncommon oxidation state. This consideration, as well as differences in the relative isomerization rates with various Pd(II) catalysts, has led to proposal of an alternative mechanism [21]. The alternative involves electrophilic attack of Pd^{2+} on the olefin to give an incipient carbonium ion **15**. This species

$$RCH_2\overset{\oplus}{C}HCH_2-Pd^+ \quad \underset{+H^+}{\overset{-H^+}{\rightleftharpoons}} \quad RHC\overset{\overset{H}{\underset{|}{C}}}{\diagup}{}_{Pd^+}\diagdown CH_2 \quad \underset{}{\overset{+H^+}{\rightleftharpoons}} \quad RHC\overset{\overset{H}{\underset{|}{C}}}{\diagup}{}_{Pd^+}\overset{\oplus}{\diagdown}CH_3$$

15   **16**   **17**

rearranges via a heterolytic cleavage of a C-H bond to form an allylic intermediate **16**, which is in equilibrium with **17**. Compound **17** may be regarded as a partial carbonium ion that is in equilibrium with the isomerized olefin. Clearly, solvent polarity will be a factor in determining whether this mechanism or that of Figure 2.2 will operate in a given situation.

A 1,3-hydrogen shift with a somewhat different intimate mechanism has been demonstrated for the [Rh(diphosphine)(solvent)$_2$]$^+$ catalyst system. This type of catalyst is valuable for the isomerization of allylic alcohols to enols [22] and of allylic amines to enamines [23]. The latter reaction proceeds via a suprafacial 1,3-shift. If done with a chiral catalyst and a prochiral allylamine, it permits synthesis of a single optical isomer of the enamine, as described below.

2.3   ENANTIOSELECTIVE ISOMERIZATIONS

As described in Chapter 1, one of the major driving forces for the practical use of soluble catalysts is the need to make specific optical isomers of biologically active compounds. One of the most striking examples is the development of enantioselective catalysts for isomerization of olefins. In principle, a metal complex, which is coordinated preferentially to one face of a prochiral olefin, should be able to effect a stereoselective 1,3-hydrogen shift to produce specifically one of the two optical isomers of the product olefin. This possibility has been realized for allylic amines and, as described below, provides the basis of an industrial synthesis of L-menthol, a major fragrance chemical [23-26].

Work at the Takasago Perfumery first showed the possibility of enantioselective isomerization of an allylic amine [24]. The catalyst, prepared by reducing a cobalt(II) compound with an organoaluminum in the presence of a chiral ligand, gave only modest enantioselectivity, but subsequent work with rhodium catalysts gave spectacular results. A key to success was use of the BINAP ligand [26], which has also been useful in enantioselective hydrogenations of olefins and ketones (Section 3.4).

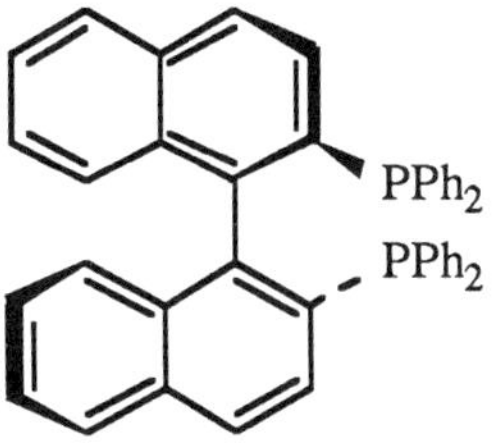

(-)-BINAP

The most important commercial application of this chemistry [26] is shown in Figure 2.3. β-Pinene, an abundant natural terpene, is pyrolyzed to form myrcene **18**. Myrcene, in turn, is treated with diethylamine in the presence of lithium diethylamide to form diethylgeranylamine **19** and its Z-isomer, diethylnerylamine. Either of these allylic amines may be isomerized to R(-)-diethyl-E-citronellalenamine **20**, but the "handedness" of the product depends on the chirality of the BINAP ligand present in the catalytic rhodium complex. For example, **19** is boiled in tetrahydrofuran containing 0.1 mole % [Rh(-)-BINAP(COD)](ClO$_4$) for 21 hours to give a 94% yield of **20**. If one starts with the isomeric nerylamine, it is necessary to use the (+)-BINAP complex [27]. Hydrolysis of **20** with cold aqueous acetic acid gives R(+)-citronellal **21** in 91% chemical yield with an optical purity of about 95% [25]. The Lewis-acid-catalyzed ring closure to form **22** accomplishes formation of two more chiral centers, the stereochemistries of which are determined by the chirality at the carbon β to the aldehyde function. Finally, hydrogenation of the remaining C=C bond in **22** over a Raney nickel catalyst generates L-menthol.

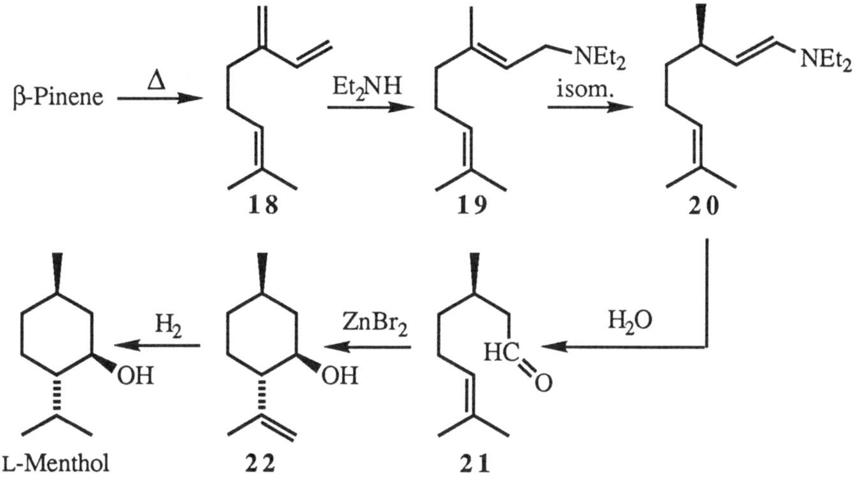

Figure 2.3 Synthesis of L-menthol by enantioselective isomerization of allylic amines.

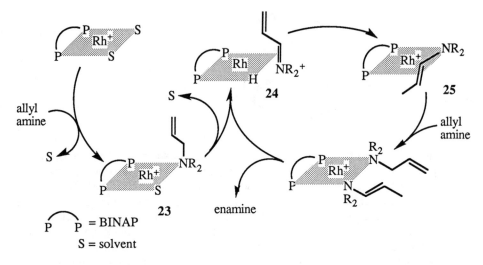

Figure 2.4 Mechanism of enantioselective isomerization of allylic amines (from ref. 23).

This menthol synthesis is remarkable in that three chiral centers are created, all of which are necessary to produce the characteristic menthol odor and local anesthetic action. Even more remarkable is the fact that this complex multistep synthesis can be economically competitive with production of menthol from natural sources. The key seems to lie in the enantioselective isomerization, which creates the first chiral center.

The mechanism of the enantioselective isomerization has been studied extensively [23]. As shown in Figure 2.4, it is critically dependent on the presence of the amine function, which provides the initial binding of the substrate to the catalyst. An allylic amine such as diethylgeranylamine coordinates to the BINAP rhodium(I) complex through N, as shown in **23**. Dissociation of a solvent molecule permits transfer of a hydrogen from the α carbon of the amine to rhodium, formally oxidizing the metal to Rh(III). The product of C-H cleavage may be viewed either as an iminium complex **24**, as written, or as an allyl complex bearing an NR_2 substituent on C-1. If viewed as a π-allyl complex, it must have *syn* stereochemistry because return of H from Rh to C-3 gives the trans enamine in compound **25**. Displacement of the enamine from **25** by fresh allylamine completes the catalytic cycle by reforming **24**. Stereochemical analysis of the transition states leading to cleavage has rationalized the role of the BINAP ligand in determining which of the two α–hydrogens migrates to rhodium.

The enantioselective isomerization reaction appears to have numerous applications in fragrance chemistry. In a reaction closely related to the L-menthol synthesis, the allylic amine **26** is isomerized to the enamine **27** using a BINAP rhodium catalyst [27]:

The enamine can be hydrolyzed to an aldehyde with a "lily of the valley" fragrance. As in the citronellal synthesis, the (-)-BINAP ligand gives rise to the desired R-stereochemistry at the new chiral center in **27**. The enamine again forms with the trans configuration.

It should be noted that the process of Figure 2.3 is not the only industrial synthesis of L-menthol that employs organometallic chemistry. In another commercial process [28], L-citronellene **28**, derived from pyrolysis of *cis*-pinane, is treated with *tris*(isobutyl) aluminum at 125-150°C in a stoichiometric reaction to form **29** and isobutylene:

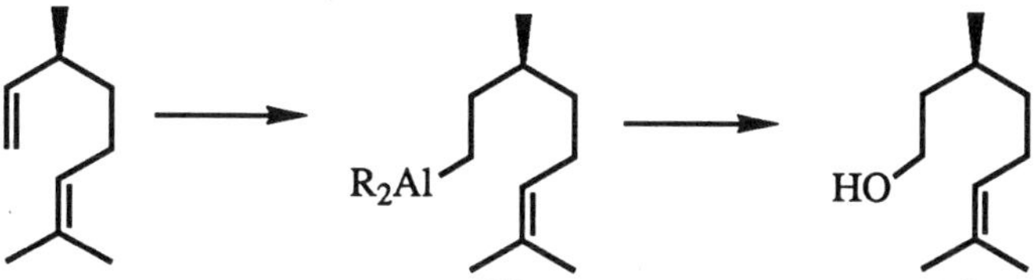

Oxidation and hydrolysis of the Al-C bond in **29** yields (+)-citronellol **30**, which is converted to L-menthol by conventional organic reactions. This oxidative hydrolysis of an organoaluminum compound closely parallels that used in the Conoco process (Section 4.3) to convert an Al-terminated ethylene oligomer to a fatty alcohol [29]. Citronellol **30**, in addition to being a menthol intermediate, is also useful as a component of rose scent.

2.4 ISOMERIZATION OF FUNCTIONAL OLEFINS

The location of a C=C bond within an organic molecule can be shifted by moving substituents other than hydrogen. As mentioned at the opening of this chapter, the isomerization of an allylic cyanide by CN migration from C-3 to C-1 in the allyl function affects olefin isomerization. This major industrial process is discussed in Section 3.6. Another major process, the isomerization of 1,4-dichloro-2-butene,

$$ClCH_2CH=CHCH_2Cl \quad \rightleftharpoons \quad ClCH_2CHCH=CH_2$$
$$\underset{\displaystyle Cl}{|}$$

involves an analogous C-1 $\rightleftharpoons$ C-3 migration of chlorine (Section 12.1).

In addition to these two isomerizations, conducted on a scale of many thousands of tons per year, there are many small-scale specialized applications of allylic isomerization. In general, they involve the use of a soluble catalyst to move a hydroxyl or carboxylate group in the production of a biologically active compound. For example, in the BASF synthesis of vitamin A [20], there is an isomerization of a diacetoxybutene exactly analogous to the dichlorobutene isomerization mentioned above.

$$AcOCH_2CH=CHCH_2OAc \rightleftharpoons AcOCH_2\underset{\underset{OAc}{|}}{C}HCH=CH_2$$

A mixture of *cis*- and *trans*-diacetoxy-2-butene is heated with a PtCl$_4$ catalyst. A slow feed of oxygen and chlorine is added, presumably to keep the catalyst in a high oxidation state, while the mixture is slowly distilled. The distillate is enriched in the lower boiling 3,4-diacetoxy-1-butene, which is formed in 95% yield with 98-99.5% purity [30]. The product is hydroformylated as described in Chapter 5 to produce part of the vitamin A side chain.

Fragrance Chemicals

Allylic isomerization is applied to a variety of terpenoid substrates in the perfume industry. These C$_{10}$ compounds, which are nominally isoprene dimers, are produced industrially both from natural products such as turpentine and from petrochemicals. Linalool **32**, a key intermediate in making a number of fragrance alcohols, is obtained both from pinenes and from the 6-methylheptenones **10** and **11** discussed earlier in this chapter. The network of chemistry involving linalool is sketched in Figure 2.5 [20,31]. Ethynylation of **11** with acetylene and a metal acetylide catalyst (Chapter 8) gives dehydrolinalool **31**, which is of interest as a precursor to linalool **32**, citral **33**, and vitamin A [20]. The conversion of **31** to **32** involves selective hydrogenation of the C≡C bond with a heterogeneous catalyst.

The isomerization of propargylic alcohols (**31** ⟶ **33**) and allylic alcohols (**32** ⇌ **34** + **35**) is catalyzed by alkyl vanadate esters. In the isomerization of dehydrolinalool **31**, the starting alcohol is heated at 140-160°C with tricyclohexyl

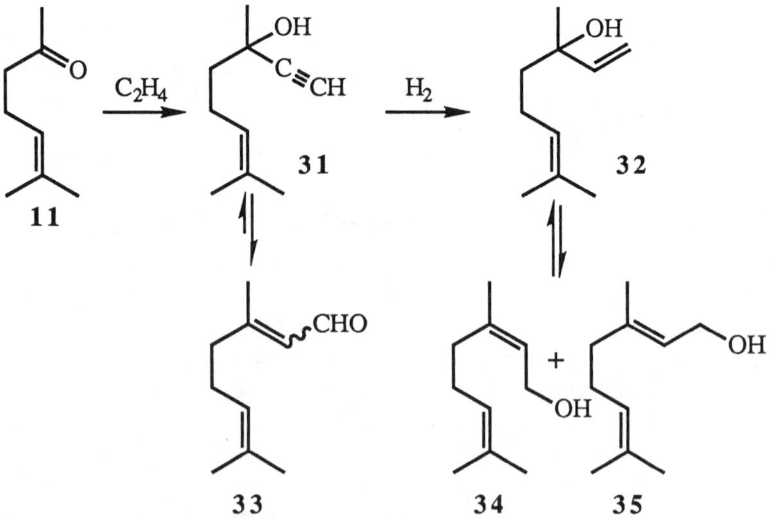

Figure 2.5 The role of linalool in fragrance chemistry.

vanadate in a paraffin solution [32]. Migration of the OH group from C-3 to C-1 presumably gives a transient enol, which spontaneously tautomerizes to citral **33**. In typical examples, the aldehyde is formed in 60-80% yield at 20-33% conversion [33]. Tris(trialkylsilyl) vanadates also catalyze this reaction [34].

The equilibrium nature of the isomerization process is more evident in the vanadium-catalyzed reaction of linalool **32** [33,35]. When linalool is heated at 160°C with *tris*(tetrahydrolinalyl) or *tris*(triphenylsilyl) vanadate, the mixture that forms contains approximately 30% of the primary alcohols, geraniol **34** and nerol **35**, along with unchanged linalool. Starting from the other side of the equation, another vanadate-based catalyst system converts either geraniol or nerol to a mixture that contains 68-70% linalool [36].

The equilibrium mixtures obtained in the linalool isomerization are not altogether satisfactory for perfumery use because separation of the desired geraniol (a rose scent constituent) is difficult. The reaction as practiced by SCM, however, employs a trick to shift the equilibrium in favor of the terminal alcohols [37]. The linalool is converted to a borate ester by ester exchange with tributyl borate. Equilibration of the borate esters yields 75-80% of the primary alcohols [38].

The vanadium-catalyzed isomerization appears to occur within the coordination sphere of the metal as sketched in Figure 2.6 [33]. The allylic alcohol such as linalool enters the coordination sphere by ester exchange with the trialkyl vanadate catalyst. The fundamental rearrangement of the allylic vanadate ester **36** resembles the classical Claisen rearrangement. The transition state may be rather like **37** in which C-1 is forming a bond to the V=O oxygen while the VO-C bond to C-3 is breaking. The rearrangement produces the new allylic vanadate ester **38** derived from the primary alcohol. Ester exchange with free tertiary alcohol (linalool) releases the isomerized product (geraniol or nerol). All steps in the process are reversible; hence, the composition of the equilibrated product mixture is determined by the relative thermodynamic stabilities of the products.

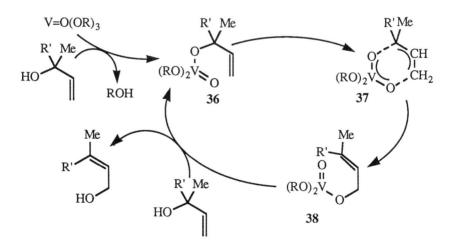

Figure 2.6 Proposed mechanism for isomerization of allylic alcohols.

The vanadate-catalyzed isomerization of allylic alcohols contrasts with the rhodium-catalyzed isomerization [22] mentioned earlier in that vanadium produces a 1,3-oxygen shift while rhodium shifts a hydrogen. This mechanistic difference leads to a difference in product. Vanadium makes an isomeric allyl alcohol, but rhodium produces an enolate of an aldehyde or ketone.

Food Additives

The 1,3-shift of a hydroxyl or acetoxyl group may also be catalyzed by copper salts or copper powder. BASF makes extensive use of this chemistry in the production of vitamin A and citranaxanthin, a food colorant [20].

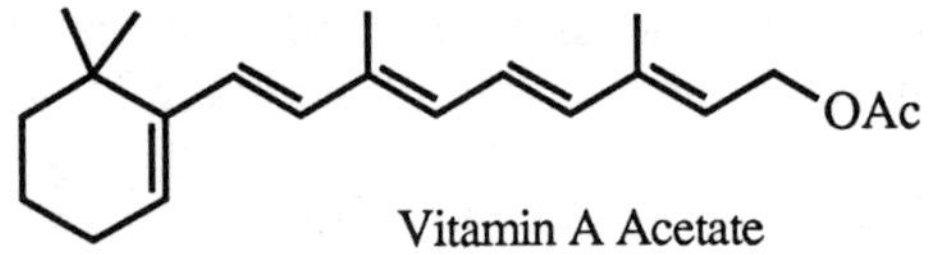

Vitamin A Acetate

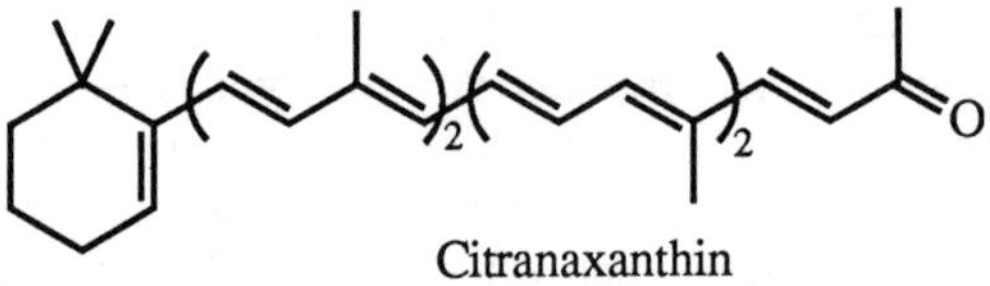

Citranaxanthin

One example of this reaction is the isomerization of the isoprenoid tertiary alcohol **39** to the primary chloride **40** [39]. (Under the acidic reaction conditions, the acetal function is hydrolyzed to the aldehyde.) The allylic chloride is then converted to a triphenylphosphonium salt **41** which, in turn, provides a Wittig reagent used in building the unsaturated side chains in vitamin A and citranaxanthin. The high chloride concentration in the reaction medium for isomerization of **39** makes it likely that the tertiary alcohol is first converted to the tertiary allylic chloride, which then isomerizes by a copper-assisted 1,3-chlorine shift like that described for the dichlorobutenes in Section 12.1.

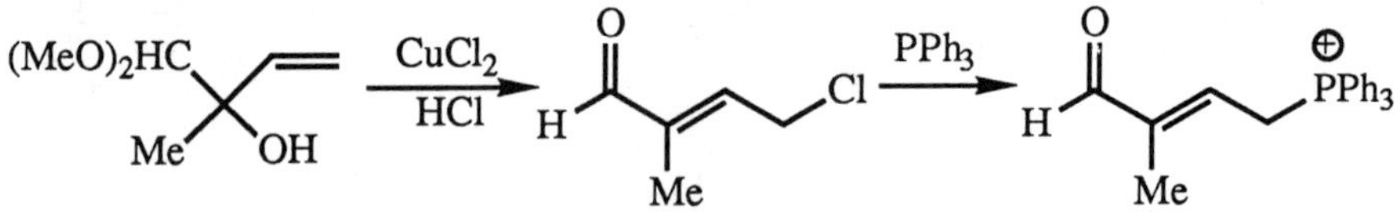

| | **39** | **40** | **41** |

The acetate corresponding to **39** undergoes an analogous isomerization to 4-acetoxy-2-methyl-2-butenal on treatment with Cu(OAc)$_2$ in acetic acid [40,41].

SPECIFIC REFERENCES

1. P. A. Chaloner, *Handbook of Coordination Catalysis in Organic Chemistry*, Butterworths, London, 1985, pp. 403-448.
2. M. Orchin, *Adv. Catal.*, **16**, 1-47 (1966).
3. G. W. Parshall and W. A. Nugent, *Chemtech*, **18**, 184-90, 314-20, 376-83 (1988).
4. P. A. Verbrugge and G. J. Heiszwolf, British Patent 1,416,317 (1975).
5. E. R. Gum and C. R. Freitas, *Chem. Eng. Prog.*, **75**, 73-6 (1979).
6. K. Bauer and D. Garbe, *Common Fragrance and Flavor Materials*, Verlag Chemie, Weinheim, 1985, p. 9.
7. L. H. Slaugh and R.D. Mullineaux, U.S. Patents 3,239,569 and 3,239,570 (1966).
8. G. Ver Strate, *Encycl. Polym. Sci.*, **6**, 522-64 (1986).
9. H. E. Fritz, K. E. Atkins, and G. L. O'Connor, U.S. Patent 3,347,944 (1967).
10. W. Schneider, U.S. Patent 3,535,395 (1970).
11. J. Andrieux, D. H. R. Barton, and H. Patin, *J. Chem. Soc. Trans. I Perkin*, 359-63 (1977).
12. A. J. Birch and G. S. R. Subba Rao, *Tetrahedron Lett.*, 3797-8 (1968).
13. E. J. Corey and J. W. Suggs, *J. Org. Chem.*, **38**, 3224 (1973).
14. D. Bingham, B. Hudson, D. Webster, and P. B. Wells, *J. Chem. Soc. Dalton Trans.*, 1521-4 (1974).
15. R. J. DePasquale, *Synth. Commun.*, **10**, 225-31 (1980).
16. C. A. Tolman, *J. Am. Chem. Soc.*, **94**, 2994-9 (1972); C. A. Tolman, R. J. McKinney, W. C. Seidel, J. D. Druliner, and W. R. Stevens, *Adv. Catal.*, **33**, 1-45 (1985).
17. D. Bingham, D. E. Webster, and P. B. Wells, *J. Chem. Soc. Dalton Trans.*, 1514-8, 1519-20 (1974).
18. R. Cramer, *J. Am. Chem. Soc.*, **88**, 2272-82 (1966); *Acc. Chem. Res.*, **1**, 186-93 (1968).
19. M. Tuner, J. V. Jouanne, H-D. Brauer, and H. Kelm, *J. Mol. Catal.*, **5**, 425-32, 433-46, 447-57 (1979).
20. H. Pommer and A. Nurrenbach, *Pure Appl. Chem.*, **43**, 527-51 (1975).
21. A. Sen and T-W. Lai, *Inorg. Chem.*, **20**, 4036-8 (1981).
22. S. H. Bergens and B. Bosnich, *J. Am. Chem. Soc.*, **113**, 958-67 (1991).
23. S. Inoue, H. Takaya, K. Tani, S. Otsuka, T. Sato, and R. Noyori, *J. Am. Chem. Soc.*, **112**, 4897-905 (1990).
24. H. Kumobayashi, S. Akutagawa, and S. Otsuka, *J. Am. Chem. Soc.*, **100**, 3949-50 (1978).
25. K. Tani, T. Yamagata, S. Otsuka, H. Kumobayashi, and S. Akutagawa, *Org. Synth.*, **67**, 33-43 (1989).
26. R. Noyori and H. Takaya, *Acc. Chem. Res.*, **23**, 345-50 (1990).
27. K. Tani, T. Yamagata, S. Akutagawa, H. Kumobayashi, T. Taketomi, H. Takaya, A. Miyashita, R. Noyori, and S. Otsuka, *J. Am. Chem. Soc.*, **106**, 5208-17 (1984).
28. B. D. Sully, *Chem. Ind.* 263-7 (1964); R. Rienacker and G. Ohloff, German Patent 1,118,775 (1962).
29. A. Lundeen and R. Poe, "Alpha-Alcohols" in J. J. McKetta and W. A. Cunningham, Eds., *Encyclopedia of Chemical Processing and Design*, Vol. 2, Marcel Dekker, New York, 1977, p.465-81.
30. J. Hartig and H-M. Weitz, German Patent 2,736,695 (1979).
31. P. Z. Bedoukian, *Am. Perfum. Cosmet.*, **86**(4), 25-36 (1971).

32. P. Chabardes and Y. Querou, British Patent 1,204,754 (1970); *Chem. Abstr.*, **72**, 43923 (1970).

33. P. Chabardes, E. Kuntz, and J. Varagnat, *Tetrahedron*, **33**, 1775-83 (1977).

34. H. Pauling, D. A. Andrews, and N. C. Hindley, *Helv. Chim. Acta*, **59**, 1233-43 (1976).

35. P. Chabardes, C. Grard, and C. Schneider, U.S. Patent 3,925,485 (1975).

36. S. Matsubara, T. Okazoe, K. Oshima, K. Takai, and H. Nozaki, *Bull. Chem. Soc. Jpn.*, **58**, 844-9 (1985).

37. *Chem. Eng. News*, 5-6 (22 Nov. 1982).

38. B. J. Kane, U.S. Patent 4,254,291 (1981).

39. W. Reif, R. Fischer and H. Pommer, U.S. Patent 3,940,445 (1976).

40. R. Fischer and H. Pommer, U.S. Patent 3,639,437 (1972).

41. W. Reif and H. Grassner, *Chem.-Ing.-Tech.*, **45**, 646-54 (1973).

3 | REACTIONS OF OLEFINS AND DIENES - HYDROGENATION AND HY ADDITIONS

Some of the best studied homogeneous catalytic processes are the additions of H_2, $HSiR_3$, and HCN to the C=C bond. In the past, hydrogenation with soluble catalysts was primarily of academic interest because heterogeneous catalysts such as Pd/C were so much more convenient. Recently, however, enantioselective hydrogenations with soluble chiral catalysts have become important to the pharmaceutical industry and to the organic synthesis chemist. The hydrosilylation and hydrocyanation reactions have been applied industrially for over 20 years. The synthesis of adiponitrile, a nylon intermediate, by hydrocyanation of butadiene is carried out on a scale of almost 500,000 tons per year. Sections of this chapter are devoted to each of these classes of reactions.

3.1 HYDROGENATION OF SIMPLE OLEFINS

As evidenced by hundreds of publications, including monographs and reviews, the homogeneous hydrogenation of olefins has been studied extensively, perhaps more so than any other reaction catalyzed by a soluble metal complex. This intensive study seems anomalous because heterogeneous catalysts are usually more active and more convenient for practical applications such as the hydrogenation of cyclododecatriene to cyclododecane or of dicyanobutene to adiponitrile. Until the mid-1980s, the sole commercial use of a soluble catalyst for olefin hydrogenation was Monsanto's reduction of an unsaturated amino acid to a precursor of the drug L-dopa. This operation, begun in 1970, pioneered the commercial use of asymmetric induction by an optically active catalyst. This rapidly burgeoning field of study is discussed in Section 3.4.

The reduction of a sterically accessible C=C bond is usually simple experimentally. One mixes hydrogen, olefin, and catalyst in an organic solvent at 25-100°C and 1-3 atmospheres pressure. The reaction is usually clean, and the products are separated from catalysts by conventional techniques such as distillation or washing with water. Dozens of soluble transition metal complexes catalyze hydrogenation of olefins, but four classes are preferred for practical hydrogenation:

- Wilkinson's catalyst, $RhCl(PPh_3)_3$, and the related $[Rh(diene)(PR_3)_2]^+$ complexes.
- Mixtures of platinum and tin chlorides.
- Anionic cyanocobalt complexes.
- Ziegler catalysts prepared from a transition metal salt and an alkylaluminum compound.

Each of these classes has particular advantages in synthesis. The first three classes are mechanistically distinct from one another, as discussed in Section 3.2, while the mechanisms of the Ziegler catalysts are not well defined.

The best studied soluble catalyst for olefin hydrogenation is Wilkinson's catalyst, $RhCl(PPh_3)_3$ [1]. This moderately stable, commercially available compound catalyzes the hydrogenation of many sorts of olefins under mild conditions (see General References). Terminal olefins such as 1-hexene are rapidly hydrogenated at room temperature and atmospheric pressure. The hydrogenations of internal olefins proceed slowly, but excellent results are often attained. The stereochemistry of the products may differ from that obtained with heterogeneous catalysts [2]. This catalyst selectively reduces C=C bonds in the presence of other easily reduced functions such as nitro and -CH=O. It adds H_2 or D_2 cleanly cis to the C=C bond and usually produces little HD scrambling when it is used to introduce deuterium. Its application in synthesis is illustrated by the *Organic Syntheses* procedure for the hydrogenation of carvone [3]:

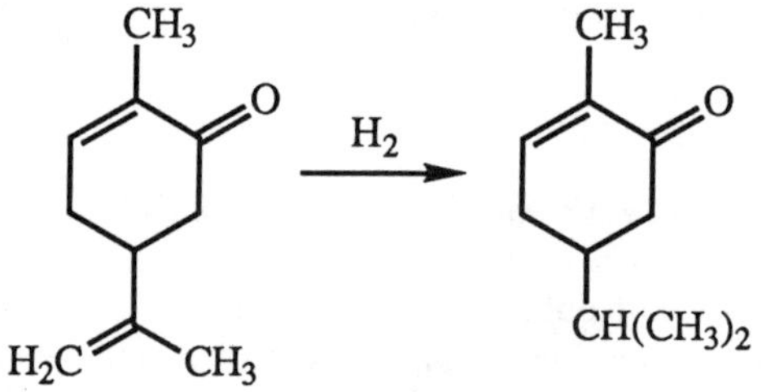

In this particular application, a 1,1-disubstituted double bond is hydrogenated in preference to a trisubstituted olefin and a ketone.

The kind of selectivity seen in the dihydrocarvone synthesis has been applied industrially in the synthesis of ivermectin, a drug valuable for the treatment of parasitic diseases such as onchocerciasis (African river blindness) [4,5]. It is also broadly useful in veterinary medicine to kill parasites such as heartworm in dogs [6]. Ivermectin is made by hydrogenation of the macrocyclic compound, avermectin-B_1, which is isolated from fermentation broths of *Streptomyces avermitilis*. The key to producing the derivative with optimal biological activity is to hydrogenate specifically one of the five C=C bonds, as indicated in the abbreviated structure below. This selectivity is achieved by hydrogenation with $RhCl(PPh_3)_3$ in toluene at 25°C and 1 atmosphere H_2 [7,8]. The targeted double bond, which is cis and disubstituted, is hydrogenated in preference to the trans and trisubstituted C=C bonds. Yields of ivermectin are reported to be 85%.

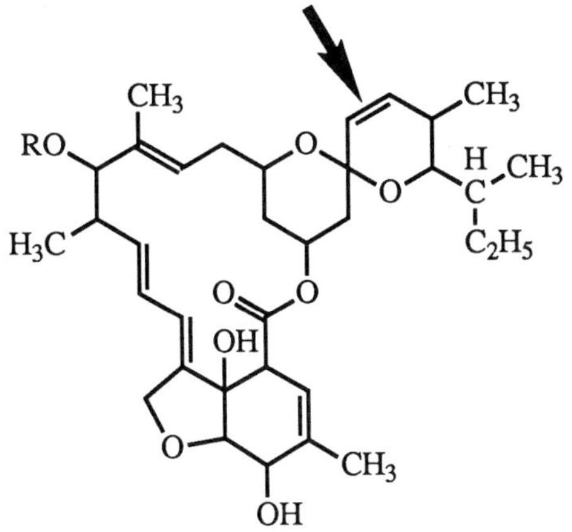

Wilkinson's catalyst is slow for the hydrogenation of internal double bonds, but as illustrated above, its selectivity can make it useful in organic synthesis. Other potential applications include steroid hydrogenation [9] and the hydrogenation of dicyanobutenes to adiponitrile [10].

$$NCCH_2CH=CHCH_2CN \xrightarrow[\text{base}]{H_2, \text{RhCl(PPh}_3)_3} NC(CH_2)_4CN$$

In contrast to Wilkinson's catalyst, the closely related catalysts $[Rh(diene)(PR_3)_2]^+$, developed by Schrock and Osborn [11,12], give respectable rates with highly substituted olefins. These cationic catalysts are employed in most systems for the asymmetric hydrogenation of olefins (Section 3.4). The results obtained vary significantly with the reaction conditions. In acidic media, the active catalytic species appears to be a cationic dihydride, $[RhH_2(PR_3)_2(\text{solvent})_2]^+$, which hydrogenates olefins with little concomitant isomerization. In nonacidic media, the dihydride appears to deprotonate to form a neutral complex $[RhH(PR_3)_2(\text{solvent})_x]$, which catalyzes both olefin isomerization and hydrogenation.

Another family of catalysts suitable for hydrogenation of C=C in the presence of C=O is generated by mixing platinum and tin chlorides [13]. The commercially available H_2PtCl_6 and $SnCl_2 \cdot 2H_2O$ react in methanol to form deep red solutions which contain species such as $[Pt(SnCl_3)_5]^{3-}$. These solutions catalyze hydrogenation of simple linear olefins and have been extensively studied for hydrogenation of vegetable oils to remove excessive unsaturation which is responsible for flavor instability [14,15]. Other ligands such as phosphines are often added to modify the catalytic activity.

For hydrogenations in water with an inexpensive catalyst, solutions containing cobalt salts and excess cyanide ion are useful [16,17]. These solutions contain complex anions such as $[Co(CN)_5]^{3-}$ and $[HCo(CN)_5]^{3-}$. The catalysts are selective for hydrogenation of C=C bonds, which are conjugated with one another or with C=O, C≡N, or phenyl groups. In contrast to other diene hydrogenation catalysts, the cobalt cyanides are relatively unreactive with unconjugated dienes such as 1,5-cyclooctadiene.

Industrial hydrogenation of unsaturated polymers is usually carried out with Ziegler-type systems like those used in arene hydrogenation (Chapter 7). These systems are prepared by mixing a hydrocarbon-soluble complex of a first-row transition metal with an alkane solution of an alkylaluminum compound. Typically, cobalt acetylacetonate or 2-ethylhexanoate is used with triethyl- or triisobutylaluminum [18,19]. The mixtures are dark, air-sensitive solutions, which may contain some colloidal metal. Because of the highly reactive alkyl-metal bonds, these catalysts are affected by functional groups such as OH, C=O, and C≡C-H. In a typical application [20], a 75:25 butadiene-styrene block copolymer is hydrogenated with a catalyst prepared from reaction of a $CoCl_2$ complex with triisobutylaluminum in hexane. The C=C bonds in the polybutadiene segments, which contain both internal double bonds and pendant vinyl groups, are completely hydrogenated. The process gives a tough, oxidation-resistant rubber of the kind ordinarily found in shoe soles and heels.

One of the virtues of soluble catalysts in the hydrogenation of polymers is that the catalyst can diffuse to the site of the C=C bond in the polymer chain. In contrast, a heterogeneous catalyst would require that the polymer unfold to gain access to the catalytic site. This virtue of the molecular catalysts is not confined to polymer solutions. It has been shown that $RhCl(PPh_3)_3$ can diffuse to amorphous regions in 1,2-polybutadiene at 60°C and affect saturation of most of the pendant vinyl groups [21]. This process is not used commercially, but it has some practical potential as a hydrogen sequestrant in nuclear applications.

In addition to the catalyst families described above, $Co_2(CO)_8$ and $[Co(CO)_3(PBu_3)]_2$ are useful hydrogenation catalysts. The latter is relatively stable and shows excellent selectivity in the hydrogenation of polyenes to monoolefins. Other carbonyls such as $Cr(CO)_6$ and $Fe(CO)_5$ are also useful. They are less active, but become effective when activated by heating or radiation to expel a carbonyl ligand and create a vacant coordination site.

3.2 MECHANISM OF OLEFIN HYDROGENATION

Most soluble catalysts add hydrogen to a C=C bond very simply. The olefin and H_2 are brought together as ligands in the coordination sphere of the metal. A rearrangement of the hydrido olefin metal complex to a metal alkyl is followed by some sort of M-C bond cleavage process. Catalysts differ in the mode of cleaving H_2 to form the metal-hydride ligand and in the mechanism of cleavage of the metal-alkyl bond to form alkane. Three different H_2 cleavage mechanisms are observed for the rhodium(I), platinum-tin, and cobalt-cyanide catalysts. The last system may differ quite fundamentally in its hydrogenation mechanism.

At least three mechanistic pathways have been demonstrated for Wilkinson's catalyst [22-24]. The kinetically dominant mechanism is shown in Figure 3.1. The Rh(I) species shown at the top of the catalyst cycle is probably a solvated three-

coordinate complex formed by dissociation of a triphenylphosp
parent compound

$$RhCl(PPh_3)_3 \rightleftharpoons RhCl(PPh_3)_2 + PPh_3$$

This Rh(I) species is very coordination-deficient, with a formal ele   .. ui 14.
It readily undergoes oxidative addition of an H_2 molecule to form a dinydride which is
formally Rh(III) if the H ligands are regarded as H^-. In the presence of
triphenylphosphine, $RhH_2Cl(PPh_3)_3$ is detected in solution by nmr [22]. When the
concentration of triphenylphosphine is limited, the dihydride coordinates an olefin to
form the complex 1. This complex, in turn, undergoes migratory insertion to produce
the alkylrhodium-hydride complex 2. The alkylrhodium-hydride complex rapidly
eliminates alkane and regenerates the catalytically active Rh(I) species to complete the
cycle.

A similar cycle probably operates for the $[Rh(diene)(PR_3)_2]^+$ catalysts which have
been studied intensively because of their use in asymmetric hydrogenation. In this
instance, the Rh(I) species is solvated $[Rh(PR_3)_2]^+$, which is formed by
hydrogenation of the diene ligand. The sequence of reaction with H_2 and olefin is
reversed with this catalyst [25]. The olefin coordinates *before* the oxidative addition
of hydrogen, which is the rate-limiting step. In the hydrogenation of α-
acetamidocinnamic acid, the alkyl hydride complex has been detected by nmr at low
temperatures [25].

The hydrogenation catalyst prepared by mixing H_2PtCl_6 or K_2PtCl_4 with stannous
chloride follows a different reaction pathway. The catalyst solutions contain several
anionic complexes of which the best characterized is $[Pt(SnCl_3)_5]^{3-}$. The $SnCl_3^-$
ligands in this anion are very labile and dissociate to give vacant coordination sites for
reaction with H_2 and with olefins. The ligands also inhibit reduction of the Pt(II) ion
to metallic platinum. The $SnCl_3^-$ ligand appears to be a weak σ-donor and a good π-
acceptor like carbon monoxide. These ligand-metal bond characteristics seem to favor
stability of low-valent metal centers.

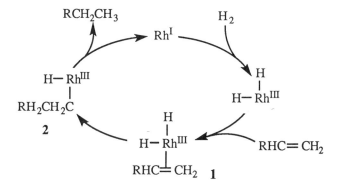

Figure 3.1 A major mechanistic pathway for olefin hydrogenation by Wilkinson's catalyst.

A major difference from Wilkinson's catalyst is the mechanism of hydrogen activation. With the anionic platinum catalyst, it occurs by heterolytic cleavage of H_2:

$$H_2 + [Pt(SnCl_3)_5]^{3-} \rightleftharpoons H^+ + [HPt(SnCl_3)_4]^{3-} + SnCl_3^-$$

The anionic platinum hydride, which can be isolated as a tetraalkylammonium salt [26], reacts with an olefin to give an alkyl complex $[RPt(SnCl3)4]^{3-}$. Presumably the coordination of the olefin and the insertion reaction with the Pt-H bond occur much as indicated for Wilkinson's catalyst. The second major difference between the two systems lies in the cleavage of the metal-alkyl bond. In the platinum system, protonolysis by the acid formed in H_2 cleavage is the most likely reaction:

$$[RPt(SnCl_3)_4]^{3-} + H^+ \longrightarrow RH + [Pt(SnCl_3)_4]^{2-}$$

The third major mechanism is based on homolytic cleavage of the dihydrogen molecule by metal-metal bonded species or by a paramagnetic complex. Two important examples are found in cobalt chemistry:

$$Co_2(CO)_8 + H_2 \rightleftharpoons 2 HCo(CO)_4$$

$$2\,[Co(CN)_5]^{3-} \rightleftharpoons [Co_2(CN)_{10}]^{6-} \xrightarrow{H_2} 2\,[HCo(CN)_5]^{3-}$$

The kinetics of the cobalt cyanide system [27] are those expected if H_2 cleavage is the slow step. It is interesting that the two hydrogen cleavage reactions shown above occur with different oxidation states of the metal. The carbonyl reaction involves a formal transition from Co(0) to Co(I), whereas the cyanide complex changes from Co(II) to Co(III).

The reaction of the cobalt hydride with an olefin such as styrene is proposed to occur without precoordination [28]:

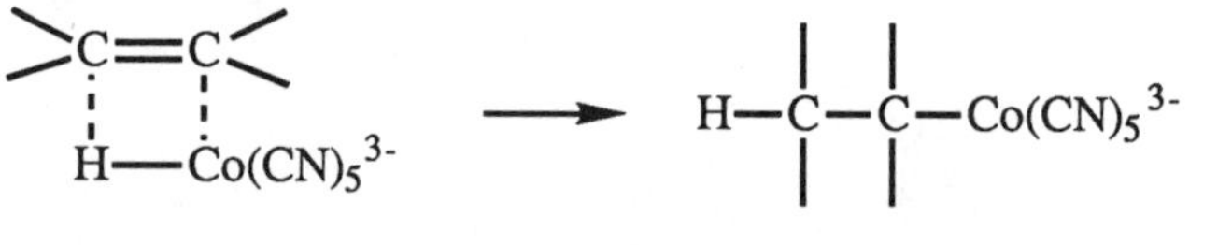

If this proposal is correct, it is an exception to the general rule that olefin insertion reactions involve a coordinated C=C bond. The rate and course of styrene hydrogenation change little with changes in cyanide ion concentration, whereas major effects would be expected if CN^- dissociation were required to free a coordination site for the olefin. Cyanide concentration does change the stereochemistry of diene hydrogenation [17], presumably by regulating formation of a π-allyl intermediate. With either alkyl or allyl intermediates, the reduced organic product is released from the metal by reaction with a second cobalt hydride anion:

$$H-\overset{|}{\underset{|}{C}}-\overset{|}{\underset{|}{C}}-Co(CN)_5^{3-} + HCo(CN)_5^{3-} \longrightarrow H-\overset{|}{\underset{|}{C}}-\overset{|}{\underset{|}{C}}-H + 2\ Co(CN)_5^{3-}$$

3.3   SELECTIVE HYDROGENATION OF POLYENES

The hydrogenation of dienes and trienes to monoolefins has several potentially useful applications in industry. Heterogeneous catalysts are used commercially for selective hydrogenation of polyunsaturated vegetable oils to give shortenings with better physical properties [29]. Typically, a linolenate ester with three C=C bonds per C_{18} chain may be reduced to a linoleate with two double bonds or an oleate with only one double bond per chain. In applications of this type, in which only moderate selectivity is required, heterogeneous catalysts excel because they are easy to separate from reaction products by decantation or filtration. Soluble catalysts are very difficult to separate from high-boiling materials such as vegetable oils except when the catalyst can be extracted into a polar solvent.

Homogeneous catalysts are preferred over their heterogeneous analogs when high selectivity is required.   One industrial example is the hydrogenation of 1,5-cyclooctadiene to cyclooctene. This selective hydrogenation converts a readily available butadiene dimer (Chapter 4) to a precursor of specialty dienes and of polyoctenamer rubber (Chapter 9). Another similar hydrogenation is not operated industrially, but may have considerable potential. This is the hydrogenation of 1,5,9-cyclododecatriene, a butadiene trimer, to cyclododecene, a precursor of dodecanedioic acid and laurolactam. Although heterogeneous catalysts such as palladium on alumina can be used for this hydrogenation, higher selectivity is available with soluble catalysts (Table 3.1). The two soluble catalysts prepared from nonprecious metals, $[Co(CO)_3(PBu_3)]_2$ and $NiI_2(PPh_3)_2$, are interesting because they are inexpensive and give very high yields of the desired cyclododecene. Similar results are obtained in the hydrogenation of 1,5-cyclooctadiene to cyclooctene with some of the same catalysts [33,34,36].

The catalysts listed in Table 3.1 effect partial hydrogenation of either conjugated or unconjugated dienes and trienes.   In contrast, a second group of catalysts hydrogenate only conjugated dienes.   Catalysts such as $[Co(CN)_5]^{3-}$ and $[Cr(CO)_3(methyl\ benzoate)]$ reduce 1,3,7-octatriene, a linear butadiene dimer (Section 4.5), to mixtures of octadienes [16,37] by selective H_2 addition to the 1,3-diene function:

The unconjugated diene products are hydrogenated slowly, if at all, under standard reaction conditions.

Table 3.1   Hydrogenation of 1,5,9-Cyclododecatriene (CDT) to Cyclododecene (CDE)[a] at Full Conversion

Catalyst	Temp. (°C)	Pressure (atm)	Time (hr)	Solvent	Yield of CDE (%)	Ref.
Pd/Al$_2$O$_3$	150	1.5		CDT	91	30
[Co(CO)$_3$(PBu$_3$)]$_2$	140	30	1.7	C$_6$H$_6$	99	31
NiI$_2$(PPh$_3$)$_2$	160	80	6	CDT	99	32
[Pt(Cl$_3$)$_5$]$^{3-}$	160	100	10	Et$_4$NSnCl$_3$	87	26
RuCl$_2$(PPh$_3$)$_3$	25	10	30	C$_6$H$_6$/EtOH	87	33
RuCl$_2$(CO)$_2$(PPh$_3$)$_2$	140	5	1	C$_6$H$_6$	98	34
RuCl$_3$(py)$_3$/NaBH$_4$	75	1	0.1	DMF	91	35

[a] Most of the hydrogenations employed commercially available cis, trans, trans-1,5,9-cyclododecatriene and yielded a mixture of cis- and trans-cyclododecene.

A major distinction between the two classes of catalysts is that those in Table 3.1 are isomerization catalysts while the cobalt cyanide and chromium carbonyl catalysts are not. The more versatile catalysts can convert unconjugated dienes or trienes to conjugated systems through double-bond migration. Since most of the catalysts in this class are hydride complexes or form hydrides when treated with H$_2$, it is likely that the isomerization occurs by an M-H addition-elimination process (Section 2.2).

The selectivity for hydrogenation of dienes in the presence of monoolefins usually arises from the exceptional stability of π-allyl complexes. Regardless of the H$_2$ cleavage mechanism, M-H addition to a conjugated diene can generate a π-allyl intermediate. In a hydrogenation mixture that contains diene, monoolefin, and a platinum-tin chloride catalyst, the following reactions are believed to be in competition [26]:

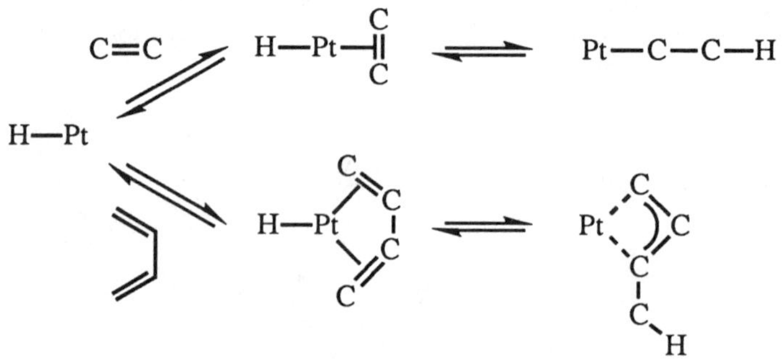

The reaction pathway involving the π-allyl intermediate is favored, especially when the olefin or diene must compete with excess ligand such as R_3P, CO, or $SnCl_3^-$ for a coordination site. Consequently, the diene in the reaction mixture is almost completely hydrogenated before the concentration of olefin increases to the point that olefin gains access to the catalyst. Similar competition for catalyst sites is believed to be responsible for selectivity in hydrogenation of dienes by heterogeneous catalysts.

3.4   ASYMMETRIC HYDROGENATION

Synthesis of optically active organic compounds from nonchiral starting materials is perhaps the most elegant application of homogeneous catalysis. This asymmetric induction can occur in many reactions catalyzed by transition metal complexes (Chapter 1), but the first commercial application [38] was the enantioselective hydrogenation involved in Monsanto's synthesis of L-dopa, a drug used in the treatment of Parkinson's disease.

One characteristic feature for selective synthesis of one optical isomer of a chiral substance is an asymmetric catalyst site that will bind a prochiral olefin preferentially in one conformation. This recognition of the preferred conformation is achieved by using a chiral ligand, which creates a "chiral hole" in the coordination sphere of the metal. In general, the ligands used in practical applications are chelating diphosphines. The chirality may reside either at the phosphorus or in the organic groups bound to the phosphorus. In general, some of the groups bound to the P atom are aryl, but recently good results have been reported [39] with aliphatic groups that impart chirality. Considerable understanding of the basis for this enantioselectivity has come from studies performed during the past 15 years, as discussed later in this section.

Realization of the dream of synthetic catalysts with enzyme-like stereoselectivity came about through the convergence of three factors in the late 1960s:

- Development of highly active and selective catalysts for olefin hydrogenation [11].
- Synthesis of asymmetrically substituted phosphines that would serve as ligands to create a chiral catalyst site.
- Evolution of a need to manufacture pharmaceuticals as specific optical isomers.

The last requirement was met through medical research that provided moderate relief for Parkinsonism patients through the administration of L-3,4-dihydroxyphenylalanine (Levodopa, 3). Subsequently, needs have developed for enantioselective commercial processes for phenylalanine and for Naproxen®, a highly effective anti-inflammatory drug.

Levodopa

The industrial synthesis of Levodopa [38,40] is shown in Figure 3.2. Vanillin 4 is converted to an oxazolidinone 5 by reaction with N-acetylglycine in the presence of sodium acetate and acetic anhydride. The heterocyclic ring in 5 is then hydrolyzed

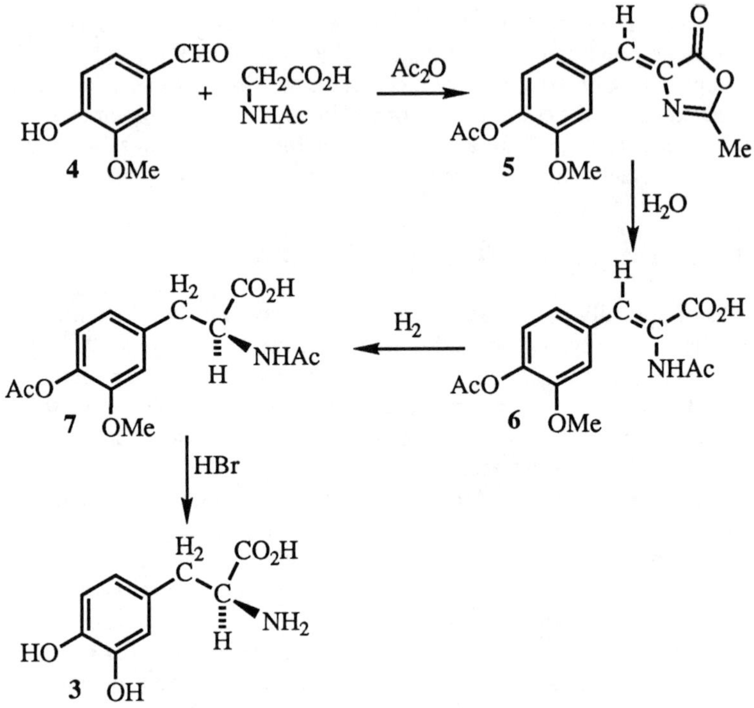

Figure 3.2   Enantioselective synthesis of Levodopa.

with aqueous acetone to give enamide **6**. This acetamidocinnamic acid is the prochiral substrate in the enantioselective hydrogenation, which is the key step in the overall process. The subsequent hydrolysis of the hydrogenation product **7** to give Levodopa **3** appears to be straightforward, although care is needed to avoid epimerization to a racemate.

The catalyst used in the hydrogenation is prepared by the Schrock-Osborn method [11] of reacting the [Rh(COD)₂]⁺ cation (COD = 1,5-cyclooctadiene) with the chiral phosphine in aqueous ethanol or isopropanol [38,41]. Initially, a monodentate phosphine was used, but subsequent research showed much better enantioselectivity with the chelating diphosphine DIPAMP:

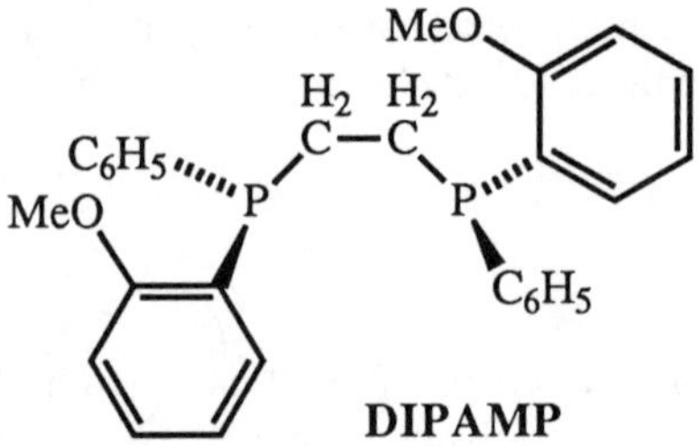

DIPAMP

and this change was incorporated in the commercial process in 1974 [38]. The hydrogenation is carried out at about 50°C and 3 atmospheres pressure by adding the solid enamide **6** to the catalyst solution. It slowly dissolves and reacts, after which the chiral aryl acetamidopropionic acid **7** crystallizes. It can be filtered away from the catalyst solution, which contains a little racemic product. The optical yield is high (ca. 95% ee) and about a 90% yield of the desired enantiomer **7** is isolated.

A wide variety of different ligands has been explored for asymmetric hydrogenations like that employed in the Levodopa synthesis [42-44]. In general, they have employed the $[Rh(ligand)_2(diene)]^+$ catalyst system, as discussed after mention of the phenylalanine process.

L-Phenylalanine

Technology somewhat similar to that employed in the L-dopa synthesis has been commercialized to make phenylalanine in Europe. The demand for this amino acid has increased substantially as a result of the commercial success of the synthetic sweetener aspartame:

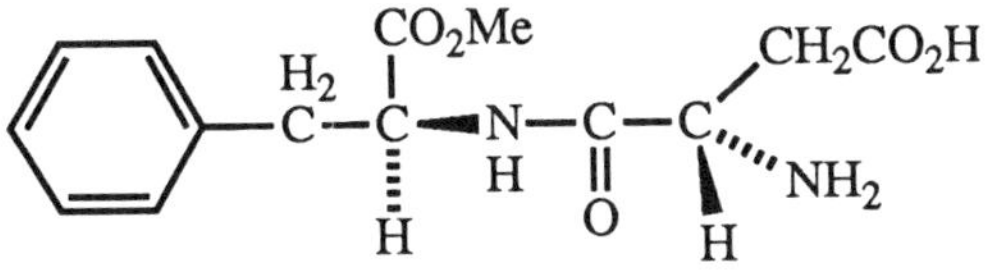

Aspartame

The dipeptide can be prepared by reaction of L-phenylalanine with N-acetylated aspartic anhydride in glacial acetic acid [45]. After removal of the acetyl protecting group, the dipeptide is converted to the methyl ester [46].

The hydrogenation of α-acetamidocinnamic acid **8** is carried out in ethanol with a cationic rhodium catalyst [47,48]:

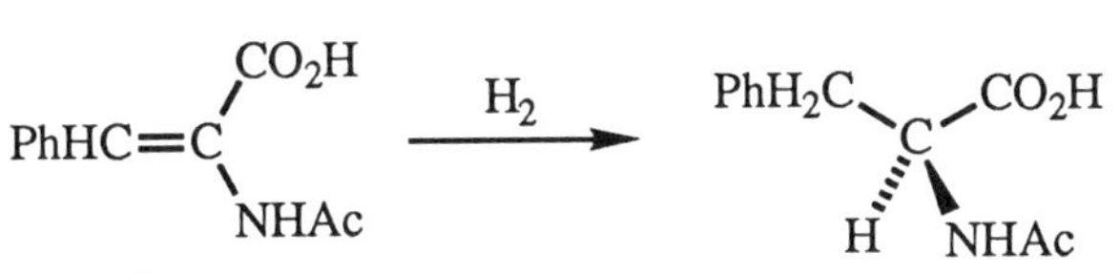

As in the L-dopa synthesis, hydrogen is transferred stereospecifically to one face of the olefin to produce N-acetyl-L-phenylalanine in high optical yield. The catalyst appears to be a [Rh(PNNP) (norbornadiene)]⁺ salt, a type of catalyst that has received extensive study [49]. The PNNP ligand (shown below), in contrast to the P-chiral DIPAMP ligand used by Monsanto, has its chirality centered on the α-phenethyl groups two atoms removed from the phosphorus atoms, which bind to the rhodium ion. Nevertheless, the enantioselectivity is good in this reaction.

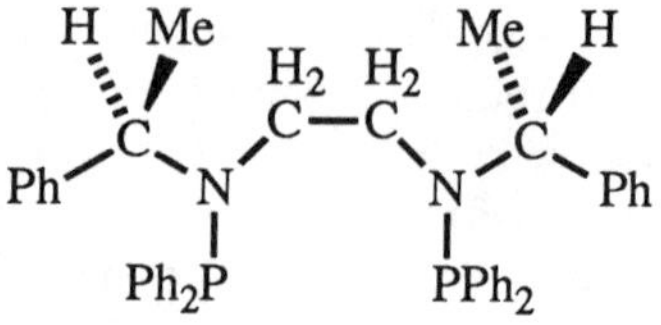

PNNP

The commercial viability of this phenylalanine process has been challenged by new fermentation technology [50] commercialized by Ajinomoto.

Other Catalyst Systems

The scientific and technological potential of enantioselective catalysis has stimulated extensive research and the discovery of many new catalyst systems that are effective. Perhaps the most important of these is the BINAP ligand:

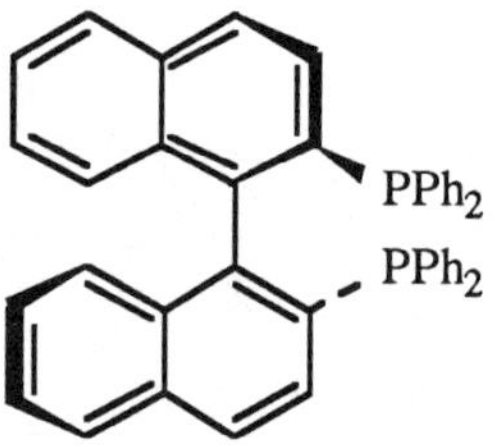

(-)-BINAP

which has been used with both Rh(I) and Ru(II) catalyst systems [51]. The rhodium system has limited utility, but complexes such as $Ru(BINAP)(O_2CR)_2$ catalyze the asymmetric hydrogenation of α-amidocinnamic acids, allylic alcohols [52], and acrylic acids. The latter capability seems likely to be applied in an industrial process for manufacture of the anti-inflammatory drug Naproxen®, as described below. A particularly useful characteristic of the ruthenium system is the ability to hydrogenate β-ketoesters stereoselectively. Substrates such as the benzamidomethyl-3-oxobutanoates **9** yield desirable precursors to the β-lactam antibiotics [53]:

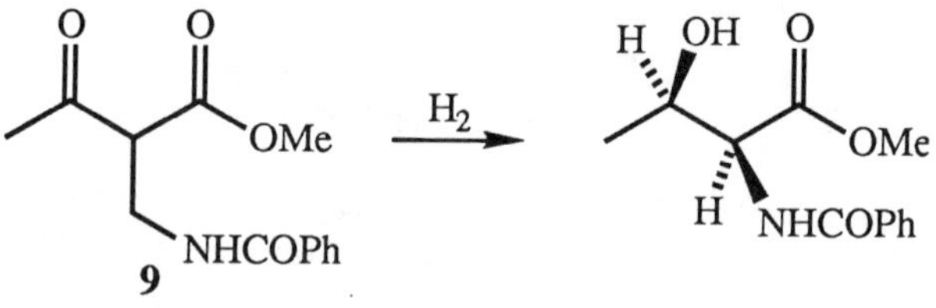

In this instance, the preferred catalyst is a cationic Ru(II) complex, $[Ru(BINAP)(arene)I]^+$. Such halide-containing complexes [54] seem most effective for hydrogenation of the C=O function, whereas the $Ru(BINAP)(O_2CR)_2$ complexes are best for allylic alcohols and amines.

Much of the other scouting for effective ligands for asymmetric hydrogenation catalysts has employed the cationic rhodium system [42-44,49]. The general procedure is illustrated by an *Organic Syntheses* example employing a substituted acrylic ester as the substrate [55]. Many give optical yields, expressed as enantiomeric excesses, over 90%. Considerable effort has been devoted to the development of catalytic systems that facilitate separation of the catalyst from the product mixture. One useful approach is attachment of the catalyst to an insoluble support such as polystyrene or silica gel [56,57] so that it may be simply filtered from the reaction mixture. Another approach involves sulfonation of the chiral phosphine ligand [58] to make it water soluble, as in the Ruhrchemie hydroformylation process (Chapter 5). This technique permits decantation of an aqueous catalyst-containing phase from a nonpolar organic phase that contains product and any unreacted starting material.

One limitation of all the catalyst systems discussed to this point is that they do not produce good optical yields with simple olefins; optimum results require the presence of a nearby functional group to help orient the C=C bond when it binds to the catalytic complex. An exception to this requirement is found with a family of Ziegler-Natta catalysts that catalyze both enantioselective polymerization (Chapter 4) and hydrogenation of olefins such as styrene and 1-pentene [59]. The optical yields are not exceptional (up to 65% ee with styrene), but the results provide an interesting research lead. The enantioselectivity results from the use of a chelating chiral ligand derived from coupling two tetrahydroindenyl groups. The ligand is coordinated to zirconium in a $Zr(CH_3)_2(ligand)$ complex used in conjunction with a methylaluminoxane catalyst modifier. The hydrogenation is carried out under mild conditions (25°C, 1-20 atmospheres hydrogen pressure).

Naproxen®

One of the world's largest-selling prescription drugs is the anti-inflammatory Naproxen® **10**. It is sold as the pure S-isomer because the R-isomer is a liver toxin. Currently the desired isomer is obtained by a conventional optical resolution of the racemate. Production by an enantioselective synthesis appears to offer a good commercial opportunity, especially since the original patent on the drug expires in 1993. Many enantioselective routes have been explored, but it seems likely that the new industrial process will employ an asymmetric hydrogenation, as outlined in Figure 3.3 [60,61].

The proposed synthesis employs an innovative electrochemical reduction (aluminum anode, lead cathode) of the acetylnaphthalene derivative **11** in the presence of CO_2. The resulting α-hydroxypropionic acid **12** is dehydrated over an acidic catalyst to produce the α-naphthylacrylic acid **13**. This compound is the substrate for enantioselective hydrogenation employing an (S)-BINAP ruthenium(II) chloride complex [60]. For optimal results, the reaction is carried out at low temperatures and high hydrogen pressure in the presence of excess triethylamine. Optical yields are reported to be in the range of 96-98% ee [54,60-62].

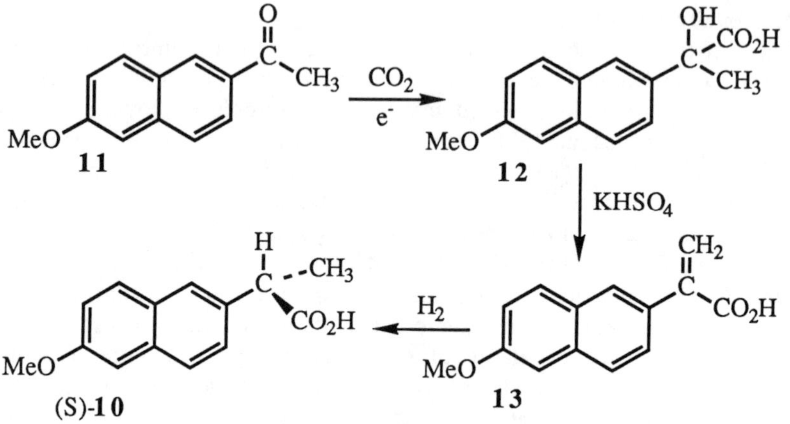

Figure 3.3   Probable future commercial synthesis of Naproxen®.

Mechanism of Asymmetric Hydrogenation

The scientific challenge associated with understanding the enzyme-like specificity of enantioselective hydrogenation has led to intensive study, particularly of the rhodium(I) catalysts bearing chelating diphosphine ligands [42,63,64]. A deep understanding of the mechanism of olefin hydrogenation has resulted.

The mechanism proposed for operation of the [Rh(DIPAMP)(solvent)$_2$]$^+$ catalyst 14 used in L-dopa synthesis is sketched in Figure 3.4. The initial step, as with most [Rh(PR$_3$)$_2$]$^+$ catalysts, is coordination of the olefinic substrate. In these studies [64], methyl Z-acetamidocinnamate 15 is the substrate rather than the actual L-dopa precursor. The chirality of the R,R-DIPAMP ligand permits two diastereomeric olefin complexes 16 and 17 to form. Because the following step, H$_2$ addition to Rh(I), has a significant activation energy and is often rate-limiting, detectable quantities of the more stable olefinic complex 16 accumulate in solution. The substrate 15 is bound to the metal by coordination of both the C=C bond and the amide oxygen [65].

Perhaps the most surprising and significant result of this study is that the major product 20 arises from the less-stable initial olefin complex 17. Evidently the rate-limiting reaction of 17 with hydrogen is sufficiently faster than that of 16 that it dominates the kinetics of the overall process. Hence, it determines the chirality of the final product. (The oxidative addition of H$_2$ and subsequent steps appear to be essentially irreversible.) The chirality of 18 determines that of the product because a hydride ligand transfers to the Rh-bound face of the coordinated olefin to produce the hydrido alkyl complex 19. The subsequent reductive elimination of Rh-H and Rh-C bonds forms the major organic product 20 and regenerates the original [Rh(DIPAMP)(S)$_2$]$^+$ catalyst 14. At low temperatures, the reductive elimination step becomes rate limiting and it has been possible to isolate a (DIPHOS)Rh(I)hydrido alkyl complex analogous to 19 [25].

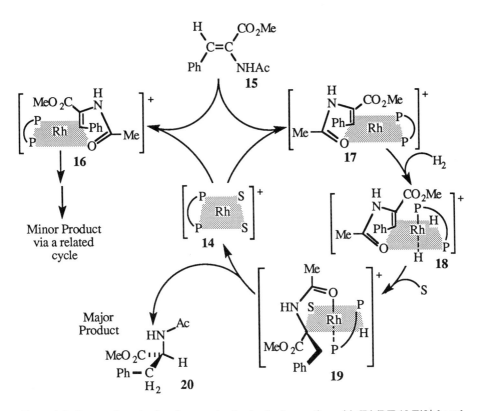

Figure 3.4 Proposed mechanism for enantioselective hydrogenation with [Rh(DIPAMP)]$^+$-based catalyst (S = solvent).

Other enantioselective catalyst systems such as the versatile BINAP Ru(II) complexes [66] are currently undergoing mechanistic studies like that applied to the DIPAMP Rh(I) system. Deeper insights into the basis for enantioselectivity should result.

3.5   HYDROSILYLATION

The addition of an Si-H bond to a C=C function has been explored intensively as a route to alkylsilanes [67,68]. In addition to its use in laboratory syntheses, the hydrosilylation reaction is used in many ways in the manufacture of silicone polymers. Probably the broadest application is the "curing" of silicone rubbers, a step that converts a syrupy polymer to a gum rubber. Similarly a putty-like polymer may be converted to a hard material such as dental cement. This toughening process is accomplished by forming crosslinks between polymer chains. Commonly, an SiH function of one chain is added to a vinyl group of another chain [69].

$$\underset{\substack{-O \\ -O}}{\overset{R}{\underset{H}{\diagdown}}}Si\overset{R}{\underset{H}{\diagup}} + \underset{H_2C=C}{\overset{R}{\underset{H}{\diagup}}}Si\overset{O-}{\underset{O-}{\diagup}} \longrightarrow \underset{-O}{\overset{R}{\diagdown}}Si\overset{R}{\underset{C-C}{\diagup}}\underset{H_2 \ H_2}{\overset{R}{\diagup}}Si\overset{O-}{\underset{O-}{\diagup}}$$

Typically the crosslinking reaction is carried out by mixing two components: (a) a syrupy vinylsilicone polymer containing a trace of a platinum complex like those described below, and (b) a small amount of an Si-H functional compound such as [CH$_3$SiH-O]$_4$ [70]. Such mixtures are stable for several hours at room temperature, but react rapidly at 50-100°C to form a crosslinked network. This reaction may be carried out in a mold to produce a tough or hard molded object directly in a process known as reaction injection molding (RIM).

The vinyl silanes that provide the crosslinking sites are often made by adding an Si-H function to acetylene [71]. Long-chain alkyl substituents are introduced by adding an Si-H group to a C=C bond of a terminal olefin. For addition of silane Si-H bonds to unactivated olefins, the usual catalyst of choice is chloroplatinic acid, H$_2$PtCl$_6$·6 H$_2$O, often designated as Speier's catalyst. Although dozens of transition metal complexes catalyze the addition, the stable, easily available platinum compound is preferred [67]. It catalyzes Si-H addition in the presence of many kinds of functional groups. A strong steric influence is noted since the silicon almost always attaches to the less-crowded end of a C=C bond. Terminal olefins are hydrosilylated in preference to internal olefins. In fact, internal olefins often isomerize to form terminal products [72] as in

$$\text{3-Heptene} \rightleftharpoons \text{1-heptene} \longrightarrow \text{n-C}_7\text{H}_{15}\text{SiR}_3$$

If the silane is optically active, it retains its configuration. Asymmetric induction has been observed in hydrosilylation of olefins with catalysts that bear chiral ligands [73,74].

In a typical hydrosilylation [72], the olefin and the silane are mixed with a solution of Speier's catalyst (10^{-5} mole Pt/mole Si) in a polar solvent such as 2-propanol. After a brief induction period, a vigorous exothermic reaction occurs. The mixture is commonly heated to complete the reaction and the product is isolated by distillation. The induction period can be reduced or eliminated by using a preformed zero-valent platinum compound such as

$$Pt_2\left(\begin{array}{c} Me_2Si\!-\!O\!-\!SiMe_2 \\ | \qquad\qquad | \\ H_2C\!=\!CH \qquad HC\!=\!CH_2 \end{array}\right)_3$$

This compound, known as Karstedt's catalyst [70,71], produces extremely rapid hydrosilylation of alkenes and alkynes. Careful study of Speier's catalyst has shown that H$_2$PtCl$_6$ and other platinum chlorides are rapidly reduced to metallic platinum colloids by SiH compounds [75].

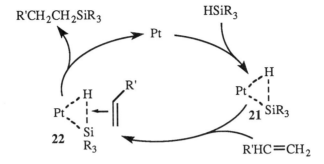

Figure 3.5 Proposed mechanism for catalytic addition of a silane Si-H bond to an olefin.

Recent work [76] suggests that an O_2-stabilized platinum colloid catalyzes the hydrosilylation of an olefin by a mechanism like that shown in Figure 3.5. A surface atom of the colloid, designated Pt in the figure, reacts with the Si-H bond of the silane to form an η_2 complex **21** or, perhaps an alternative product R_3Si-Pt-H, formed by formal oxidative addition of the Si-H bond. Insertion of the olefin into the coordinated Si-H may be a concerted step, as indicated by **22**, or may involve discrete insertion into an H-Pt bond to give a species such as $R'CH_2CH_2$-Pt-SiR_3. Either way, elimination of the $R'CH_2CH_2SiR_3$ product regenerates zero-valent Pt for another catalytic cycle. Most truly homogeneous hydrosilylation catalysts such as $RhCl(PPh_3)_3$ are believed to function via discrete oxidative addition, insertion, and reductive elimination mechanisms [68].

Another application of hydrosilylation is for the synthesis of siloxane precursors bearing polar substituents such as CN or CF_3. These materials are used to synthesize silicone gums and rubbers with good resistance to gasoline and other hydrocarbon solvents. The common reactions are:

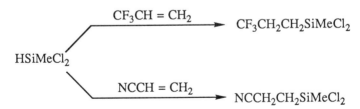

The resulting dichlorosilanes may be hydrolyzed alone or with Me_2SiCl_2 to give silicones. The reaction with trifluoropropene may be carried out with one of the platinum catalysts described above [77]. The reaction of acrylonitrile with the silane is catalyzed by a mixture of cuprous chloride, tetramethylethylenediamine, and a trialkylamine [78]. The chelating diamine probably solubilizes the copper(I) salt. When all three components are present, 75-80% yields of the β-cyanoethylsilane result.

3.6 HYDROCYANATION OF OLEFINS AND DIENES
(by R. J. McKinney)

The ease with which C≡N groups may be converted to other functional groups has made organic nitriles commercially important intermediates. For example, acrylonitrile, the largest volume organonitrile, is an important monomer both for plastics and synthetic fibers. Hydrogenation of nitriles to amines provides important intermediates both for polyurethanes (by way of isocyanates) and polyamides (nylons). Adiponitrile, after hydrogenation to the diamine and reaction with adipic acid, accounts for more than 4 billion pounds per year of nylon-6,6. Further examples include cyanohydrins and aminonitriles, important intermediates in the production of acrylates, chelating agents such as ethylenediaminetetraacetate (EDTA) and nitrilotriacetate (NTA), and a variety of water treatment products based on hydantoin derivatives. Many smaller volume commercial processes use nitrile intermediates for the production of pharmaceuticals, agrichemicals, cosmetics, and other consumer products.

The addition of HCN to an unsaturated substrate is not the only method of producing an organonitrile, but it is often the easiest and most economical. The addition of HCN to aldehydes and ketones is readily accomplished with simple base catalysis, as is the addition to activated olefins (Michael addition). However, the addition of HCN to unactivated olefins and the regioselective addition to dienes is best accomplished with a transition metal catalyst.

Safety in HCN Use

The lack of reported incidents despite the widespread commercial use of volatile and toxic HCN points to the respect and care with which it is treated within the industry. In a laboratory, HCN should be used only in small quantities in a well-ventilated fume hood. Vapors may be scrubbed or small samples (up to a few milliliters) may be disposed of by treatment with an equimolar mixture of NaOH and hypochlorite (NaOCl, bleach). Larger samples or vapor streams should be burned. The use of a "buddy system" and "first response" training is essential when using HCN [79].

Commercial Use

The Du Pont adiponitrile (ADN) process [80,81] involves nickel-catalyzed anti-Markovnikov hydrocyanation of butadiene according to eqs. 1 to 4. All the reactions are catalyzed by air- and moisture-sensitive triarylphosphite-nickel(0) complexes.

The first HCN addition (eq. 1) occurs at practical rates above 70°C under sufficient pressure to keep butadiene condensed in solution and produces the 1,4- and 1,2-addition products (3-pentenenitrile, **3PN**, and 2-methyl-3-butenenitrile, **2M3BN**) in a 2 to 1 ratio. Fortunately, thermodynamics favors **3PN** (about 9:1) and **2M3BN** may be isomerized to **3PN** (eq. 2) in the presence of the nickel catalyst. The isomerization of **2M3BN** goes through a relatively rare C-C cleavage reaction.

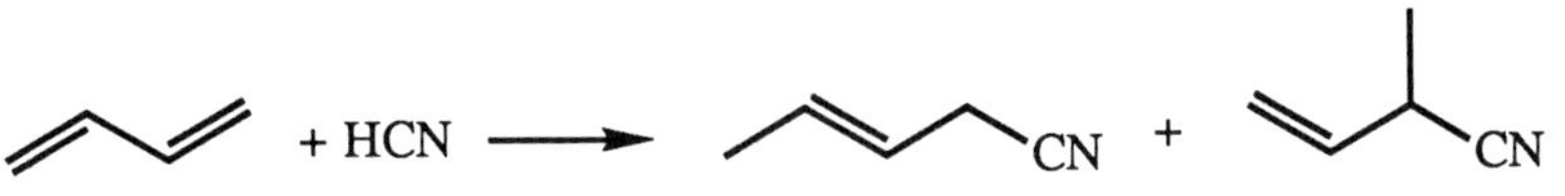

The selective anti-Markovnikov addition of the second HCN to provide **ADN** requires the concurrent isomerization of **3PN** to 4-pentenenitrile (**4PN**) (eq. 3), and HCN addition to **4PN** (eq. 4). A Lewis acid promoter is added to control selectivity and increase rate in these latter steps. Temperatures in the second addition are significantly lower and practical rates are achieved above 20°C at atmospheric pressure. 2-Methylglutaronitrile (**MGN**), ethylsuccinonitrile (**ESN**), and 2-pentenenitrile (**2PN**) are byproducts of this process.

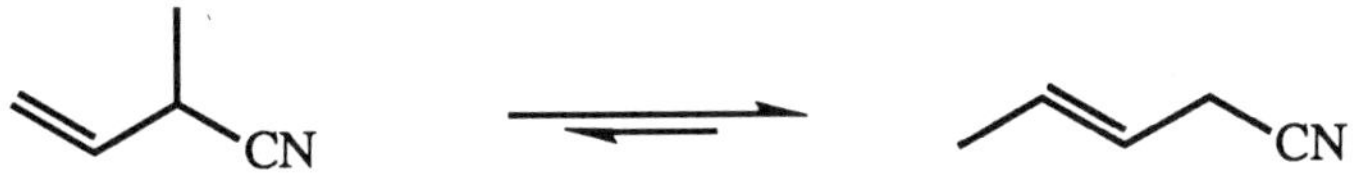

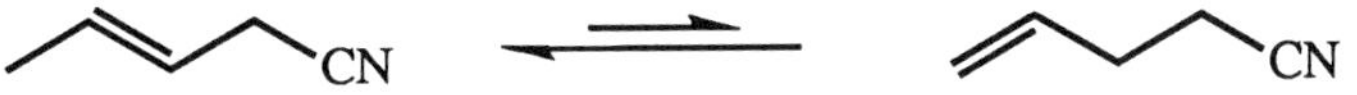

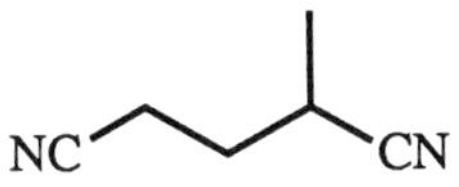

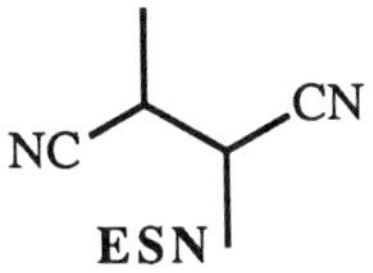

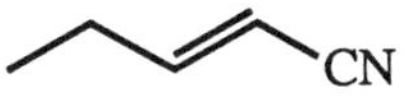

Similar catalyst systems permit selective HCN addition to many other unsaturated substrates. α-Olefins such as those obtained from Shell's Higher Olefin Process (SHOP) are readily hydrocyanated under conditions similar to that described for eq. 4. The products may be converted to primary amines, amides, and acids containing odd numbers of carbons. Addition of HCN to styrene and other vinylarenes gives high yields of 2-aryl-propionitriles [82], which are excellent intermediates for 2-arylpropionic acids, a class of nonsteroidal anti-inflammatory drugs, which include ibuprofen and Naproxen®. The hydrocyanation of dienes, trienes [83,84], and other polyenes leads to polyfunctional systems that have many potential applications including polymers and lubricants.

Mechanism

Though most chemical and mechanistic studies related to the 20-year-old Du Pont ADN process have been published only recently, many of the fundamental studies that advanced the understanding of organometallic catalysis occurred during the development of this process. Some of the identifiable concepts to arise from this work include the proposal that many catalytic processes occur through discrete 16- and 18-electron intermediates [85], the elucidation and quantification of important steric effects (ligand cone-angle) in ligand coordination and dissociation [86], and the use of "Tolman cycles" in describing catalytic sequences. Added to these are pioneering studies [87] on the mechanism of olefin isomerization in the presence of metal hydride species (hydride addition and elimination mechanism, Section 2.2).

Studies suggest that ligand dissociation from NiL_4 complexes precedes addition of HCN to generate hydrido nickel cyanide complexes [88,89]. Butadiene complexation and insertion into the metal-hydride bond affords π-allylnickel cyanide species, observable by NMR spectroscopy at ambient temperature, which may reductively eliminate **3PN** (or **2M3BN**) according to eq. 5. The reaction is reversible and provides the mechanism for the isomerization of **2M3BN** to **3PN** [90].

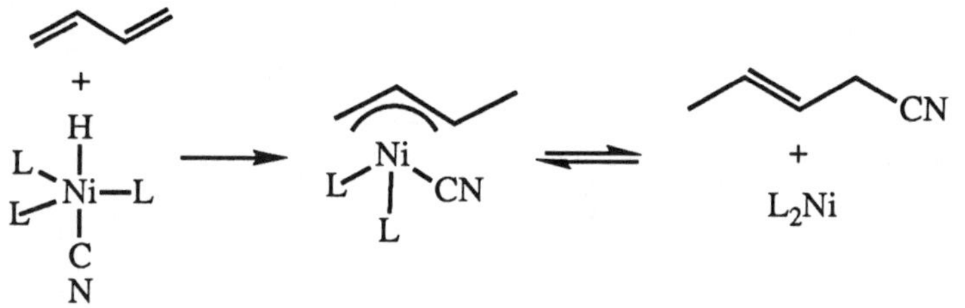

$$(5)$$

Consistent with this picture, the addition of DCN to cyclohexadiene has been shown to be stereospecifically cis [91].

The isomerization of **3PN** to **4PN** is catalyzed by a cationic nickel hydride generated by removal of cyanide by a Lewis acid promoter (designated A in eq. 6). A remarkable

$$HNi(CN \cdot A)L_3 \quad \rightleftharpoons \quad HNiL_3^+ \; CN \cdot A^- \qquad (6)$$

feature of this isomerization is the kinetic control that allows **4PN** to be made more than 70 times faster than the thermodynamically favored **2PN**! The ability of the nitrile group to coordinate to the catalyst and direct the insertion (Figure 3.6) is believed responsible [92].

The insertion of an olefin (**3PN**) into the Ni-H bond, followed by reductive elimination from the resulting alkylnickel cyanide complex, are the principle features of the second HCN addition in the ADN process. It is clear that the Lewis acid binds preferentially to the nickel-cyanide moiety and that the size of the Lewis acid has direct bearing on product selectivity. Bulkier Lewis acids favor the production of linear nitriles as shown in Figure 3.7 [93,94].

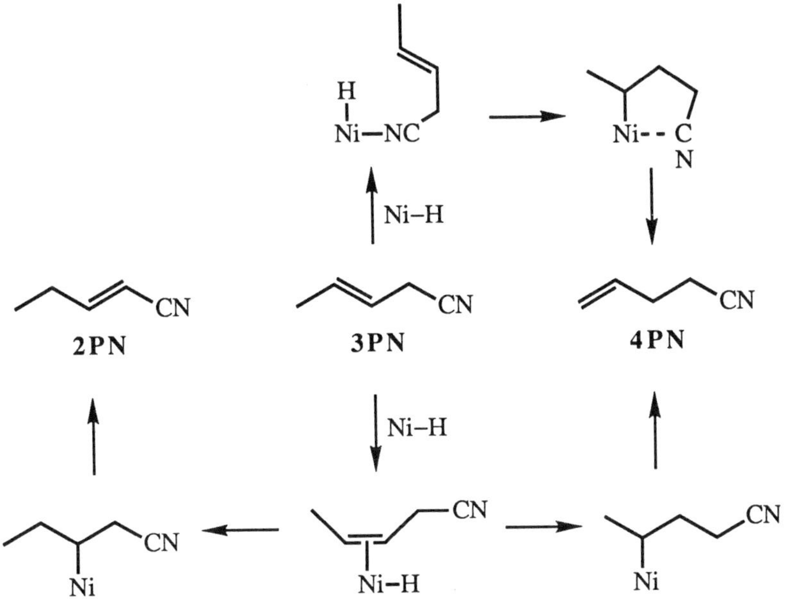

Figure 3.6 Isomerization of 3-pentenenitrile by a nickel hydride catalyst.

The exact nature of the complex from which the final reductive elimination occurs is subject to speculation. The involvement of a five-coordinate nickel complex appears likely based on the observation that reductive elimination of benzonitrile from $(Et_3P)_2Ni(CN)Ph$ is promoted by triethyl phosphite [95] and that reductive elimination of propionitrile from $[(o\text{-tolyl})_3P](C_2H_4)Ni(CN)Et$ is first order in phosphite concentration [96].

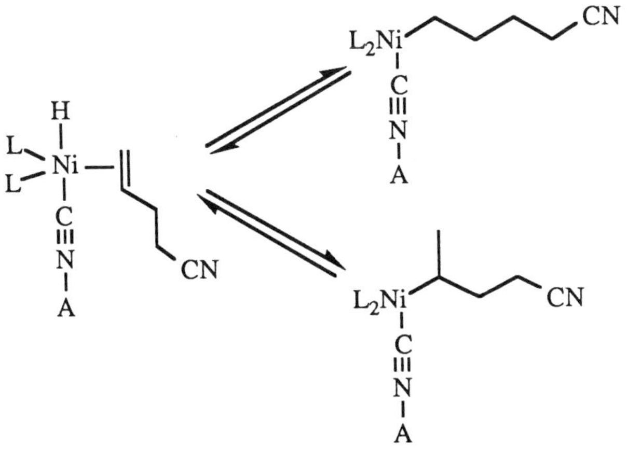

Figure 3.7 Stereochemistry of 4-PN insertion into an Ni-H bond (A = Lewis acid).

Other metal complexes are known to catalyze hydrocyanation of olefins, among the best being some palladium and cobalt complexes. Copper(I) compounds catalyze the addition of HCN to butadiene [97] and the displacement of allylic halide by cyanide ion [98].

Asymmetric Hydrocyanation

Catalytic asymmetric carbon-carbon forming reactions are potentially valuable. Three groups have reported asymmetric hydrocyanation of norbornene (or derivatives) with enantiomeric excesses up to 40%. These additions are catalyzed by nickel or palladium complexes containing chiral chelating diphosphine or diphosphinite ligands [99-101]. High optical yields have been attained in the hydrocyanation of a vinylnaphthalene to a Naproxen® precursor [102] as shown below.

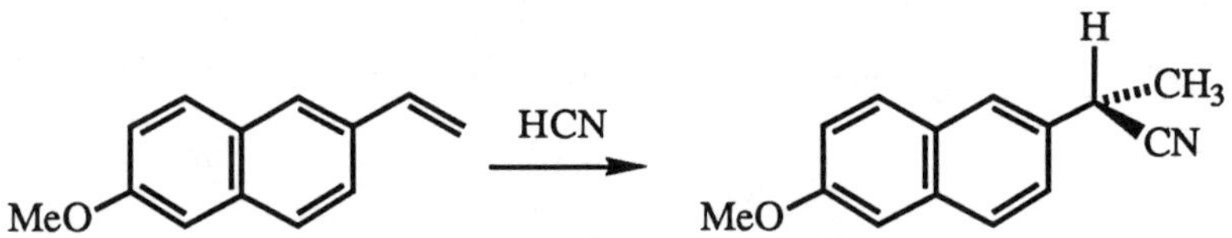

The versatility of HCN as a building block for complex organic molecules coupled with the growing need for optically pure pharmaceuticals and agrichemicals makes this a rewarding area for research.

GENERAL REFERENCES

Hydrogenation

B. R. James, *Homogeneous Hydrogenation*, Wiley, New York, 1973; also "Hydrogenation Reactions Catalyzed by Transition Metal Complexes," *Adv. Catal.*, **17**, 319-405 (1979); also in G. Wilkinson, Ed., *Comprehensive Organometallic Chemistry*, Vol. 8, Pergamon Press, New York, 1982.

F. J. McQuillen, *Homogeneous Hydrogenation in Organic Chemistry*, D. Reidel, Dordrecht, 1976.

P. A. Chaloner, *Handbook of Coordination Catalysis in Organic Chemistry*, Butterworths, London, 1986.

P. Rylander, *Catalytic Hydrogenation in Organic Syntheses*, Academic Press, New York, 1979; *Hydrogenation Methods*, Academic Press, New York, 1985.

Hydrosilation

J. L. Speier, "Homogeneous Catalysis of Hydrosilation by Transition Metals," in *Adv. Organomet. Chem.*, **17**, 407-47 (1979).

I. Ojima, "The Hydrosilylation Reaction," in S. Patai and Z. Rappoport, Eds., *The Chemistry of Organic Silicon Compounds*, Wiley, New York, 1989, pp. 1479-1526.

Hydrocyanation

E. S. Brown, "Addition of Hydrogen Cyanide to Olefins," in I. Wender and P. Pino, Eds., *Organic Synthesis via Metal Carbonyls*, Vol. 2, Wiley-Interscience, New York, 1977, p. 655; also *Aspects of Homogeneous Catalysis*, **2**, 57-78 (1974).

C. A. Tolman, R. J. McKinney, W. C. Seidel, J. D. Druliner, and W. R. Stevens, "Homogeneous Nickel-Catalyzed Olefin Hydrocyanation," in *Adv. Catal.*, **33**, 1-45 (1985).

SPECIFIC REFERENCES

1. J. A. Osborn, F. H. Jardine, J. F. Young, and G. Wilkinson, *J. Chem. Soc. A*, 1711-32 (1966).
2. A. Yanagawa, Y. Suzuki, I. Anazawa, Y. Takagi, and S. Yada, *J. Mol. Catal.*, **29**, 41-54 (1985).
3. R. E. Ireland and P. Bey, *Org. Syn.*, Coll. Vol. **6**, 459-60 (1988).
4. H. R. Taylor, M. Pacque, B. Muno, and B. M. Greene, *Science*, **250**, 116-8 (1990).
5. R. Baker and C. J. Swain, *Chem. Br.*, 692-6 (1989).
6. S. C. Stinson, *Chem. Eng. News*, 51 (5 Oct. 1987).
7. J. C. Chabala and M. H. Fisher, U.S. Patent 4,199,569 (1980).
8. J. C. Chabala, H. Mrozik, R. L. Tolman, P. Eskola, A. Lusi, L. H. Peterson, M. F. Woods, M. H. Fisher, W. C. Campbell, J. R. Egerton, and D. A. Ostlind, *J. Med. Chem.*, **23**, 1134-6 (1980).
9. W. Voelter and C. Djerassi, *Chem. Ber.*, **101**, 58-68 (1968).
10. F. N. Jones, U.S. Patent 3,459,785 (1969).
11. R. R. Schrock and J. A. Osborn, *J. Am. Chem. Soc.*, **98**, 2134-43, 2143-7 (1976).
12. R. R. Schrock and J. A. Osborn, *J. Am. Chem. Soc.*, **98**, 4450-5 (1976).
13. R. D. Cramer, E. L. Jenner, R. V. Lindsey, and U. G. Stolberg, *J. Am. Chem. Soc.*, **85**, 1691-2 (1963); I. Yasumori and K. Hirabayashi, *Trans. Faraday Soc.*, **67**, 3283-96 (1971); W. Strohmeier and L. Weigelt, *Z. Naturforsch.*, **31b**, 387-9 (1976).
14. J. C. Bailar, *Plat. Met. Rev.* **15**, 2 (1971); *J. Am. Oil Chem. Soc.*, **47**, 475 (1970).
15. D. H. Goldsworthy, F. R. Hartley, and S. G. Murray, *J. Mol. Catal.*, **19**, 257-68 (1983).
16. J. Kwiatek, I. L. Mador, and J. K. Seyler, "Reactions of Coordinated Ligands and Homogeneous Catalysis," in *Advances in Chemistry Series*, Vol. 37, American Chemical Society, Washington, DC, **37**, pp. 201-15, 1963; J. Kwiatek, *Catal. Rev.*, **1**, 37-71 (1967).
17. M. S. Spencer and D. A. Dowden, U.S. Patent 3,009,969 (1961).

18. M. F. Sloan, A. S. Matlack, and D. S. Breslow, *J. Am. Chem. Soc.*, **85**, 4014-8 (1963) .

19. F. K. Shmidt, S. M. Krasnopolskaya, V. G. Lipovich, and B. A. Bazhenov, *Kinet. Katal.*, **15**, 86-91 (1974).

20. A. F. Halasa, U. S. Patent 3,872,072 (1975).

21. L. R. Gilliom, *Macromolecules*, **22**, 662-5 (1989).

22. C. A. Tolman, P. Z. Meakin, D. L. Lindner, and J. P. Jesson, *J. Am. Chem. Soc.*, **96**, 2762-74 (1974).

23. J. Halpern, *Inorg. Chim. Acta*, **50**, 11-19 (1981).

24. Y. Ohtani, M. Fujimoto, and A. Yamagishi, *Bull. Chem. Soc. Jpn.*, **50**, 1453-9 (1977); *ibid.*, **51**, 2562-5 (1978).

25. A. S. C. Chan and J. Halpern, *J. Am. Chem. Soc.*, **102**, 838-40 (1980).

26. G . W. Parshall, *J. Am. Chem. Soc.*, **94**, 8716-9 (1972).

27. J. Halpern and M. Pribanic, *Inorg. Chem.*, **9**, 2616-8 (1970).

28. J. Halpern and L. Y. Wong, *J. Am. Chem. Soc.*, **90**, 6665-9 (1968).

29. *Bailey's Industrial Oil and Fat Products*, D. Swern, Ed., Wiley-Interscience, New York, 1964.

30. K. Katsuragawa and K. Yoshimitsu, *Jap. Kokai*, 7, 408,481 (1974); *Chem. Abstr.*, **81**, 17216 (1974) .

31. A. Misono and I. Ogata, *Bull. Chem. Soc. Jpn.*, **40**, 2718-9 (1967); U.S. Patent 3,715,405 (1973).

32. L. W. Gosser, U.S. Patent 3,499,050 (1970).

33. J. Tsuji and H. Suzuki, *Chem. Lett.*, 1083-4 (1977).

34. D. R. Fahey, *J. Org. Chem.*, **38**, 80-7, 3343-8 (1973); "Selective Hydrogenations," in *Catalysis in Organic Syntheses*, Academic Press, New York, 1976, pp. 287-303.

35. A. D. Shebaldova, V. I . Bystrenina, V. N. Kravtsova, and M. L. Khidekel, *Izv. Akad. Nauk SSSR, Ser. Khim.*, 2101-5 (1975).

36. D. R. Fahey, U.S. Patent 3,717,585 (1973).

37. L. W. Gosser, U.S. Patent 3,673,270 (1972).

38. W. S. Knowles, M. J. Sabacky, and B. D. Vineyard, *Ann. N. Y. Acad. Sci.*, **295**, 274-82 (1977); W. S. Knowles, *Acc. Chem. Res.*, **16**, 106-12 (1983); *J. Chem. Ed.*, **63**, 222-5 (1986).

39. M. J. Burk, J. E. Feaster, and R. L. Harlow, *Tetrahedron: Asymmetry*, **2**, 569-92 (1991); M. J. Burk, *J. Am. Chem. Soc.*, **113**, 8518-9 (1991).

40. W. S. Knowles, M. J. Sabacky, and B. D. Vineyard, U.S. Patent 4,005,127 (1977).

41. B. D. Vineyard, W. S . Knowles, M. J. Sabacky, G. L. Bachman, and D. J. Weinkauff, *J. Am. Chem. Soc.*, **99**, 5946-52 (1977).

42. B. Bosnich, Ed., *Asymmetric Catalyses*, Martinius Nijhoff Publishers, Dordrecht, 1986.

43. J. D. Morrison, Ed., *Asymmetric Catalysis*, Vol. 5, Academic Press, New York, 1985.

44. H. Brunner, *Top. Stereochem.*, **18**, 129-247 (1988).

45. G. L. Bachman, M. L. Oftedahl, and B. D. Vineyard, U.S. Patent 3,933,781 (1976).

46. G. L. Bachman and B. D. Vineyard, U.S. Patent 4,173,562 (1979).

47. J. W. Scott, *Ind. Chem. News*, **7**(3), 32-5 (1986).

48. M. Fiorini, M. Riocci, and G. M. Giongo, European Patent 77,099 (1983); *Chem Abstr.*, **99**, 122916 (1983).

49. M. Fiorini and G. M. Giongo, *J. Mol. Catal.*, **5**, 303-10 (1979); *ibid.*, **7**, 411-3 (1980).

50. E. Goto, M. Ishihara, S. Sakurai, H. Enei, and K. Takinami, U.S. Patent 4,403,033 (1983).

51. R. Noyori and H. Takaya, *Acc. Chem. Res.*, **23**, 345-50 (1990).

52. H. Takaya, T. Ohta, N. Sayo, H. Kumobayashi, S. Akutagawa, S. Inoue, I. Kasahara, and R. Noyori, *J. Am. Chem. Soc.*, **109**, 1596-7, 4129 (1987).

53. K. Mashima Y. Matsumura, K. Kusano, H. Kumobayashi, N. Sayo, Y. Hori, T. Ishizaki, S. Akutagawa, and H. Takaya, *J. Chem. Soc., Chem. Commun.*, 609-10 (1991).

54. K. Mashima, K. Kusano, T. Ohta, R. Noyori, and H. Takaya, *J. Chem. Soc., Chem. Commun.*, 1208-10 (1989); *Pure Appl. Chem.*, **62**, 1135-8 (1990).

55. J. M. Brown, P. L. Evans, and A. P. James, *Org. Synth.*, **68**, 64-75 (1989).

56. R. Selke, H. Burneleit, and M. Capka, East German Patents 280,527-280,529 (1990); *Chem. Abstr.*, **114**, 82546-8 (1991).

57. U. Nagel and E. Kinzel, *J. Chem. Soc. Chem. Commun.*, 1098-9 (1986).

58. Y. Amrani, L. Lecomte, D. Sinou, J. Bakos, I. Toth, and B. Heil, *Organometallics*, **8**, 542-7 (1989).

59. R. Waymouth and P. Pino, *J. Am. Chem. Soc.*, **112**, 4911-4 (1990).

60. Monsanto, World Patent WO9,015,790A (1991).

61. A. S. C. Chan, National Meeting, American Chemical Society, Boston, MA, 23 April 1990.

62. T. Ohta, H. Takaya, M. Kitamura, K. Nagai, and R. Noyori, *J. Org. Chem.*, **52**, 3174-6 (1987).

63. J. Halpern, *Science*, **217**, 401-7 (1982); "Asymmetric Catalytic Hydrogenation," in J. D. Morrison, Ed., *Asymmetric Synthesis*, Vol. 5, Academic Press, New York, 1985, pp. 41-70.

64. C. R. Landis and J. Halpern, *J. Am. Chem. Soc.*, **109**, 1746-54 (1987).

65. B. McCulloch, J. Halpern, M. R. Thompson, and C. R. Landis, *Organometallics*, **9**, 1392-5 (1990).

66. M. T. Ashby and J. Halpern, *J. Am. Chem. Soc.*, **113**, 589-94 (1991).

67. E. Lukevics, Z. V. Belyakova, M. G. Pomerantseva, and M. G. Voronkov, *Organomet. Chem. Rev.*, **5**, 1-179 (1977).

68. J. F. Harrod and A. J. Chalk, "Hydrosilation Catalyzed by Group VIII Complexes," in I. Wender and P. Pino, Eds., *Organic Syntheses via Metal Carbonyls*, Vol. 2, Wiley-Interscience, New York, 1977, p. 673-704.

69. B. Hardman and A. Torkelson, "Silicones," in *Encyclopedia of Polymer Science and Engineering*, Vol. 15, Wiley, New York, 1989, pp. 204-308.

70. B. D. Karstedt, U.S. Patent 3,775,452 (1973).

71. L. N. Lewis, K. G. Sly, G. L. Bryant, and P. E. Donahue, *Organometallics*, **10**, 3750-9 (1991).

72. J. C. Saam and J. L. Speier, *J. Am. Chem. Soc.*, **80**, 4104-6 (1958).

73. I. Ojima, K. Yamamoto, and M. Kumada, "Asymmetric Hydrosilylation by Means of Homogeneous Catalysts with Chiral Ligands," in *Aspects of Homogeneous Catalysis*, Vol. 3, D. Reidel, Dordrecht, 1977, pp. 186-227.

74. H. Brunner, *Angew. Chem. Int. Ed.*, **22**, 897-907 (1983).

75. L. N. Lewis and N. Lewis, *J. Am. Chem. Soc.*, **108**, 7228-31 (1986); *Chem. Mater.*, **1**, 106-14 (1989).

76. L. N. Lewis, *J. Am. Chem. Soc.*, **112**, 5998-6004 (1990).

77. S. Smith, "Fluoroelastomers," in R. E. Banks, Ed., *Preparation, Properties and Industrial Applications of Organofluorine Compounds*, Ellis Horwood Ltd., Chichester, 1982, pp. 271-7.

78. B. A. Bluestein, *J. Am. Chem. Soc.*, **83**, 1000-1 (1961).

79. For additional details on using HCN safely see *Prudent Practices for Handling Hazardous Chemicals in Laboratories*, National Academy Press, Washington, DC, 1981, pp. 45-47. Safety literature is also delivered with commercial samples from Fumico, Inc.

80. W. C. Drinkard and R. V. Lindsey, U.S. Patent 3,496,215 (1970).

81. C. A. Tolman, R. J. McKinney, W. C. Seidel, J. D. Druliner, and W. R. Stevens, *Adv. Catal.*, **33**, 1-46 (1985).

82. W. A. Nugent and R. J. McKinney, *J. Org. Chem.*, **50**, 5370-2 (1985).

83. W. Keim, A. Behr, H.-O. Luhr, and J. Weisser, *J. Catal.*, **78**, 209-16 (1982).

84. E. M. Campi, P. Elmes, W. R. Jackson, C. G. Lovel, and M. K. S. Probert, *Aust. J. Chem.*, **40**, 1053-61 (1987).

85. C. A. Tolman, *Chem. Soc. Rev.*, **1**, 337-53 (1972).

86. C. A. Tolman, *Chem. Rev.*, **77**, 313-48 (1977).

87. C. A. Tolman, *J. Am. Chem. Soc.*, **94**, 2994-9 (1972).

88. J. D. Druliner, A. D. English, J. P. Jesson, P. Meakin, and C. A. Tolman, *J. Am. Chem. Soc.*, **98**, 2156-60 (1976).

89. C. A. Tolman, W. C. Seidel, J. D. Druliner, and P. J. Domaille, *Organometallics*, 3, 33-8 (1984).

90. J. D. Druliner, *Organometallics*, **3**, 205-8 (1984).

91. J.-E. Backvall and O. S. Andell, *Organometallics*, **5**, 2350-5 (1986).

92. R. J. McKinney, *Organometallics*, **4**, 1142-3 (1985).

93. B. W. Taylor and H. E. Swift, *J. Catal.*, **26**, 254-60 (1972).

94. R. J. McKinney and W. A. Nugent, *Organometallics*, **8**, 2871-5 (1989).

95. G. Favero, M. Gaddi, A. Morvillo, and A. Turco, *J. Organomet. Chem.*, **149**, 395-400 (1978).

96. R. J. McKinney and D. C. Roe, *J. Am. Chem. Soc.*, **108**, 5167-73 (1986).

97. E. Puentes, A. F. Noels, R. Warin, A. J. Hubert, Ph. Teyssie, and D. Y. Waddan, *J. Mol. Catal.*, **31**, 183-90 (1985).

98. J. V. Supniewski and P. L. Saltzberg, *Org. Synth.*, *Coll. Vol. 1*, 46-7 (1932).

99. P. S. Elmes and W. R. Jackson, *Aust. J. Chem.*, **35**, 2041-51 (1982).

100. M. Hodgson, D. Parker, R. J. Taylor, and G. Ferguson, *Organometallics*, **7**, 1761-6 (1988).

101. M. J. Baker and P. G. Pringle, *J. Chem. Soc., Chem. Commun.*, 1292-3 (1991).

102. T. V. RajanBabu and A. Casalnuovo, *J. Am. Chem. Soc.*, **113**, 000 (1992).

4 | POLYMERIZATION AND OLIGOMERIZATION OF OLEFINS AND DIENES

Organometallic catalysts, both soluble and insoluble, find wide practical application in C-C bond-forming reactions of olefins and dienes. The largest of these are the polymerizations and copolymerizations of ethylene, propylene, butadiene, isoprene, and higher α-olefins. Polymerizations of these hydrocarbons with catalysts based on transition metal complexes yield ordered polymers with physical properties different from those of free-radical polymers. Similarly, oligomerization of olefins and dienes with soluble metal catalysts is used extensively to produce dimers, trimers, and other "low polymers." The mechanisms of these reactions are based on the same elementary steps as those of Chapter 3. The major difference is that polymerization and oligomerization involve olefin insertion into a M-C bond in addition to the M-H insertion involved in hydrogenation.

4.1 α-OLEFIN POLYMERIZATION

The major applications of organometallic catalysts in the polymerization of olefins and dienes are listed in Table 4.1. Both homogeneous and heterogeneous catalysts are used commercially, but the solid catalysts are increasingly preferred because they have technical advantages in many processes. Even the nominally homogeneous catalysts are probably not soluble under reaction conditions. For example, catalytic solutions of $VOCl_3$ and i-Bu_2AlCl are clear to the naked eye, but light-scattering experiments suggest the presence of aggregates.

The largest volume plastic in the United States is polyethylene. Three major varieties are produced. High-density polyethylene (HDPE) is a linear polymer with a density approaching 0.97 and a melting point of about 136°C. It is made by coordination polymerization, as discussed below. Low-density polyethylene (LDPE), which has a density near 0.92 and a wide melting range, is a highly branched polymer prepared by free-radical polymerization of ethylene at high

Table 4.1 Polyolefin Production with Coordination Catalysts

Polymer	Major Catalysts	1978 U.S. Production[a]	1988 U.S. Production[a]
High-density polyethylene	$TiCl_x/AlR_3$, Cr/silica	1900	3800
Linear low-density polyethylene	$TiCl_x/AlR_3$, Cr/silica	<500	2000
Polypropylene	$TiCl_3/AlR_2Cl$, $MgCl_2/TiCl_4$	1600	3600
Ethylene/propylene/ diene rubbers	$VOCl_3/AlR_2Cl$, $TiCl_4/MgCl_2/AlR_2Cl$, $ZrCp_2Cl_2/MAO(1990)$	180	270
cis-1,4-Polybutadiene	TiI_4/AlR_3, $Co(O_2CR)_2/Al_2R_3Cl_3$, $Ni(O_2CR)_2/AlR_3/BF_3$	350	360

[a] Thousand tons.

pressure (approximately 2000 atmospheres). Linear low-density polyethylene (LLDPE) refers to a class of polyethylenes intermediate between high- and low-density materials prepared for specific applications by modification of the conditions used for manufacture of the high-density product.

Both heterogeneous and homogeneous catalysts are discussed here because they are closely related mechanistically. The treatment is necessarily brief, but recent books and reviews which cover most aspects of olefin polymerization are listed at the end of the chapter. The references cited in these reviews represent many hundreds of research-years in the past quarter century. Curiously, despite this massive effort, there remain questions about the mechanism of coordination polymerization. Fundamental facts such as the oxidation state and structure of the catalytic metal site and the nature of the insertion mechanism are not well established in many cases. The mode of action of the many additives to these reactions are often explained only in empirical terms.

Polyethylene

Two families of ethylene polymerization catalysts were developed in the early 1950s. Ziegler catalysts [1] are prepared by reaction of an alkylaluminum compound with $TiCl_4$ or $TiCl_3$ to give compositions that sometimes appear to be soluble in hydrocarbon solvents. The Phillips catalysts [2,3] are clearly insoluble materials prepared by deposition of chromium oxides on silica. Despite early acceptance of the Ziegler systems, chromium-based catalysts lead the production of polyethylene in the United States. The chromium systems now include chromacene-on-silica catalysts developed by Union Carbide [4,5].

Colloidal Ziegler catalysts for solution-based polymerizations (like those of Dow and Du Pont Canada) are prepared by reaction of $TiCl_4$ with trialkylaluminum compounds in cyclohexane or heptane. The titanium is alkylated and reduced to Ti(III). The catalyst can also include vanadium compounds such as $VOCl_3$. Other Ziegler catalysts which are clearly insoluble are prepared from crystalline $TiCl_3$ or $TiCl_3 \cdot 1/3 AlCl_3$. In laboratory experiments, these materials polymerize ethylene vigorously at room temperature and atmospheric pressure [6]. In commercial practice, the catalyst slurry is fed to a reactor along with ethylene at pressures that vary from 10 to 160 atmospheres. Rapid polymerization occurs at 130-270°C to give a viscous solution of a highly linear polymer with a relatively narrow molecular weight range. The combination of short reaction times and small, versatile reactors, and the desirable polymer properties, is the major advantage of this process and compensates to some extent for the need to remove the corrosive chloride-containing catalyst from the viscous polymer solution. New generations of these catalysts using siloxane-modified or other aluminum alkyl activators have achieved activities sufficiently high that the catalyst residues can be left in the polymer [7]. Another very significant advance has been the development of a new generation of $MgCl_2$-supported catalysts for slurry polymerizations at lower temperatures where the polyethylene is precipitated as it is produced. Productivities of these catalysts are very high, again allowing the catalyst residues to be left in the polymer. In addition, important factors such as molecular weight, molecular weight distribution, and long- and short-chain branching can be controlled by careful catalyst design.

The fundamental processes for chain growth and termination for the titanium-based catalysts appear to be similar to those sketched below for chromium-based catalysts. The titanium catalysts are discussed in more detail in connection with their use in polypropylene production where they enjoy more widespread acceptance.

Heterogeneous chromium-containing catalysts give polyethylene with properties somewhat different from those obtained with Ziegler systems. The heterogeneous catalysts are generally noncorrosive and are left in the product, affording a substantial process advantage. The catalysts are used in both lower-temperature slurry reactions and in "gas-phase" processes in which polyethylene is grown directly on the surface of the catalyst in the absence of solvent. Polymerization pressures are fairly low (10-30 atmospheres).

The chromium-on-silica catalysts are obtained in two very different ways. A silica gel may be impregnated with aqueous chromate solution or with a hydrocarbon solution of $(Ph_3SiO)_2CrO_2$ to give a dispersion of Cr(VI) oxide sites on the surface of the support. These catalyst sites appear to be inactive before contact with a reducing agent such as an organoaluminum compound, CO, H_2, or ethylene. Reduction yields Cr sites with an oxidation state of +2 to +3 (a subject of discussion) [8-10]. Alternatively, in a Union Carbide process, reaction of silica with $Cr(C_5H_5)_2$ gives low valent sites in which a cyclopentadienyl ligand has been liberated by protonation by the surface OH groups:

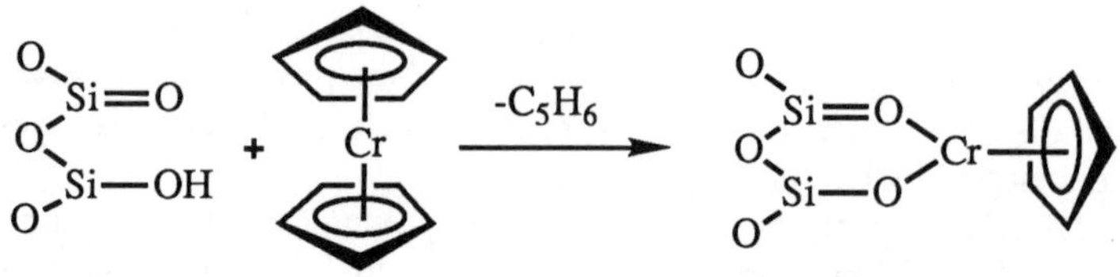

It seems likely that chromium-on-silica catalysts prepared by both methods contain Cr-H sites which react with ethylene by a repetitive insertion process to give a high molecular weight polymer:

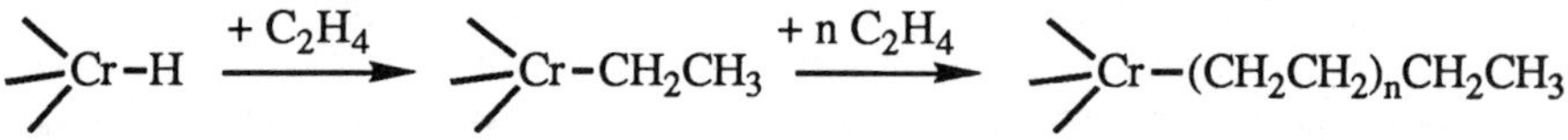

Polymer grown in this manner is chemically bonded to the catalyst surface through a labile Cr-C bond. The polymer molecules are released from the surface by a molecular weight control agent such as hydrogen or by thermal cleavage. In either situation, a metal hydride site is regenerated and can initiate growth of a new polymer chain.

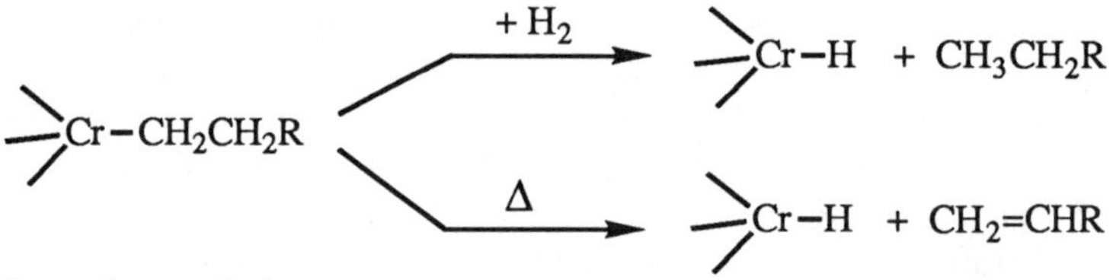

R = polymer chain

The elimination of β-hydrogen in the thermal process generates a high molecular weight olefin. This olefin can copolymerize with ethylene to form a branch in a polyethylene chain. If olefin formation by β-elimination takes place and some of these high molecular weight α-olefins are incorporated into the growing polymer chain, a highly branched, low-density polyethylene can be produced. This subsequent incorporation is generally rare because the concentration of olefinic chain ends is small in comparison with the monomer concentration and incorporation of any α-olefin is usually much slower than ethylene incorporation, but even if branching occurs only rarely, it can have an impact on end-use applications.

In the last decade, there has been a minor revolution in the preparation of polyethylene. Advances in catalysts and process conditions have allowed the customization of polymers to specific applications ranging from blow-molded bottles, injection-molded parts, and blown films. The vast majority of polyethylene is produced using Phillips' loop reactor technology [11,12] or Union Carbide's UNIPOL® fluidized-bed polymerization technology [13,14]. Carbide's process will be discussed in more detail to illustrate the flexibility which can be achieved. BP's fluidized-bed process provides growing competition to Carbide.

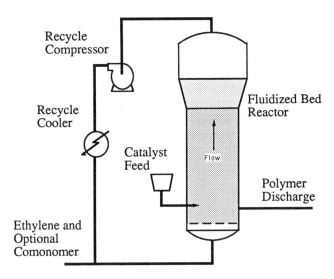

Recycle
Compressor

Recycle
Cooler

Fluidized Bed
Reactor

Catalyst
Feed

Flow

Polymer
Discharge

Ethylene and
Optional
Comonomer

Figure 4.1 Schematic representation of Union Carbide's Unipol process.

The heart of the technology is a large, fluidized-bed reactor, represented schematically in Figure 4.1, in which supported catalysts are reacted with gas-phase ethylene. The process eliminates solvents and the associated recycle streams. Conversion per pass is low, allowing the considerable heat of polymerization to be removed by the fluidizing ethylene. The catalyst is introduced in the form of small spheres of a relatively uniform size. Ideally, each catalyst particle grows to a single polymer particle. The polymer particles are then removed from the reactor through a rotary valve. Degassing and introduction of the stabilizer additives can be accomplished without pelletizing, allowing considerable savings in energy for the process.

Higher α-olefin monomers can be copolymerized with the ethylene to introduce short-chain branches off the main backbone of the polymer chain, thereby yielding linear low-density polyethylene. This LLDPE process is economically advantageous because it operates at a much lower pressure than the radical-initiated LDPE processes. LLDPE can often be prepared in HDPE plants, allowing the capacity to be swung from HDPE to LLDPE products as the market dictates. The comonomers range from 1-butene to 1-octene. Incorporation of comonomer was originally limited to a few percent, but Union Carbide has been pushing densities downward through "low-density" polyethylene products (0.925-0.900) into the range (0.900-0.885) considered to be "ultralow." The resulting FLEXOMER® polymers have elastomeric properties between LLDPE and those of ethylene propylene rubbers discussed below [15]. As densities are lowered, the tackiness of the polymers increases. Maintaining fluidization becomes more difficult so temperatures and polymerization rates are lowered. Under some conditions, a liquid coating may form on the polymer particles and reactor walls; this condition is generally avoided in fluidization technology but seems to be essential to the current process. Isolation also becomes an increasingly difficult engineering problem [16] because the polymer particles agglomerate if allowed to remain in contact with each other. As the limits of the technology are being pushed, it is becoming increasingly difficult to draw a clear distinction between low-density polyethylenes, the polypropylene copolymers, and the ethylene-propylene-diene rubbers discussed below.

As mentioned above, incorporation of 1-butene and higher α-olefins into the growing polymer chain can have profound impact on the properties of the polymers [17]. For instance, despite the high inherent strength of high-density polyethylene, it creeps under load, resulting in rapid failure under a slight but constant stress. This creep is a result of polymer chains reptating (snaking) past each other in the polymer crystal at temperatures well below the melting point. This reptation can be inhibited by incorporating a low percent of C_6-side chains on the main backbone by copolymerizing ethylene with octene.

As copolymerization has became more sophisticated, greater control over polymer architecture has developed. For some applications, a homogeneous copolymer with very even and random incorporation of comonomer is desired [18]. Alternatively, comonomer can be incorporated into the high or low molecular weight fraction of the polymer, can be randomly incorporated, or can be incorporated into blocks. Diluents can be used to increase the randomness of incorporation. For some applications, specifically tailored bimodal distributions are desirable [19]. With the recognition that such control was possible and desirable, catalysts and polymerization conditions have been designed to meet each of these needs.

A new generation of catalysts, sometimes referred to as "single-site" catalysts, is based upon modifications of $(Cp)_2ZrCl_2$ combined with methyl aluminoxanes [20,21] which are complex mixtures containing the elements represented by

The methyl aluminoxanes, that are the products of controlled partial hydrolysis of $AlMe_3$ with one equivalent of water, are often represented as simple linear or cyclic species with a single methyl group on each aluminum. In fact, they are complex mixtures of materials having many $Al(-O-)_3$ and $Al(-O-)_4$ groups, as well as $AlMe_2$ species; free $AlMe_3$ is required to keep these "homogeneous species" in solution [22] and is probably coordinated to the bridging oxygen atoms of the aluminoxane.

The simple complex $(Cp)_2ZrCl_2$ is an extremely active ethylene polymerization catalyst. Because there is a single type of active species in solution, a high degree of control over polymer properties is possible. Narrow molecular weight distributions are possible, or the molecular weight distribution can be controlled by ligand and process variations. Greatly improved control of comonomer incorporation is also possible; for instance, the comonomer can be incorporated specifically into one region of the molecular weight distribution. Exxon, Dow and others are bringing these custom-tailored polymers to the higher end of the polyolefin market [23].

Polypropylene will be covered in the next section of this chapter, but single-site catalysts give interesting results when applied to propylene polymerization. The complex $(Cp)_2ZrCl_2$ is not chiral and as a result, gives polypropylene that is purely atactic. The first generation polypropylene processes described below gave more than enough atactic polymer to meet commercial needs, but with new catalysts, this byproduct has been eliminated from many processes and atactic polypropylene is now made on purpose. When substituents are added to the cyclopentadienyl rings, and ring rotation is stopped by bridging the two rings, the resulting "single-site constrained-geometry" [23] catalysts are chiral:

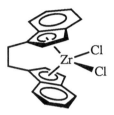

Generally, the surface of a heterogeneous catalyst is required to generate a site with suitable geometric constraints, but variations of these catalysts can produce isotactic polypropylene [24-26]. It is even possible to resolve the optical isomers [27]; propylene and higher oligomers resulting from hydrooligomerization using the resolved catalyst are optically active [28-30]. Syndiotactic polymers have been prepared using another version of this catalyst [31]. Remarkably it has been demonstrated that the methylaluminoxane is not required as a cocatalyst if a base-free or weakly ligated cationic complex can be generated [32,33]; compounds such as the zwitterionic $Cp^*_2Zr(C_6H_4BPh_3)$ polymerize ethylene or propylene in the absence of aluminum alkyl activators.

Polypropylene

Commercial polypropylene is a very regular polymer with properties that vary with the amount of crystallization that occurs during processing. Densities range from about 0.85 for amorphous material to 0.93 for crystalline isotactic polymer. Other properties change correspondingly. The basic polymer chain is ordered with respect to placement of the methyl groups at the chiral centers in each polymer unit. In the common isotactic polymer, each polymer chain will have all of the methyl groups on a single side of the extended chain:

$$R \diagup \overset{CH_2}{\diagup} \diagup \overset{CH_2}{\diagup} \diagup \overset{CH_2}{\diagup} \diagup \overset{CH_2}{\diagdown} R'$$
$$CH_3 \; H \quad CH_3 \; H \quad CH_3 \; H$$

This selectivity for ordering all the methyls on one side of the chain rather than randomly (atactic) is related to the symmetry of the catalyst site.

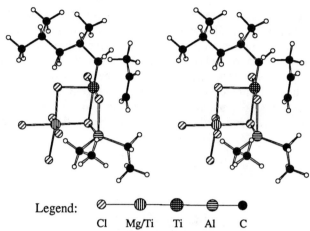

Legend:

| Cl | Mg/Ti | Ti | Al | C |

Figure 4.2   Stereo view of the active site of a Ziegler-Natta catalyst showing incoming propylene; the growing isotactic polypropylene chain; part of the support; and the residue of the AlEt₃ promoter.

Older commercial polypropylene catalysts are modifications of the $TiCl_3/AlR_3$ systems developed by Ziegler and Natta. In one catalyst preparation which is used in laboratory polymerizations [34], a slurry of violet crystalline $3TiCl_3 \cdot AlCl_3$ in heptane is treated with diethylaluminum chloride in the presence of propylene. Rapid polymerization occurs at room temperature and 3-4 atmospheres pressure to give highly isotactic polypropylene. The polymer slurry is treated with ethanol to kill the catalyst and the polymer is purified by washing. Similar procedures and catalysts are used in laboratory preparation of isotactic poly-1-butene [35] (see Polybutene below) and crystalline polystyrene [36].   The older-generation commercial production of polypropylene resembles the laboratory preparation to some extent. Propylene is reacted with a hydrocarbon slurry of the alkylated $TiCl_3$ catalyst at 50-85°C and 20-40 atmospheres pressure.

The chemistry at the catalyst surface is not well known despite intensive study [37]. The treatment with AlR_2Cl is thought to alkylate surface titanium sites. The active site pictured in Figure 4.2 is assumed to be a monoalkylated titanium (+3) ion attached to the crystal by Ti-Cl-M (M = Mg or Ti) bridges. It seems likely that alkylaluminum groups are attached at or near the site by Al-Cl-Ti coordination. The growing polymer chain is attached to the crystal by a Ti-C σ-bond. As shown in the figure, propylene coordinates to the titanium adjacent to the Ti-C bond [38]. The polymer grows by insertion of the coordinated olefin into the Ti-C bond. The propylene inserts with great regularity with respect to both head-to-tail orientation and to placement of all the methyl groups on the same side of the polymer chain. The regularity must arise from the stereochemistry of the coordination sphere in which the entering olefin molecule coordinates. To generate purely isotactic polypropylene, this coordination site must emerge unchanged from each insertion sequence. There is growing evidence that this process is aided by coordination of a second olefin molecule - a process which has been referred to as a "trigger" for insertion [39]. The movements of atoms and repopulation of orbitals have been calculated by both CNDO and ab initio methods [40,41].

The new generation of catalysts uses $MgCl_2$ as a support for the $TiCl_4$ catalyst and incorporates Lewis base electron donors. Anhydrous $MgCl_2$ of very fine particle size is prepared by grinding $MgCl_2$ in the presence of a Lewis base, commonly ethyl benzoate, or by reacting magnesium compounds with chlorine donors in the presence of $TiCl_4$ and electron donors. The catalyst is activated with aluminum alkyls in combination with additional electron donors. There has been a great deal of study devoted to characterization of these catalysts and the role of each of the components. A useful functional picture has emerged, but a molecular model remains elusive.

In the Himont Spheripol® process [42,43], polypropylene is made in a large loop reactor similar to that used in the Phillips polyethylene process, but which is completely filled with liquid propylene which serves as the reactant and the slurrying medium. The process schematic, pictured in Figure 4.3, indicates that the heat of polymerization is removed by cooling jackets around the loops. The $MgCl_2$-supported catalyst is prepared in the shape of small, uniform spheres. It is impregnated with $TiCl_4$ and an internal electron donor. It is then activated with an aluminum alkyl combined with external electron donors. These electron donors include Lewis bases such as ethyl benzoate or related esters, silyl ethers such as phenyltrimethoxysilane, and hindered amines such as 2,2,6,6-tetramethylpiperidine-derived light stabilizers. The resulting catalysts have a number of advantages which make this a versatile and highly economical process. Catalyst activities are very high so that catalyst residues do not have to be removed from the polymer. The electron donor molecules selectively poison the catalyst sites which make atactic polymer; the difficult removal of an atactic fraction is no longer required. Because the liquid propylene is the reaction medium, a separate solvent recycle is eliminated.

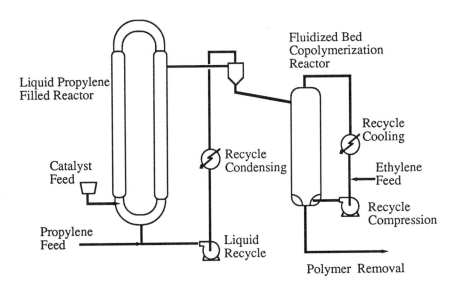

Figure 4.3 Schematic representation of the Himont Spheripol polypropylene process.

Finally, because the polymer is grown in the shape of uniform spheres having good bulk density, there is no need to go through a final energy-intensive pelletizing step.

Polymer is removed from the loop reactor as a slurry in liquid propylene. The liquid propylene is flashed off, recondensed, and returned to the loop reactor. If a simple homopolymer is desired, the residual catalyst in the polymer is deactivated and the uniform spheres are treated with the chemical stabilizers and dried, and are then ready for shipment. The final steps are facilitated by the small residual open-pore structure in the polymer particles. Alternatively, the spheres are fed to the second reactor in the figure where an ethylene-propylene copolymerization is carried out in the gas phase. The rubbery EP-copolymer is formed within the pore structure of the preformed sphere, so there is no sticking problem. This intimate combination of polypropylene and an EP rubber is particularly useful in automotive or other applications requiring impact resistance.

More recent versions of the process imbibe free-radical initiators and polar comonomers such as maleic or itaconic anhydrides into the pore structure of the preformed particles. After wetting the internal surface structures, the particles are warmed and the resulting free-radical polymerization grafts short polar chains onto the polypropylene, providing functionality needed for specialty applications.

Ethylene-Propylene-Diene Rubbers

While the homopolymers of ethylene and propylene are usually plastics, a family of ethylene-propylene copolymers are elastic and are used in place of rubber in many industrial applications. The simple copolymers are often blended with more crystalline polymers as toughening agents. More commonly, though, a small amount of an unsymmetrical diene is copolymerized with ethylene and propylene to give an elastomer which can be cured by conventional rubber technology [44,45]. Common diene comonomers are ethylidenenorbornene, dicyclopentadiene, and *trans*-1,4-hexadiene (Section 4.4). Each of these dienes has a highly reactive double bond which readily copolymerizes to affix the diene to the polymer chain. The less reactive C=C bond in the diene remains intact to provide a site for crosslinking polymer chains in the "cure" of the final product. The cured products are essentially saturated and any unsaturation is located in pendant groups rather than the main backbone. As a result, these rubbers are more ozone-resistant than natural rubber or the synthetic polydienes discussed below.

Most of the ethylene-propylene copolymers and terpolymers are prepared with Ziegler catalysts based on a soluble vanadium compound such as $VOCl_3$ or VCl_4, though high-activity $MgCl_2$-supported $TiCl_4$ catalysts are being used for polymers where highly random incorporation of the monomers is not required. Empirically, vanadium seems unique in its ability to incorporate the comonomers into the polymer chain in a random sequence, an important characteristic to produce an amorphous, elastomeric product. In laboratory experiments that simulate commercial practice [46], a reaction solvent such as hexane or chlorobenzene is chilled to 15°C and saturated with an ethylene-propylene mixture at one atmosphere pressure. The diene (e.g., dicyclopentadiene) is added, followed by the catalyst components, in this instance $Al_2Et_3Cl_3$ and $VOCl_3$. After a short induction period during which the active catalyst is formed, vigorous polymerization begins and the three comonomers are added at rates to maintain the desired proportion in the polymer. The relative monomer reactivity is usually ethylene>propylene>diene. When the polymer solution becomes too viscous for effective stirring, the catalyst is killed by addition of alcohol. The catalyst residues are extracted with water and the polymer is isolated by precipitation or evaporation.

The chemistry of catalyst formation has been studied extensively [47]. At the ratios of Al to V used commercially, VCl_4 and $VOCl_3$ are reduced to V(III) complexes which bear alkyl substituents [48,49]. These V(III) compounds are relatively unstable and decompose to catalytically inactive V(II) species in less than an hour at 15-40°C. Fortunately, polymerizations are usually complete in minutes. The presence of extra chloride ion seems essential for the stability of the active catalyst. Commercial polymerizations often contain an organic compound with active chlorine atoms to regenerate the catalyst in situ. Useful catalyst promoters include benzotrichloride [50,51], ethyl phenyldichloroacetate, butyl perchloro-vinylacetate [52], and even chloroform [53], but these species often require the use of increased quantities of aluminum alkyls.

It is believed that the catalytically active vanadium complexes (either soluble or insoluble) are very similar to the Ti-based Ziegler catalyst depicted in Figure 4.2. Chloride bridges to aluminum and to other vanadium ions are believed to occupy four coordination sites. Olefin and alkyl ligands are situated in adjacent positions to facilitate the repetitive insertion process by which the polymer chain grows. In contrast to the well-defined sites on the surface of a $TiCl_3$ crystal which constrain olefin molecules to insert in a very regular fashion, the vanadium catalyst sites must be quite flexible. Propylene enters the chain with random stereoregularity, random placement with respect to ethylene units, and random placement with respect to end-to-end alignment under most process conditions used commercially.

While vanadium is noted for yielding very random, amorphous polymers, zirconium has also been observed to give relatively random copolymers. Additionally, there are some applications where it is desirable to have nonrandom distribution of the ethylene/propylene monomers to provide a low degree of crystallinity. This low crystallinity allows the uncured polymer to retain its shape during fabrication prior to the chemical crosslinking. As a result, there is a growing interest in the use of the zirconocene/aluminoxane catalysts discussed above [54].

Specialty Polymers

While polyethylene, polypropylene, and EPDM rubbers dominate the polyolefin industry, several other polymers find niche applications. Values-in-use of these polymers are necessarily high because they do not enjoy the economies of scale of their larger cousins.

Poly-4-methyl-1-pentene (PMP) has the lowest density of all commercial synthetic resins, 0.83 g/cm^3. It is highly transparent, transmitting 90% of visible light; this is superior to transparent resins like polystyrene and poly(methyl methacrylate), and even glass. Its melting point is 240°C. The monomer, 4-methyl-1-pentene (4M1P), is prepared by the dimerization of propylene; this is normally carried out with a heterogeneous catalyst rather than the homogeneous species discussed below. As a result of these properties and others, it finds greatest application in medical instruments, laboratory ware, and industrial applications. PMP is synthesized in a slurry polymerization using the catalysts discussed above which give stereoregular polymers. Many grades of PMP include comonomers such as octene which lower melting point, but dramatically improve impact resistance.

High molecular weight poly(1-butene) is used mostly in flexible plastic pipe; it will expand to accommodate the expansion of freezing water. It is prepared as the isotactic polymer, though atactic grades are even more elastomeric. The catalysts used are like those discussed for poly(propylene). There has been considerable work on a gas-phase process which eliminates many of the difficulties of solution processes. A major area of research is the development of additives such as nucleating agents to shorten the curing or crystallization time of the polymer from the melt, which can otherwise be days.

Very high molecular weight poly(1-decene)s or poly(1-dodecene)s are used as flow improvers for oil pipelines. These remarkable, viscoelastic materials have helped push oil through the Trans-Alaska pipeline at rates in excess of 120% of design capacity. They function at part per million concentrations by reducing turbulence in the pipeline at high flows; the microscopic aspects of this phenomenon are unclear. Solution polymerizations are initiated using Ziegler-Natta or other catalysts at low temperatures. The temperature of the polymerization is determined by the adiabatic temperature rise because the mixtures cannot be stirred - molecular weights are so high that stirring actually breaks the long polymer chains and lowers effectiveness. The polymer solutions are then removed from the reactor and shipped to the user without isolation. Related species having lower molecular weights are used in adhesive applications. Other related crosslinked species have found utility as vibration dampeners for heavy machinery and even for seismic dampening.

Ultrahigh molecular weight polyethylene (UHMWPE) is a specialty product simply because it cannot be processed using the methods developed for high-density polyethylene or other thermoplastics. By accepted definition, the molecular weight is above 3,000,000. It is generally prepared in a low-pressure, low-temperature slurry process using modified Ziegler-Natta or Phillips catalysts. It generally leaves the reactor as a fine powder, which is a useful form for the processing methods required. Fabrication is normally carried out by compression molding, ram extrusion, or thermoforming and annealing. It is then formed into useful parts by machining. The obvious expense of fabrication is offset by the properties of UHMWPE. It has the highest abrasion resistance of all thermoplastics, and it has a low coefficient of friction. It also has excellent resistance to impact and to environmental stress cracking. As a result, it finds use in applications as diverse as artificial hip sockets and chutes for mining and mineral handling. It is also used for wear-resistant parts for machines and the bottom surfaces of skis and snowmobile skids.

Although the most common propylene polymers discussed above are crystalline thermoplastics, propylene is also homopolymerized to an elastomeric material [55-57]. Elastomeric polypropylene has a stereoblock structure with long amorphous atactic segments separated by short crystalline isotactic blocks. These isotactic blocks provide crystalline crosslinks between the flexible amorphous chains [58,59], yielding a rubbery material. The crosslinks in normal rubbers are chemical in nature, but in this case, they are physical crosslinks which are lost during melt processing and then reappear upon cooling. This material can be used to compatibilize normal isotactic polypropylene with EP or EPDM rubbers, yielding materials useful in automotive and other impact-sensitive applications. It can also be used to replace highly plasticized poly(vinyl chloride) in medical applications where the plasticizers are of concern. The catalysts for this polymerization are based upon tetraalkylzirconium compounds supported on high-surface-area minerals [60].

Generally, commercial olefin polymerization has been limited to α-olefins, except in those applications where cyclic olefins are used as minor constituents in a polymer consisting primarily of α-olefins. An exception is Mitsui Petrochemical's high-value, amorphous polyolefin "APO" [61], which is used in compact disk substrates and other optical applications such as aspherical lenses [62]. It is based upon a copolymer of ethylene and a Diels-Alder adduct which are copolymerized in toluene using a vanadium catalyst to give a highly random polymer with a narrow molecular weight distribution near 10,000.

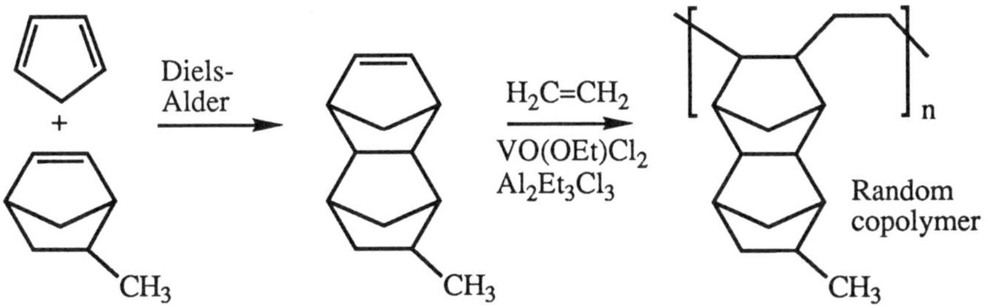

Its polyolefin nature means that it is unaffected by humidity; it is optically clear and has low birefringence; and it has a high softening temperature for thermal resistance. The polymer has several clear advantages over polymethacrylates and polycarbonates used for the same purpose, but the current low volume requires a premium price.

Polymerization of cyclopentene and norbornene to isotactic polymers with no ring opening has been reported [63].

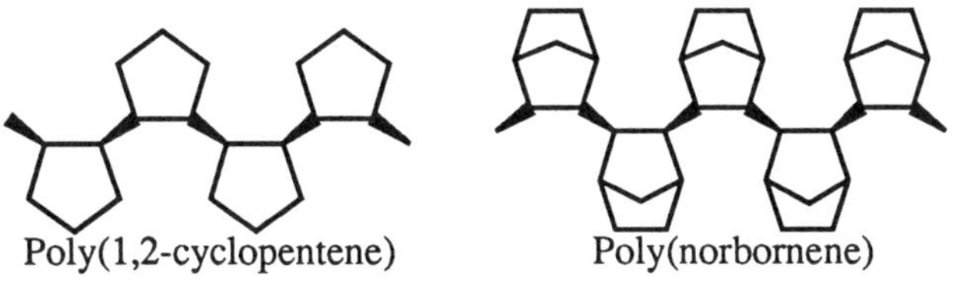

Poly(1,2-cyclopentene)   Poly(norbornene)

These unusual polymers have extremely high glass transition temperatures and melting points (300-500°C). They exhibit the oxidative instability expected for tertiary hydrocarbons, but if this limitation can be overcome, they might find utility in a number of specialty high-temperature applications. Polymerization of cyclopentene in the presence of hydrogen provides oligomers as well as polymer. The trimer has been characterized as the 1,3-isomer stereoregular isomer, calling into question the proposed 1,2-structure of the cyclopentene-based polymer [64].

Poly(1,3-cyclopentene)

Apparently, steric congestion at the metal center is relieved by the 2- to 3-isomerization before the next incoming monomer can be inserted. Poly(norbornene) is also obtained by polymerizations using labile Pd^{II} complexes [65] in what is close to a living polymerization under proper conditions [66]. The more usual polymerization using cyclic olefins is ring-opening olefin metathesis (ROMP), which is used in a number of specialty applications including the polymerization of dicyclopentadiene and cyclooctene (Section 9.2).

Most α-olefins polymerize by α-β-incorporation into the chain, producing polymer chains bearing alkyl branches two carbon atoms shorter than the olefin. One polymerization using a nickel catalyst provides β-ω-incorporation, giving only methyl branches [67-69]. The catalyst is formed by reaction of "Ni(0)" [e.g., Ni(1,5-cyclooctadiene)$_2$], and a phosphorane ligand.

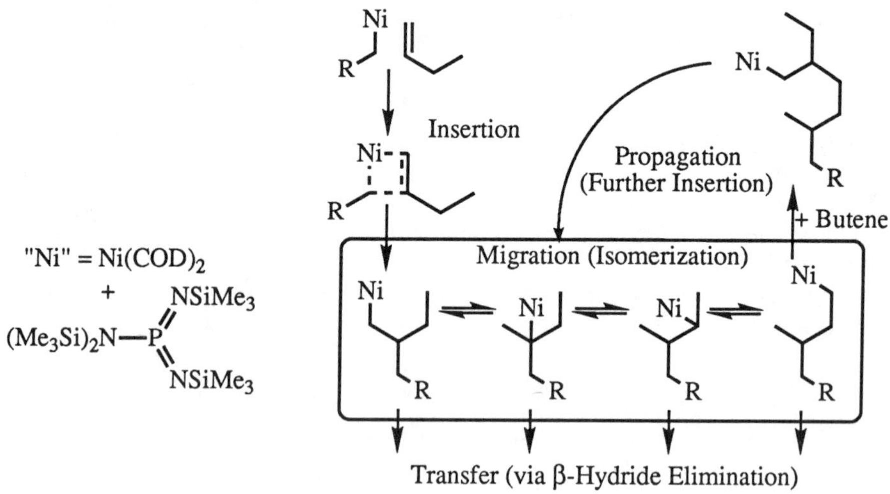

Figure 4.4 Mechanism of the β-ω-enchainment via insertion-migration polymerization of α-olefins.

The polymerization shown in Figure 4.4 is accomplished by (a) insertion of the olefin into the nickel-alkyl bond forming Ni-C_α and C_β-C_ω bonds; (b) migration of the metal center up and down the length of the alkyl chain until it reaches the end of the polymer chain; and (c) insertion of the next olefin only when the metal is at the end of the chain. Chain transfer can occur at any point during the nickel migration. No effect would be seen during ethylene or propene polymerizations. Pentene polymerization would result in a polymer equivalent to a rigorously alternating polymerization of ethylene and propylene.

4.2 POLYBUTADIENE AND POLYISOPRENE

Many of the desirable physical properties of natural rubber are due to the structure of the polymer chain which arises from a *cis*-1,4 polymerization of isoprene (n ≈ 20,000). Synthetic elastomers with similar structures and physical properties are produced by coordination polymerization of butadiene and isoprene. As applications of these rubbers become more specialized, the control of polymer microstructure has become more important. Isoprene can be inserted into the growing polymer backbone in one of four different configurations:

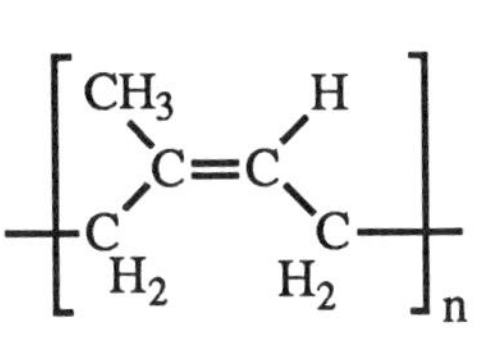

cis-1,4-addition

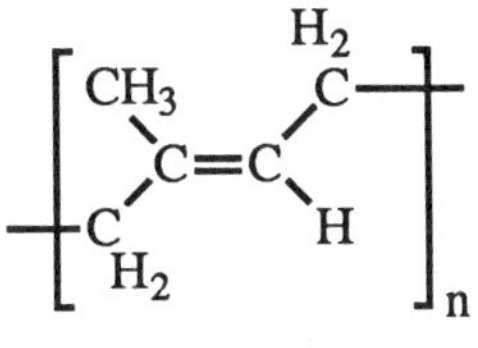

trans-1,4-addition

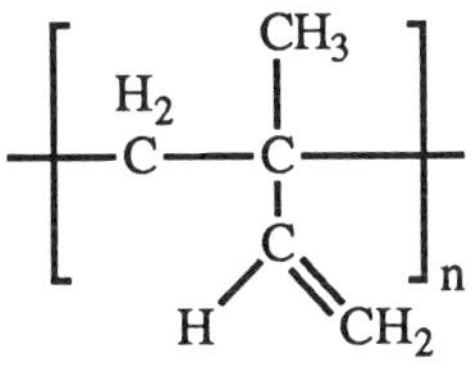

1,2-addition

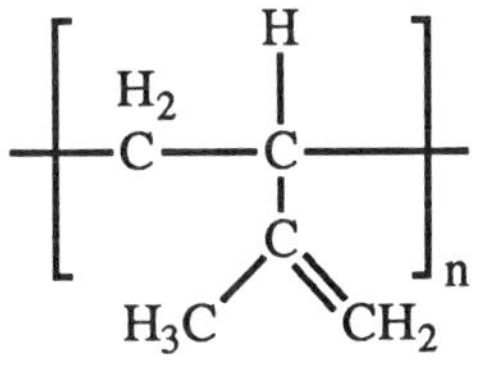

3,4-addition

Butadiene affords only the first three of these additions, 1,2-addition being indistinguishable from 3,4-addition. The 1,2- and 3,4-additions can be isotactic, syndiotactic, or atactic (see description above under polypropylene), giving further control of the polymer properties.

The two major commercial catalyst systems for the polymerization of isoprene [70] are alkyllithiums, usually BuLi for *cis*-1,4-polyisoprene, and vanadium-based Ziegler catalysts for *trans*-1,4-polyisoprene. The butyllithium is sometimes referred to as an initiator rather than a catalyst because, discounting the effects of impurities in the system, each lithium initiates one polymer chain [71]; the resulting molecular weight distribution is very narrow. Titanium-centered Ziegler catalysts, and alane (aluminum hydride) with TiCl$_4$ have been used, but yield mixed microstructures. Commercial catalysts for polybutadiene include titanium-based Ziegler catalysts, cobalt-centered Ziegler catalysts, and allylnickel complexes. When high-vinyl-content polybutadiene is desired, butyllithium initiators combined with chelating ligands such as diglyme or tetramethylenediamine are utilized.

Titanium-based Ziegler catalysts like those used to polymerize ethylene and propylene are also effective for the polymerization of butadiene and isoprene. In a preparative experiment [72], addition of TiCl$_4$ and triisobutylaluminum to an isoprene solution in pentane produces 1,4-polyisoprene. The polymer has better than 80% cis conformation. The selectivity for *cis*-1,4-polybutadiene is enhanced by iodide modification of the standard Ziegler catalyst. Catalysts prepared by reaction of TiI$_4$ with an alkylaluminum compound have been studied extensively [73]. Other catalysts have been formed by a three-component reaction such as TiCl$_4$ + AlI$_3$ + AlR$_3$. Such catalysts are suspensions of TiI$_3$ crystals which have been surface-alkylated like the polypropylene catalysts discussed earlier. They give polybutadiene which is 90-93% *cis*-1,4 conformer.

Polymerization of isoprene with a clay-supported VCl$_3$/Al(i-Bu)$_3$ catalyst gives the thermodynamically preferred *trans*-1,4-polyisoprene [74]. This modified selectivity with a vanadium catalyst is consistent with the flexible catalyst site suggested to explain its randomness in copolymerization of ethylene and propylene. With butadiene, VCl$_3$ yields extremely high stereospecificity polybutadiene with up to 99% *trans*-1,4 content [75]. Interestingly, halogen-free cobalt catalysts can yield syndiotactic 1,2-polybutadiene with high specificity [76]. This material finds application as a packaging film rather than an elastomeric product [77].

Organocobalt catalysts based on hydrocarbon-soluble cobalt salts seem to be true homogeneous catalysts. These catalysts have achieved commercial importance because they produce 96-98% *cis*-1,4-polybutadiene when used with a halide-containing aluminum compound. The soluble commercial catalysts are prepared by reacting a cobalt(II) carboxylate with Al$_2$R$_3$Cl$_3$ [78]. These conditions are nicely illustrated by a laboratory preparation of *cis*-1,4-polybutadiene [79]. Diethyl-aluminum chloride and cobalt(II) octoate react in a benzene solution of butadiene in the presence of a small amount of water to produce a very active catalyst. Polymerization at 5°C gives a high molecular weight 1,4-polybutadiene with about 98% *cis*-1,4 units. It is interesting to note that early reports of polymerizations with cobalt were often difficult to reproduce; later it was discovered that water or other "impurities" such as alcohols, reactive organic halides, or organic hydroperoxides were essential components of the catalyst preparation [80,81].

The nickel-based catalysts are used less extensively than the cobalt systems, but much more is known about their chemistry as a result of extensive study [82-84]. It seems likely that the intermediates in the cobalt and nickel systems are similar. The nickel catalysts can be prepared in several different ways, but all methods appear to give relatively stable π-allyl derivatives in the presence of butadiene. In the commercial process, alkylation of a nickel carboxylate probably proceeds through several steps:

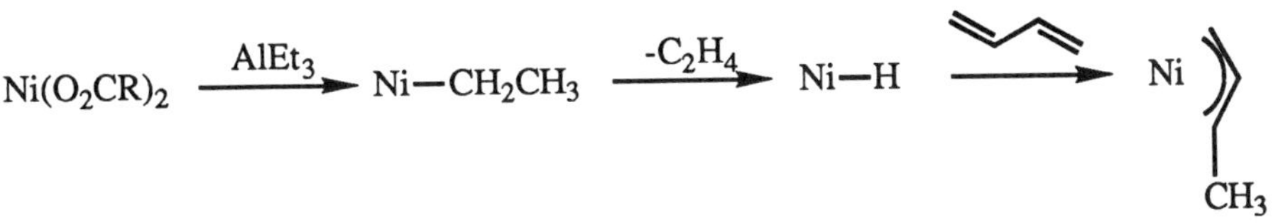

$$Ni(O_2CR)_2 \xrightarrow{\text{AlEt}_3} Ni-CH_2CH_3 \xrightarrow{-C_2H_4} Ni-H \longrightarrow Ni$$

Similar π-crotyl derivatives are obtained by protonation of zero-valent nickel complexes in the presence of butadiene [85].

Like cobalt, organonickel systems generally give high percentages of *cis*-1,4-polybutadiene. In the presence of iodide ion, though, the nickel catalysts give predominantly *trans*-1,4 product, the more stable isomer. This variation in product stereochemistry in butadiene polymerization closely parallels that in the codimerization of butadiene and ethylene to give *cis*- or *trans*-1,4-hexadiene (Section 4.4). The stereochemistry of the polybutadiene can be controlled by the concentration of potential ligands in the system. This effect can be used to prepare a "block" polymer that is half *cis*- and half *trans*-1,4-polybutadiene [86]. The polymerization is begun with π-allylnickel trifluoroacetate as the catalyst, giving rise to a segment of cis polymer. Before polymerization is complete, a trialkyl phosphite ligand is added; the ligand changes the course of the polymerization so that the second half of the growing polymer chain has the trans configuration. The stereoselectivity of the catalyst can be explained on the basis of chelate vs. monodentate coordination of the butadiene. This explanation accounts for most of the experimental observations [87], though alternative mechanisms have been proposed [88]. Figure 4.5 illustrates the chemistry involved in the preparation of a cis-trans block copolymer:

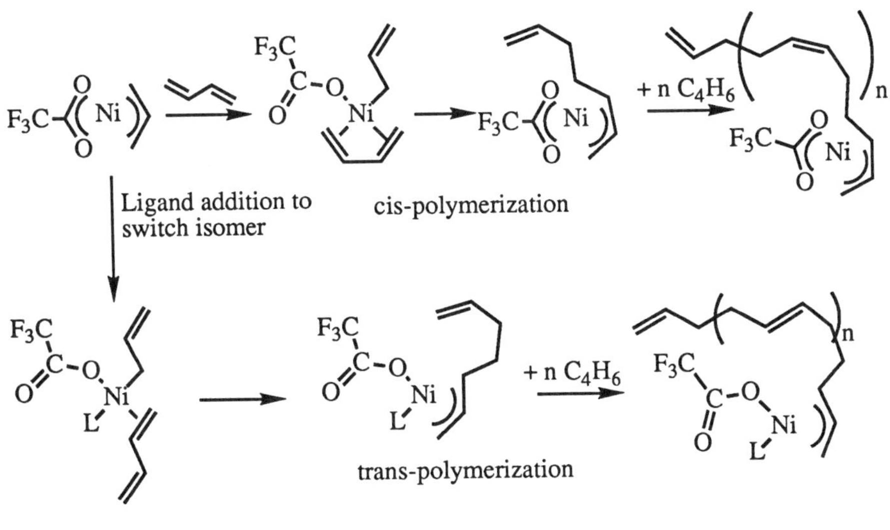

Figure 4.5 Mechanisms for production of *cis*- and *trans*-1,4-polybutadiene.

The π-allyl nickel catalyst coordinates a butadiene through both double bonds in chelate fashion to form an η^4-complex shown in the upper pathway. Insertion of the diene into the Ni-C bond of the σ-allyl ligand forms a new π-allylic ligand. The conformation of this allylic ligand ensures that the double bonds in the growing polymer chain have the cis configuration. However, if a strongly bonding ligand such as L = P(OPh)₃ is added to the system, it occupies a coordination site on nickel and forces the chemistry to follow the lower pathway. The butadiene complexes through only one C=C bond. Insertion of this single C=C bond gives a π-allyl which rearranges to a σ-allylic structure. However, the configuration of the allylic ligand formed in this way leads to a *trans*-C=C bond in the growing polymer chain. Thus the presence or absence of ligand, L, determines the stereochemistry of the growing polymer chain.

4.3 OLIGOMERIZATION OF OLEFINS

The self-addition of olefins to form dimers, trimers, and low polymers is called oligomerization. This process can be identical in mechanism to the ethylene polymerization described earlier except that chain termination occurs much more frequently. In ethylene dimerization to 1-butene, chain transfer by β-hydrogen abstraction follows every insertion into an M-C bond:

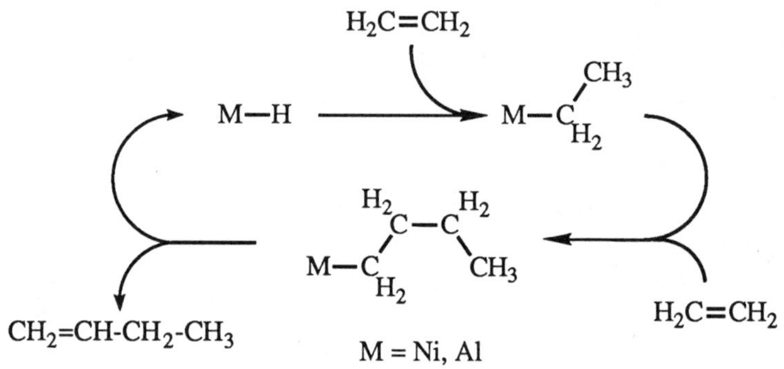

$$
\begin{array}{c}
\text{H}_2\text{C}=\text{CH}_2 \\
\text{M}-\text{H} \longrightarrow \text{M}-\text{C}\overset{\text{CH}_3}{\underset{\text{H}_2}{}} \\
\text{M}-\text{C}\underset{\text{H}_2}{}\ \overset{\text{H}_2}{}\text{C}-\text{C}\underset{\text{CH}_3}{\overset{\text{H}_2}{}} \\
\text{CH}_2=\text{CH-CH}_2\text{-CH}_3 \qquad \text{H}_2\text{C}=\text{CH}_2 \\
\text{M} = \text{Ni, Al}
\end{array}
$$

These reactions find practical applications in (a) oligomerization of ethylene by organoaluminum compounds to give linear α-olefins or α-alcohols; (b) nickel-catalyzed oligomerization of ethylene to C_{10}-C_{20} α-olefins by the Shell Higher-Olefins Process (SHOP); and (c) dimerization of propylene to branched C_6 olefins useful as octane-enhancers in motor fuel. The codimerization of ethylene and butadiene and the oligomerization reactions of dienes are discussed in Sections 4.4 and 4.5.

Aluminum-Catalyzed Oligomerization of Ethylene

Ethylene readily inserts into Al-H and Al-C bonds to form C_2-C_{40} alkylaluminum compounds. These compounds are intermediates in the commercial production of linear α-olefins and α-alcohols [89,90] according to the equations

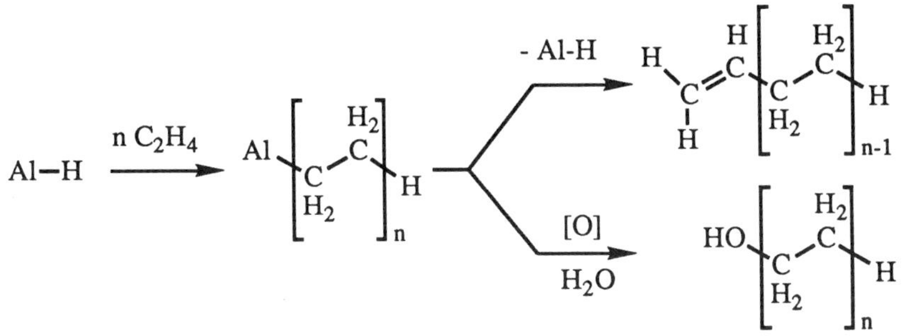

The synthesis of olefins is catalytic because β-hydrogen abstraction from the growing alkyl chain regenerates an Al-H bond that can start growth of a new alkyl chain. The linear olefins produced in this way are intermediates in the synthesis of fatty acid esters, aldehydes, and alcohols by the carbonylation reactions described in Chapter 5.

In contrast to the catalytic α-olefin synthesis, the alcohol synthesis uses the alkylaluminum compound stoichiometrically. An AlR_3 compound prepared by ethylene oligomerization is oxidized with air to $Al(OR)_3$ which is hydrolyzed to produce the linear alcohol. Such "fatty alcohols" are biodegradable and are used in the manufacture of detergents. The process competes economically with the hydroformylation of olefins even though it uses the aluminum compound stoichiometrically rather than as a catalyst. The coproduced alumina is of a high grade and adds value to the process.

The chemistry of ethylene addition to Al-C and Al-H bonds [91] dictates the manner in which ethylene oligomerization is carried out commercially. Trialkylaluminum compounds catalyze the reaction of ethylene, aluminum, and hydrogen to form triethylaluminum (which is extensively dimerized in the liquid state):

$$2\,Al + 3H_2 + 6\,C_2H_4 \xrightarrow{AlR_3} 2\,AlEt_3 \rightleftharpoons Al_2Et_6$$

The insertion of ethylene does not stop when Al-C_2H_5 groups are formed. Continued insertion into the Al-C bonds produces alkylaluminum compounds in which the alkyl groups are polyethylene chains. The length of the chains is governed by reaction conditions, but generally represents a statistical distribution of sizes based on the amount of ethylene available. Practical syntheses of terminal olefins employ reaction temperatures at which β-hydrogen elimination is frequent. Thus, chain growth and termination occur at comparable rates. The reaction is carried out typically at 200-250°C and 130-250 atmospheres pressure [89]. The high pressure of ethylene prevents the α-olefin products from reinserting into the growing alkyl chains by a simple mass-action effect. The predominance of ethylene insertion gives linear olefins with 4-8 carbon atoms as the major products.

The alcohol synthesis process developed by Conoco [92] employs a temperature of only 115-130°C in the chain-growth step in order to avoid olefin formation by β-hydrogen abstraction. Ethylene and triethylaluminum are reacted at about 120°C and 135 atmospheres to form a statistical mixture of AlRR'R" compounds with most alkyl chain lengths in the C_6-C_{14} range (all even-numbered). This mixture is oxidized with dry air at about 35°C to form the corresponding Al(OR)(OR')(OR") mixture. Hydrolysis with sulfuric acid yields a mixture of fatty alcohols. In laboratory preparations of alcohols from AlR_3 compounds, trimethylamine oxide is a convenient reagent for oxidation of the Al-R bond [93].

Shell Higher-Olefins Process

Shell uses a nickel-catalyzed oligomerization of ethylene to prepare linear α-olefins on a large scale [94,95]. Despite much research in the area, specifics of the commercial catalyst and system are still incomplete. As in the aluminum-catalyzed oligomerization, insertion of ethylene into M-H and M-C bonds forms metal alkyls with a statistical distribution of chain lengths. β-Hydrogen abstraction produces olefin and regenerates Ni-H species to repeat the chain-growth sequence. The mechanism of this process is like that discussed for olefin dimerization below. A nickel hydride catalyst is generated by reduction of a nickel salt in the presence of a chelating ligand such as diphenylphosphinobenzoic acid or nickel(0) is reacted with a phosphorus ylide [96].

In practice, a catalyst of this sort is allowed to react with ethylene in a glycol solvent such as 1,4-butanediol at about 100°C and 80 atmospheres pressure, the pressure required to attain high linearity. A rapid reaction occurs to form a mixture of linear α-olefins. The α-olefin layer is immiscible in the catalyst-containing glycol layer and is decanted. The olefin layer is then washed with additional glycol to remove traces of catalyst. The olefin chain lengths typically are distributed with about 40% in the C_4-C_8 range, 40% in the C_{10}-C_{18} range, and 20% above C_{20} [97]. Originally, the C_{10}-C_{18} olefins were the more marketable fraction, but octene is now a valuable product. The higher and lower boiling products are then utilized in a complex sequence of catalytic reactions (Figure 4.6). The low- and high-boiling olefins are isomerized separately over heterogeneous catalysts to produce internal olefins. This step is necessary because the next step is olefin metathesis, which often does not work well with terminal olefins (Section 9.1).

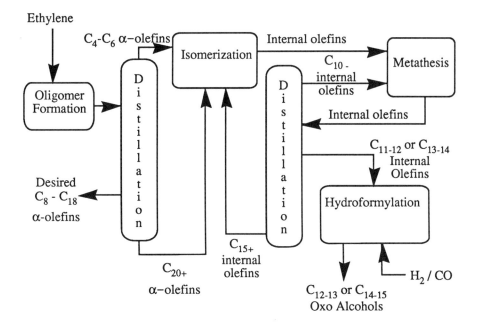

Figure 4.6 Schematic representation of the Shell Higher-Olefins Process (SHOP). Metathesis and hydroformylation are covered in Chapters 9 and 5 respectively.

In the process variant shown in the figure, the low- and high-boiling internal olefins react over a heterogeneous catalyst such as MoO_3 on Al_2O_3 to produce a broad range of internal olefins. Due to the prevalence of reactions such as

much of the product is in the useful C_{10}-C_{18} range. Hydroformylation is carried out with a cobalt catalyst that converts internal olefins to terminal alcohols (Section 5.5). In this way, a high proportion of the ethylene is converted to useful fatty alcohols.

Olefin Dimerization

The dimerization of ethylene to butenes is the simplest and one of the best studied oligomerization reactions [98,99]. Industrially, it is not very significant because butenes are generally cheaper than ethylene. However, in situ formation of l-butene during ethylene polymerization may be a convenient way to produce branched polyethylene with properties approaching those of low-density polyethylene. Another potentially useful reaction is the dimerization of propylene in mixed C_3 streams which converts this inexpensive olefin to useful C_6 compounds such as dimethylbutenes and methylpentenes [100].

The first simple olefin dimerization was discovered in the course of Ziegler's study of ethylene oligomerization to long-chain α-olefins by alkylaluminum compounds [99]. It was observed that traces of nickel from reactor corrosion diverted the ethylene oligomerization to the production of l-butene. This chance discovery led to the invention of the Ziegler catalysts for polymerization, hydrogenation, and isomerization of olefins. As discussed above, this chemistry provides the basis for the Shell α-olefins process.

The dimerization of simple olefins has been studied extensively as a model for 1,4-hexadiene synthesis (Section 4.4). Greatest attention has been given to the nickel [99] and rhodium [101] complexes that are commercially attractive for preparation of trans-1,4-hexadiene. However, very effective catalysts for olefin dimerization also arise from reaction of alkylaluminum compounds with cobalt salts and with titanium(IV) complexes.

The dimerization of propylene with nickel-based Ziegler catalysts has been explored extensively [102], and provides the basis for the IFP Dimersol® process discussed later in this section. In large-scale operations, the catalyst is prepared as it is used by mixing $NiCl_2$, triethylaluminum, and butadiene in chlorobenzene to give a π-allyl complex of nickel (Section 4.4). A phosphine is added to the solution and the mixture is fed to a continuous reactor along with liquid propylene at 15 atmospheres pressure. Rapid dimerization occurs at 30-40°C to give a mixture of n-hexenes, 2-methylpentenes, and 2,3-dimethylbutenes in 85-90% yield. The distribution of the isomeric C_6 products depends on the nature of the phosphine used in catalyst preparation. At low temperatures, the proportion of methylpentenes can be as high as 80% if PMe_3 or $Ph_2PCH_2PPh_2$ is used. However, with sterically bulky phosphines such as $PEt(t\text{-}Bu)_2$, 70-80% of the mixture is dimethylbutenes.

Effective catalysts can be prepared from combinations of π-allylnickel halides with phosphines and Lewis acids such as $AlCl_3$ or $EtAlCl_2$ [102]. Similarly, zero-valent nickel complexes such as bis(1,5-cyclooctadiene)nickel and $Ni[P(OPh)_3]_4$ react with Lewis acids to give catalysts for dimerization of propylene. The same catalyst systems also bring about the dimerization of ethylene to butenes and codimerization of ethylene and propylene to pentenes. The products are generally the thermodynamically favored internal isomers because the dimerization catalysts also catalyze double-bond migration (Section 2.1). The active catalyst for olefin dimerization in all these systems is probably a nickel hydride complex. The Ni-H function can be formed by β-hydrogen abstraction from alkyl intermediates formed by interaction of an alkyl or allylnickel complex with propylene:

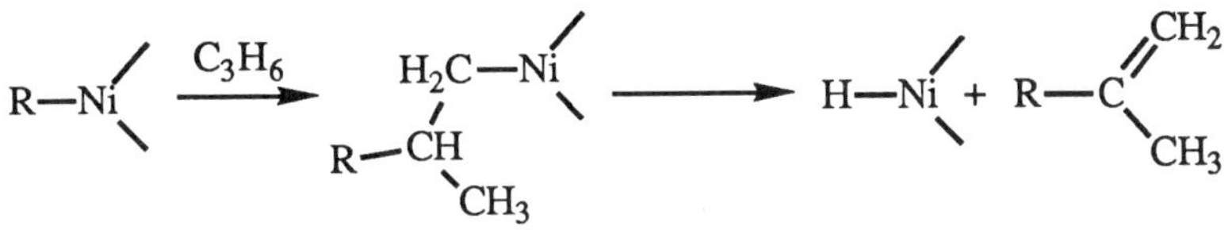

$$R-Ni \xrightarrow{C_3H_6} \begin{matrix} H_2C-Ni \\ | \\ R-CH \\ \quad CH_3 \end{matrix} \longrightarrow H-Ni + R-C \begin{matrix} CH_2 \\ \diagup\diagup \\ \diagdown CH_3 \end{matrix}$$

The other coordination sites about the nickel are occupied by phosphine, olefin, and halide ligands. The active catalyst in the well-studied rhodium system [103] may be similarly coordinated. The formation of the rhodium complex is discussed in the section on hexadiene synthesis (Section 4.4).

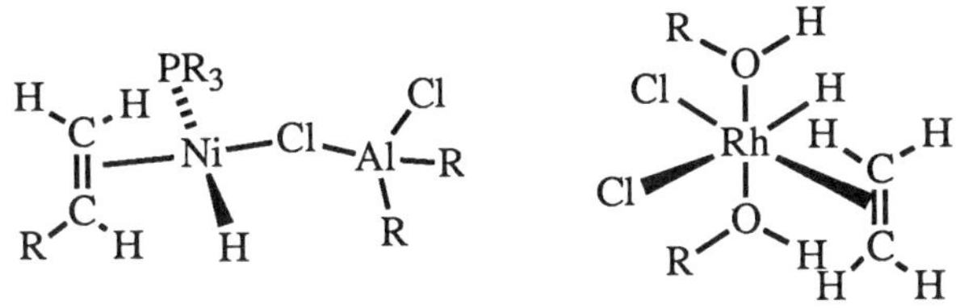

These complexes correspond to the olefin hydride complexes shown in the two o'clock position of the catalytic cycle.

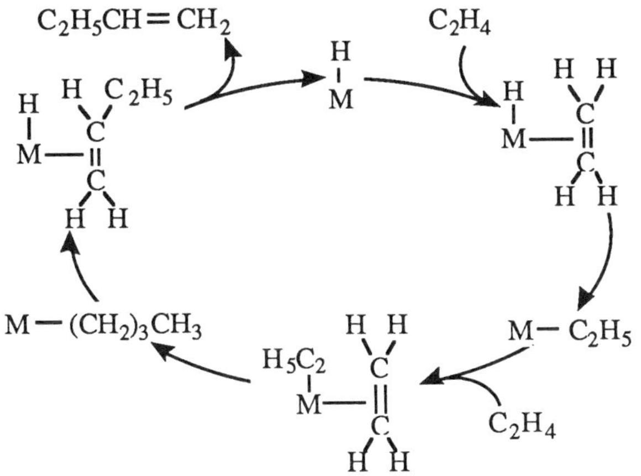

This mechanism employs the familiar steps of olefin coordination to a metal hydride, insertion into the M-H bond, and subsequent insertion into an M-C bond just as in olefin polymerization catalysts. However, the olefin dimerization catalysts effect β-hydrogen abstraction from the growing alkyl chain after almost every M-C insertion. As a result, olefin dimers predominate in the product. Only small quantities of trimers and tetramers form under normal operating conditions.

The ethylene dimerization shown above can produce only a single product because ethylene insertion into an M-H or M-C bond has no regioselective aspect. With propylene, however, both insertion steps can produce isomers:

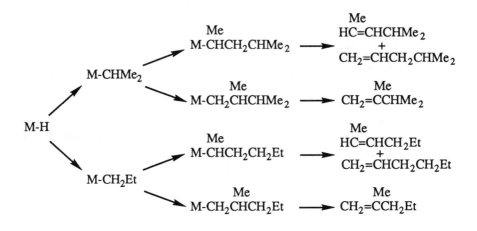

The relative frequencies of the various insertion modes determine the product distribution. A very crowded catalyst site produced by bulky ligands favors formation of alkyl groups with RCH_2 joined to the metal. This effect enhances production of the branched chain olefins. While 4-methyl-1-pentene is normally produced for the polyolefin industry using heterogeneous catalysts, a homogeneous uranium catalyst shows good selectivity [104]. BP Chemicals is commercializing a homogeneous process which yields a mixture of branched hexene isomers [105]. The mixture is treated with an isomerization catalyst to yield the internal olefin, 2,3-dimethyl-2-butene. Extraction from the mixture by distillation is followed by a reisomerization to the desired 2,3-dimethyl-1-butene. This specialty product is then used by the fragrance industry to yield a synthetic musk.

IFP's Dimersol Process and variations of that process are widely used to upgrade the value of olefin-containing streams [106]. Using Ziegler catalysts, a stream containing n-butenes is dimerized to isooctenes useful as plasticizer precursors [107]. This reaction is often carried out on a C_4 stream from a refinery cracker after it has been depleted of isobutene by reaction with methanol to form methyl-t-butylether for a gasoline additive; saturates are unreactive and are recycled to the refinery operation. The reaction can also be carried out on propylene or mixed propylene/butene streams to yield branched C_6-C_8 products useful as automotive fuels. Each of these processes use nickel-based Ziegler-Natta catalysts. IFP also licenses a zirconium-based ethylene dimerization to 1-butene. Under normal economic conditions, ethylene is the more valuable product, but there are circumstances where dimerization of polymer grade ethylene to polymer grade butene would be useful.

An entirely different mechanism for olefin dimerization has been proposed for catalysis by a stable alkylidenetantalum complex which is extremely selective for the dimerization of propylene to 2,3-dimethyl-1-butene with over 90% selectivity [108]. The alkylidene group is removed and an unstable metallocycle forms in the catalyst formation step.

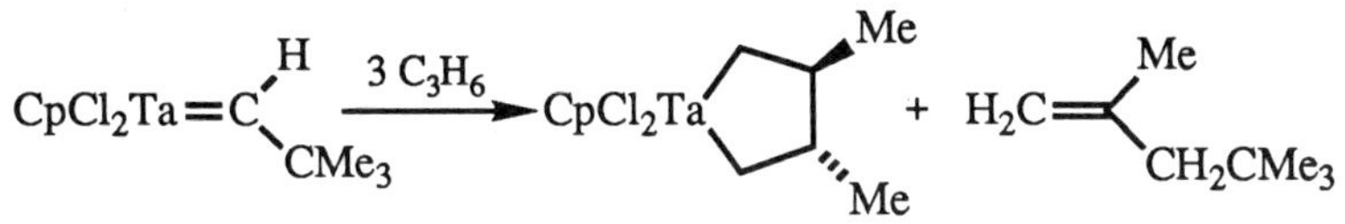

CpCl₂Ta=C
H
CMe₃
3 C₃H₆
CpCl₂Ta
Me
Me
+ H₂C=
Me
CH₂CMe₃

In the catalyst cycle, a 1,3-hydrogen shift across the face of the ring converts the metallocycle to an olefin complex. Dissociation of the dimeric olefin and coordination of propylene gives the bis(propylene) complex. Dissociation and association of olefinic ligands may occur simultaneously. Reorganization of bonds regenerates the metallocycle.

CpCl₂Ta
Me
Me
CpCl₂Ta
Me
Me
H₃C
Me
CpCl₂Ta
Me
CpCl₂Ta
2 MeCH=CH₂
Me
H₂C=CCHMe₂

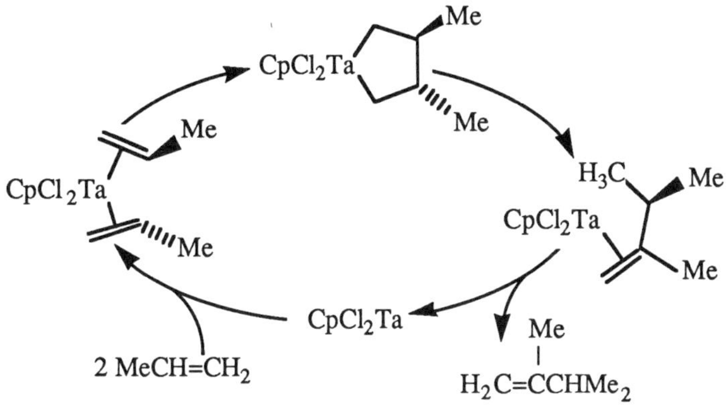

The tantalum catalyst converts ethylene to l-butene efficiently but is less effective for dimerization of l-butene and higher olefins. Because it does not involve hydride intermediates that might catalyze double-bond migration, the terminal olefin products are stable in the reaction mixture.

This metallacyclic mechanism is probably common for olefin dimerization, but its significance has not been appreciated until recently. The dimerization of ethylene to l-butene or cyclobutane has been observed with metallacyclic nickel and titanium complexes [109]. The formation of a metallacyclopentane from two moles of ethylene closely resembles the coupling of two moles of butadiene to a metallacycle, which is the first step in oligomerization of the diene by nickel or titanium catalysts (Section 4.5).

There is an interesting case where the formation of a metallacyclopentane leads to the formation of 1-hexene rather than the expected 1-butene [110]. The metallacyclopentane is formed by coupling of two ethylenes as above, but this intermediate is relatively stable. When a third ethylene is inserted, the complex is less stable, undergoing a β-elimination on one end and then reductive elimination to yield 1-hexene with high selectivity. This catalytic cycle presumably involves cycling through Cr(III) and Cr(IV).

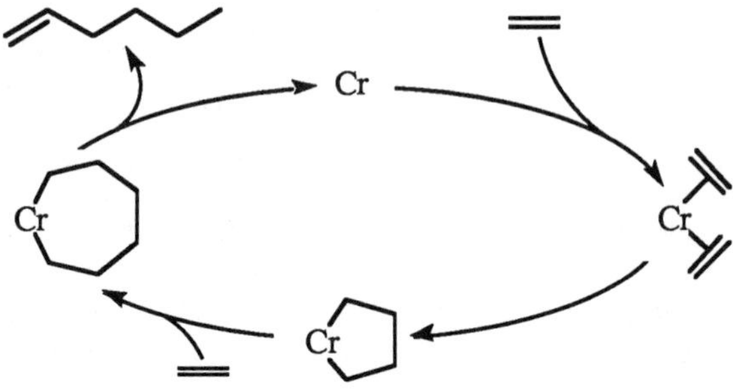

4.4 1,4-HEXADIENE

Codimerization of ethylene and butadiene with a rhodium catalyst is used to produce *trans*-1,4-hexadiene, which is a comonomer in ethylene-propylene-diene (EPDM) elastomers. Ziegler catalysts based on nickel, cobalt, and iron salts are also very effective for this codimerization [111]. The nickel catalysts produce largely the industrially interesting trans isomer while the cobalt and iron catalysts give the cis isomer cleanly. Since many of these catalysts also bring about olefin isomerization, care is required to remove the product from the reaction mixture before it isomerizes to the more stable conjugated 2,4-hexadiene. Similar nickel-based Ziegler catalysts can be used to prepare ethylidenenorbornene [112], which is also widely used in EPDM elastomers. Ethylene reacts with norbornadiene in two metal-catalyzed steps:

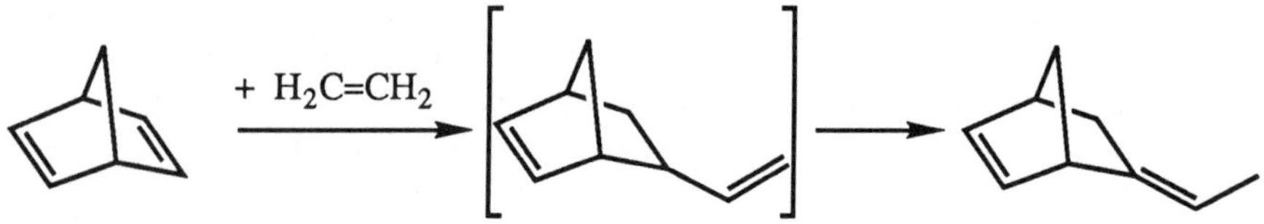

The industrial synthesis of *trans*-1,4-hexadiene is carried out under pressure, but the codimerization of ethylene and butadiene can be studied conveniently in the laboratory [113]. Addition of an equimolar mixture of ethylene and butadiene to a methanol solution of commercial $RhCl_3 \cdot 3H_2O$ leads to a slow reaction which accelerates with time. The initial induction period during which Rh(III) is reduced to Rh(I) can be eliminated by adding a preformed rhodium(I) catalyst such as $[RhCl(C_2H_4)_2]_2$. The product from the 1:1 mole ratio of ethylene and butadiene is a mixture of hexadienes which forms by isomerization of the initially formed *trans*-1,4-isomer. However, isomerization can be suppressed by maintaining high concentrations of butadiene [111], that is, by operating at low conversion.

Most hexadiene syntheses with the nickel- and cobalt-based Ziegler catalysts are carried out in metal pressure reactors. Typically, $NiCl_2(PBu_3)_2$ and $Al_2Cl_2(i\text{-}Bu)_4$ are mixed in tetrachloroethylene which has been presaturated with butadiene and ethylene [114]. The two olefins react rapidly at 65°C to give mixed hexadienes (mostly *trans*-1,4) and some 3-methyl-1,4-pentadiene, which are isolated by distillation. Instead of using a nickel(II) complex with a reducing agent, nickel(0) complexes can be used to codimerize ethylene and butadiene if acidic cocatalysts are supplied [115]. A mixture of bis(1,5-cyclooctadiene)nickel, $C_6F_5PPh_2$, $EtAlCl_2$, and Et_2AlOEt gives a high yield of *trans*-1,4-hexadiene [116]. The aluminum cocatalyst can be replaced by a protonic acid. A solution of $Ni[P(OEt)_3]_4$ and H_2SO_4 in methanol effects hexadiene synthesis under conditions in which several intermediates can be observed [117]. As noted in Chapter 3.6, such combinations of acid and Ni(0) generate a nickel hydride which is believed to be the true catalyst.

All the nickel and rhodium catalysts are based on metal hydride complexes. In Du Pont's rhodium system, several steps are involved in catalyst formation during the induction period [102]:

$$RhCl_3 + C_2H_5OH \longrightarrow \text{"RhCl"} + HCl + CH_3CHO$$

$$\text{"RhCl"} + HCl \longrightarrow \text{"HRhCl}_2\text{"} \xrightarrow{3L} HRhCl_2(L)_3$$

The rhodium hydride enters the catalytic cycle at the top. Reaction with butadiene gives a *syn*-π-crotyl complex which coordinates a molecule of ethylene to form an ethylene crotyl complex. Insertion of ethylene at the less hindered end of the crotyl ligand forms a *trans*-4-hexenyl rhodium complex. β-Hydrogen elimination yields a transient Rh-H complex of *trans*-1,4-hexadiene which dissociates the product [113].

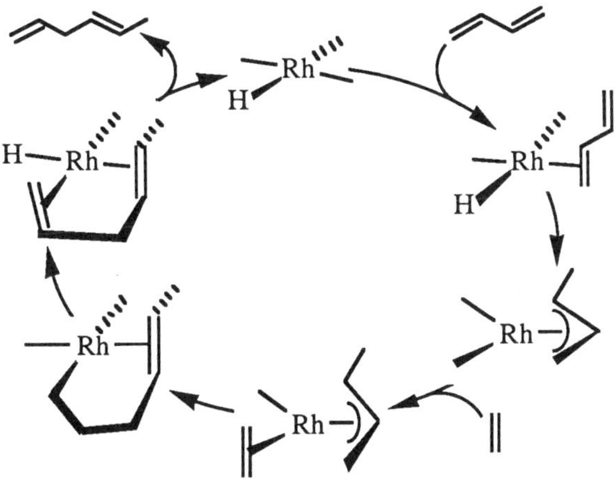

With the rhodium and the $[HNiL_4]^+$ catalysts, it seems likely that the stereochemistry about the double bond in the product is determined by an isomerization of the π-crotyl intermediate [111]. The initial insertion of butadiene into the metal-hydrogen bond is believed to give an *anti*-crotyl intermediate:

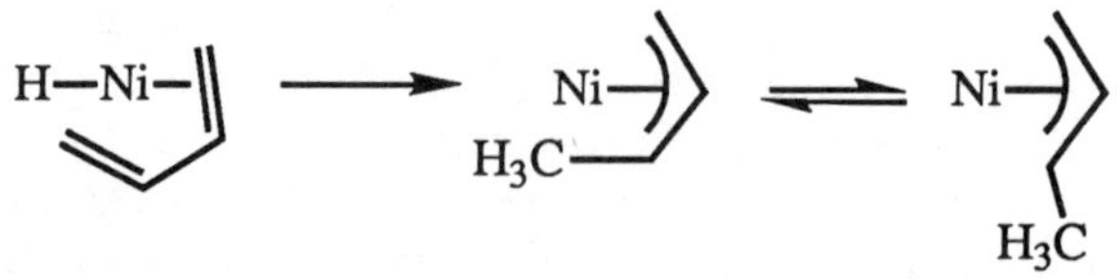

This isomer would yield *cis*-1,4-hexadiene as the final product, but isomerization to the *syn*-crotyl complex is faster than ethylene insertion [118].

In the Ziegler systems, product stereochemistry is believed to arise from mono- or bidentate coordination of the incoming butadiene ligand just as in butadiene polymerization. With nickel catalysts in the presence of excess phosphine ligand, monodentate coordination gives the *syn*-crotyl intermediate which yields *trans*-1,4-hexadiene:

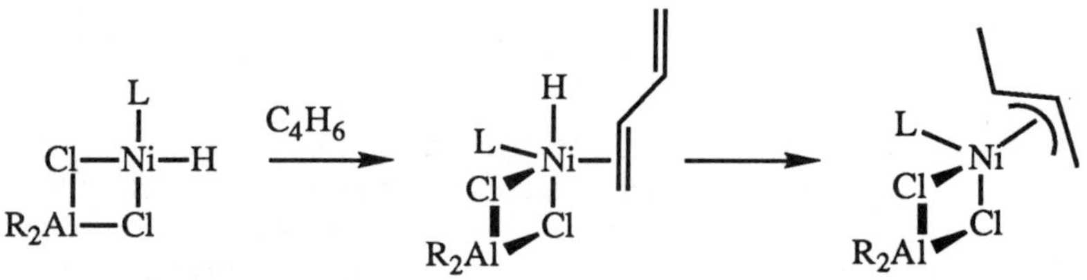

However, with a cobalt catalyst or with nickel in the absence of phosphorus ligands, the butadiene is chelated to the metal:

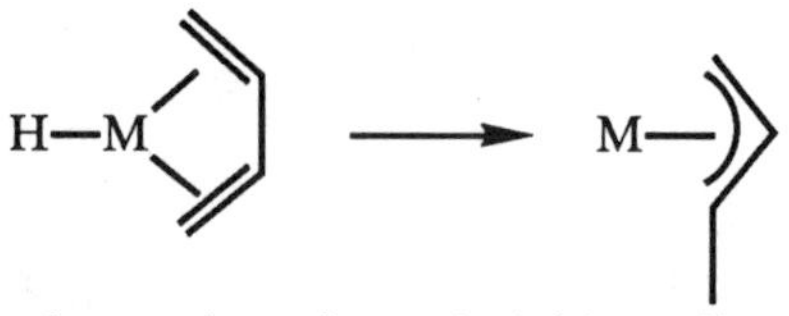

The resulting *anti*-crotyl complex gives *cis*-1,4-hexadiene very cleanly with a catalyst system such as $CoH(Ph_2PCH_2CH_2PPh_2)_2$ [119,120].

4.5 DIMERIZATION AND TRIMERIZATION OF DIENES

Butadiene and other conjugated dienes undergo a variety of oligomerization reactions to give both linear and cyclic products. Two are used commercially: (1) cyclodimerization of butadiene with a nickel catalyst to produce 1,5-cyclooctadiene, which is used in the preparation of flame retardants such as tetrabromocyclooctane; and (2) cyclotrimerization of butadiene to give 1,5,9-cyclododecatriene, an intermediate in the manufacture of dodecanedioic acid (Chapter 10), 1,13-tetradecadiene, and lauryllactam [121]. Other potential industrial applications include dimerization of butadiene to linear octatrienes and cyclodimerization of isoprene to dimethylcyclooctadienes. Some of these hydrocarbons would be difficult to synthesize by conventional methods.

Diene Dimerization

Ziegler catalysts like those used for synthesis of polybutadiene and 1,4-hexadiene can be modified to produce both linear and cyclic dimers and trimers of butadiene. Nickel catalysts are the most versatile and can be used to make almost any of the products by a suitable choice of ligands and reaction conditions [99]. Emphasis here is centered on 1,5-cyclooctadiene (COD) and 1,3,7-octatriene (OT). Other accessible dimers include 4-vinylcyclohexene (VCH) and 1,2-divinylcyclobutane (DVCB).

| COD | OT | VCH | DVCB |

Many substituted butadienes also dimerize to give analogous cyclooctadienes and octatrienes [122]. The isoprene dimers and trimers are especially interesting for synthesis of terpenoid and sesquiterpenoid compounds of biological interest. The dimethylcyclooctadienes can be used as intermediates in the synthesis of fragrances [123]. Low-valent palladium catalysts are often advantageous for preparation of the linear dimers such as 2,7-dimethyl-2,4,6-octatriene [124].

The cyclodimerization of butadiene to give 1,5-cyclooctadiene is catalyzed by a zero-valent nickel complex which contains one mole of a triaryl phosphite ligand [125]. In large-scale syntheses, nickel(II) acetylacetonate is reduced with an organoaluminum compound in the presence of the phosphite. This catalyst differs from that used in hexadiene synthesis or butadiene polymerization in that the aluminum compound need not be a Lewis acid. An equally useful catalyst is obtained by reacting the triaryl phosphite with Ni(COD)₂ in liquid butadiene. The most effective catalysts are based on bulky ligands such as tris(o-phenylphenyl) phosphite. The catalyst containing this ligand rapidly dimerizes butadiene at 80°C and 1 atmosphere pressure in a hydrocarbon solvent such as cyclooctadiene. The product is 96% 1,5-cyclooctadiene along with small amounts of trimers and 4-vinylcyclohexene. The latter product is the normal thermal dimer of butadiene although it is not commonly formed below 150°C. When the dimerization is carried out at low temperatures and low conversions, 1,2-divinylcyclobutane may be isolated in yields up to 40%. It readily isomerizes to COD and vinylcyclohexene in the presence of the catalyst.

The same catalysts that are used to cyclodimerize butadiene produce linear dimers when a slightly acidic coreactant is added to the reaction mixture [99]. The coreactant may be ROH, R₂NH, HCN, or active methylene compounds. It may be incorporated in the product or it may promote formation of an octatriene. For example, phenols can give either octatrienes [126] or phenoxyoctadienes [127] as major products:

$$2 \quad \diagup\!\!\!\diagdown\!\!\!\diagup \quad + \text{ PhOH} \quad \xrightarrow{\text{Ni}[\text{P}(\text{OPh})_3]_4} \quad \text{PhO} \diagup\!\!\!\diagdown\!\!\!\diagup\!\!\!\diagdown\!\!\!\diagup$$

In general, however, zero-valent palladium catalysts are more effective for production of these linear dimers. With Pd(PPh3)2 (maleic anhydride) as a catalyst, isoprene gives 2,7-dimethyl-1,3,7-octatriene exclusively [124]. Kuraray has developed a diol synthesis based upon the hydrodimerization of butadiene followed by hydroformylation (see Section 5.5). They have started a semiworks to hydrodimerize butadiene using a homogeneous palladium catalyst which employs a monosulfonated triphenylphosphine to solubilize the catalyst in a sulfolane/H_2O phase. Triethylamine is used to stabilize the catalyst, and the reaction is carried out under an atmosphere of CO_2.

$$2 \quad \diagup\!\!\!\diagdown\!\!\!\diagup \quad + H_2O \quad \xrightarrow[\text{Ph}_2\text{P}(m\text{-C}_6\text{H}_4\text{SO}_3\text{Na})]{\text{Pd}(\text{OAc})_2} \quad \diagup\!\!\!\diagdown\!\!\!\diagup\!\!\!\diagdown\!\!\!\diagup\!\!\!\diagdown \text{OH}$$

The hydrodimer is isomerized, hydroformylated, and hydrogenated to yield 1,9-nonanediol. Additional details are given in Chapter 5.

The activity of diene dimerization catalysts is modified by the presence of carbon dioxide. The CO_2-promoted reaction of Pt(PPh3)3 gives octatrienes [128]. With Pd(diphos)2, it gives both octatrienes and octadiene lactones [129]. The presence of CO_2 can also induce formation of octadienols from butadiene and water [130]. It has been speculated that CO_2 and water combine under pressure to give carbonic acid which adds to an octatriene precursor to give a water-sensitive octadienyl carbonate. In the nickel system, a CO_2 insertion product of an isoprene dimer complex has been characterized [131]. The mechanism of butadiene dimerization has been studied thoroughly [99,132,133]. It is likely that the same intermediates are involved in formation of the linear and cyclic dimers. Some of these intermediates also occur in the pathway to 1,5,9-cyclododecatriene, a cyclic trimer of butadiene. It appears that two molecules of butadiene coordinate to a zero-valent nickel atom which bears one phosphorus ligand, L. The critical step in dimer formation is coupling the two monodentate butadiene ligands to an octadienyl ligand:

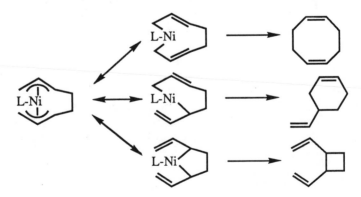

On the basis of its proton nmr spectrum, the dimer-ligand is believed to bond to the metal through a σ-allyl function as well as a π-allyl. However, the spectroscopically detectable σ,π-form is tautomeric with the three isomeric σ-allylic structures shown. Each isomer can undergo reductive elimination of two Ni-C bonds to generate a different cyclic dimer and reform the "L-Ni" catalyst. In practice, all three processes occur and each can predominate with proper choice of reaction conditions and the ligand, L. The formation of 1,3,7-octatriene can be rationalized by β-hydrogen elimination from the tautomer that ordinarily gives rise to divinylcyclobutane:

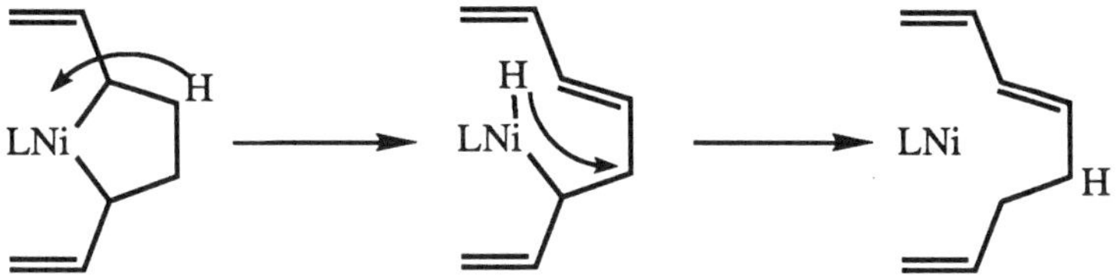

This rationalization oversimplifies the situation because it ignores the role of the weak acid cocatalyst. It has been suggested [99] that the cocatalyst protonates one allyl function to achieve the Ni-C bond cleavage shown as the second step above.

Cyclododecatriene Synthesis

Butadiene and substituted dienes trimerize readily under the influence of soluble transition metal catalysts [134], forming C_{12} molecules useful as intermediates to nylon-12's (see Section 10.4, Oxidation of Cyclododecane). The catalyst used for commercial synthesis of cyclododecatriene is almost identical to that used for polymerization of ethylene [135]. In the industrial process [136], the catalyst is prepared by mixing $TiCl_4$ with excess $Al_2Cl_3Et_3$ in benzene. Butadiene and freshly mixed catalyst solution are fed to a reactor in which trimerization occurs in a few minutes at about 70°C and 1 atmosphere pressure. After catalyst deactivation, distillation gives a 75-90% yield of 1,5,9-cyclododecatriene which is almost entirely the cis-trans-trans isomer. Minor amounts of the all-*trans* isomer, polybutadiene, 1,5-cyclooctadiene, and 4-vinylcyclohexene also form.

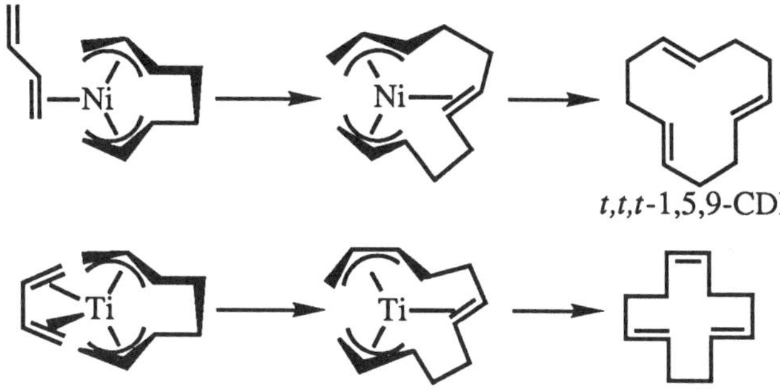

t,t,t-1,5,9-CDDT

c,t,t-1,5,9-CDDT

The cyclotrimerization of butadiene with nickel catalysts has been studied more thoroughly [137] even though it is not used industrially. When nickel(II) acetylacetonate is reduced with $Al(C_2H_5)_2(OC_2H_5)$ in the presence of butadiene, a highly reactive complex is formed. This compound, sometimes dubbed "naked nickel," catalyzes trimerization of butadiene to *trans,trans,trans*-1,5,9-cyclododecatriene. Zero-valent nickel complexes, for example, $Ni(COD)_2$, which lack phosphorus ligands, are also very effective for this purpose. The course of the nickel-catalyzed reaction is very sensitive to the nature of the ligands on nickel since, as described earlier, it can produce polymer or either of two butadiene dimers as major products. In addition, it can catalyze cotrimerization of butadiene and ethylene to form either *cis, trans*-1,5-cyclodecadiene or 1,4,9-decatriene [138]. It seems likely that the initial steps in trimerization are identical to those in diene dimerization. Two butadiene ligands couple to form an open dimer in which the double bond has a trans configuration. In the absence of phosphorus ligands, another molecule of butadiene inserts into an Ni-C bond to give a nickel complex which bears a bis(allylic)-C_{12} ligand. On warming in the presence of butadiene, the ends of the allyl functions couple to form the cyclic triene, initially as a Ni(0) complex. The trans configurations of the C=C bonds in the triene derive from syn conformations of the bis(π-allyl) intermediates. It is tempting to speculate that the monodentate bonding of the butadiene ligand dictates formation of the syn allyl intermediate. In contrast, the additional vacant orbitals of a titanium catalyst permit bidentate coordination of the diene as shown, yielding an *anti*-allyl ligand, which in turn generates a cis double bond in the final product (ctt).

The active species in Du Pont's titanium catalyst system are not well characterized. Chemical evidence [139] and analogy to olefin polymerization catalysts suggest that the Ti species are trivalent. However, bis(benzene)titanium(0) catalyzes the trimerization when used in combination with $Et_2Al_2Cl_4$ [140], suggesting a Ti(0)-Ti(II) cycle exactly analogous to that observed with nickel. The $Ti(C_6H_6)_2$ catalyst ordinarily gives *cis,trans,trans*-1,5,9-cyclododecatriene, but addition of triphenylphosphine yields the all-trans isomer, supporting the proposal that the stereochemistry of the product is dictated by the number of orbitals available for coordination of the third butadiene molecule.

4.6   FUNCTIONAL OLEFIN DIMERIZATION

The dimerization of functional olefins would be an attractive route to monomers for a variety of polymers. As a result, it has been investigated extensively. To date, the only functional olefin dimerization of commercial significance is Monsanto's electrolytic hydrodimerization of acrylonitrile to adiponitrile, the precursor to hexamethylenediamine [141,142]:

$$2\,CH_2{=}CHCN \;+\; 2e^- \;+\; 2H^+ \longrightarrow NC(CH_2)_4CN$$

This process, originally using lead alloy cathodes, was commercialized in 1965. Further improvements in the process - removal of membranes, improved electrode materials, and high-conductivity electrolytes - have made this the low-cost method of producing this important nylon intermediate on a small to medium scale.

The catalytic dimerization of acrylonitrile to 1,4-dicyanobutene has been investigated as a synthetic route to hexamethylenediamine for nylon production [143,144]. Linear dimerization of acrylates by $RuCl_3$, $RhCl_3$ [145,146], or $PdCl_2$ [147] gives unsaturated esters which can be hydrogenated directly to the corresponding ester of adipic acid.

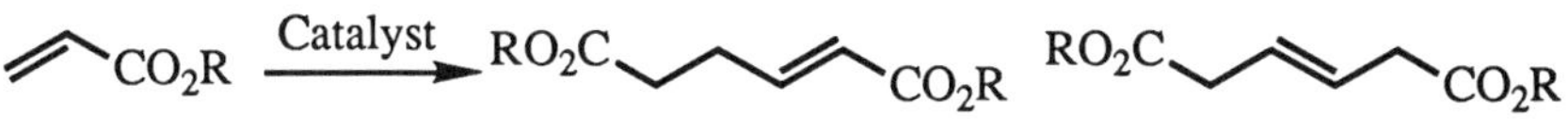

CO2R
Catalyst
RO2C
CO2R
RO2C
CO2R

The reaction is also useful in organic synthesis [148]. The preformed organometallics, $[(C_2H_4)_2RhCl]_2$ and $PdCl_2(NCPh)_2$ are more tractable starting materials and the rates of dimerization are increased markedly by addition of a Lewis acid and a proton source [149]. The dimerization is also catalyzed by the zero-valent species $Ru(\eta\text{-}C_6H_6)(CH_2{=}CHCO_2Me)_2$ in the presence of sodium naphthalenide [150] or by $Ru_3(CO)_{12}$ with PPh_3 [151].

Several mechanisms have been proposed for functional olefin dimerization, and it is not clear that all reactions go by the same mechanism. The first possible mechanism is through the intermediacy of a metallacyclopentane formed by addition of the two olefins early in the process.

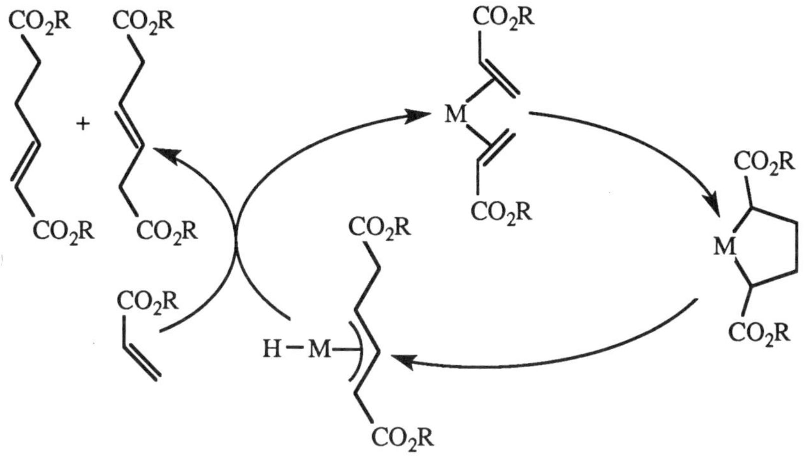

CO2R
CO2R
+
CO2R
CO2R
CO2R
M
CO2R
CO2R
M
CO2R
CO2R
CO2R
H−M
CO2R

β-Hydride elimination forms a π-allyl intermediate which transfers hydrogen back to either of two positions, giving the two products which are observed as cis or trans isomers. The second mechanism involves addition of a single olefin to the metal hydride, resulting in the formation of a functionalized ethyl group. Addition of the second olefin is followed by a second insertion to give the difunctional alkyl complex. β-Hydride elimination gives a complexed olefin which is displaced by incoming monomer. A deuterium-labelling study provides strong support for this second mechanism in acrylonitrile dimerization on ruthenium under hydrogen or deuterium [152]. It is clear that much work remains to be done to sort out the details of these reactions with different substrates and different metal catalysts.

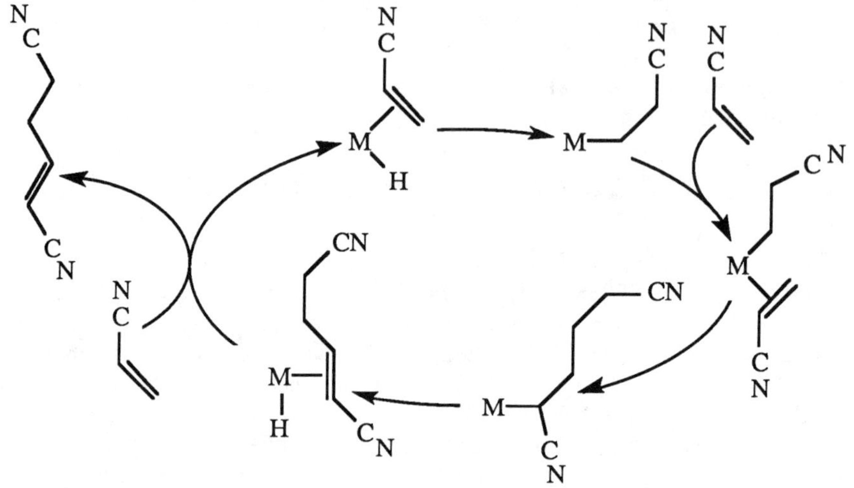

The relationship between agostic interactions in coordinatively unsaturated metal complexes and the ability to polymerize ethylene has been pointed out [153]. The agostic M-H-C interaction stabilizes the complex but is easily displaced by an incoming olefin. One result of this relationship was the discovery of cobalt-based catalysts for ethylene polymerization.

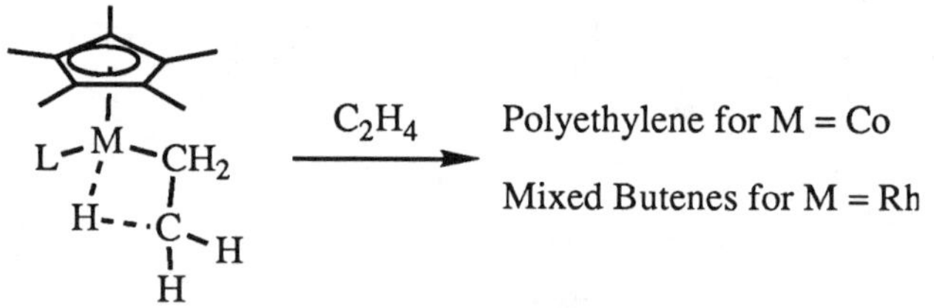

The rhodium analog is active for the dimerization of ethylene to mixed butenes [154]. When the rhodium analog is employed for methyl acrylate dimerization, an atmosphere of hydrogen prolongs the lifetime of the catalyst [155].

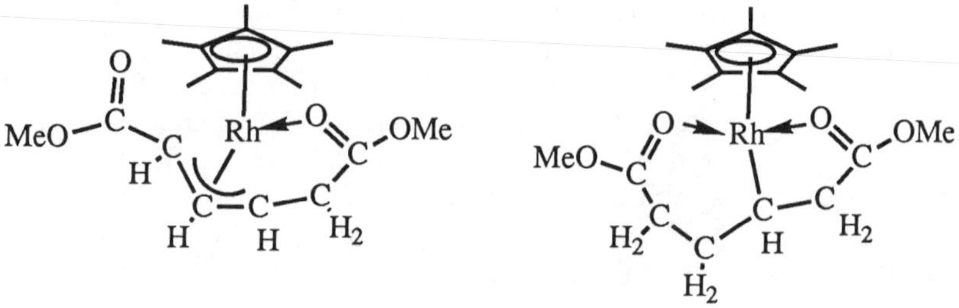

The two products isolated from the reaction proved to be a degradation product which could be reactivated by the hydrogen atmosphere and an intermediate apparently on the catalytic cycle. Both complexes display coordination of the ester carbonyl group to decrease coordinative unsaturation at the metal center. This coordination serves to highlight the role that the functional group can play in these reactions and why each monomer may generate a new reaction mechanism.

4.7 CHAIN TRANSFER CATALYSIS

Early in this chapter, the free-radical polymerization of olefins was mentioned. These reactions do not normally involve transition metal complexes except occasionally as redox couples for the generation of odd-electron species. Molecular weight is usually controlled by adjusting the concentration of initiator to reactant, sometimes with the addition of a thiol to terminate growing chains. For example, in methyl methacrylate polymerizations, the reaction is initiated by the thermal decomposition of azobis(isobutyronitrile) (AIBN or Vazo-57®) with phenyl mercaptan to chain terminate. To achieve the very low molecular weight polymers which are desired for high-solids automotive finishes or other applications, the high concentrations of initiator and terminator cause problems with toxicity and odor.

Cobalt(II) complexes of tetra-aza-ligands such as porphyrins [156,157] or dimethylglyoximates [158-161] are extremely effective catalysts for intercepting the growing polymer chains.

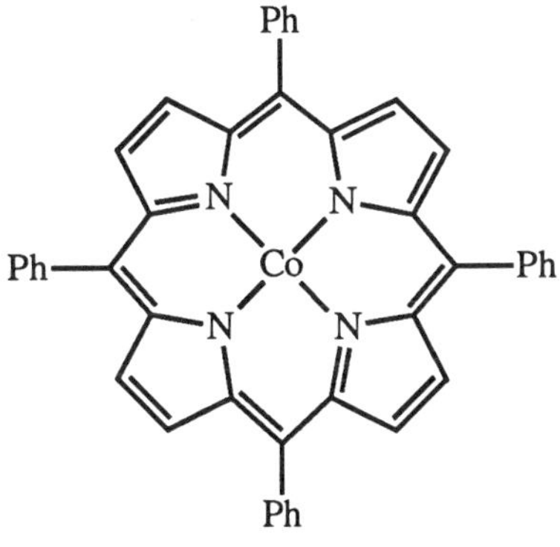

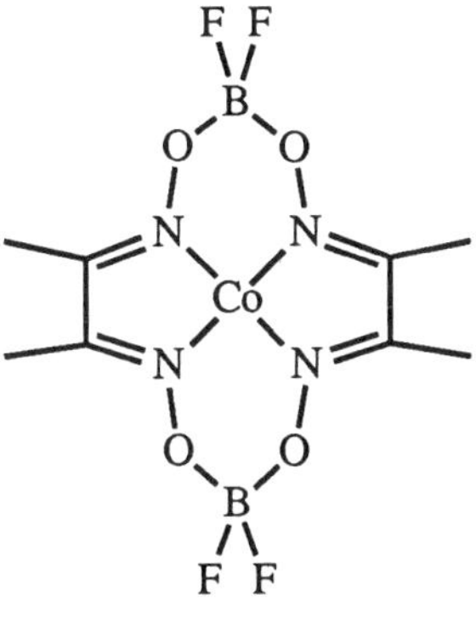

The mechanism of action is shown in Figure 4.7. Because they are essentially radical species themselves, having an odd electron count, they intercept the radical end of the growing polymer chain at diffusion-controlled rates to form a cobalt(III) alkyl. This alkyl complex then undergoes β-elimination to generate a polymer chain having an unsaturated end-group and a Co(III)-H which initiates a new free-radical chain. At parts per million concentrations of chain-transfer catalyst, molecular weights of the polymer are dropped from tens of thousands down to thousands. At higher concentrations of catalyst, the resulting products can be macromonomers incorporating just a few methacrylate units [162]. The hydrogen atom eliminated

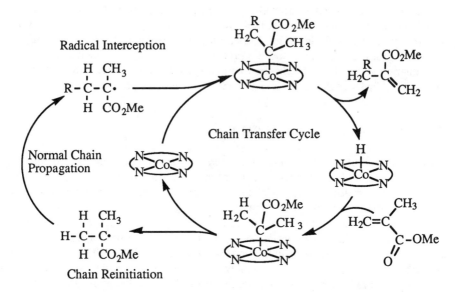

Figure 4.7 Mechanism of chain transfer catalysis in free-radical initiated methacrylate polymerizations.

from the polymer chain is almost always derived from the methyl group of the incoming monomer giving a terminal methylene group rather than an internal double bond. The catalysts are equally effective for methacrylonitrile polymers and related species, and any free-radical copolymers which incorporate these groups. The lower molecular weight polymers are useful in low-solvent automotive finishes. At higher concentrations of catalyst, the products of the reaction are oligomers which find use as macromonomers. These macromonomers are interesting in that they do not homopolymerize, but they will copolymerize with a variety of other monomers, giving comb-like copolymers.

GENERAL REFERENCES

J. Boor, *Ziegler-Natta Catalysts and Polymerizations*, Academic Press, New York, 1978.

M. Sittig, *Polyolefin Production Processes*, Noyes Data Corp., Park Ridge, New Jersey, 1976.

History of Polyolefins - The World's Most Widely Used Polymers, R. B. Seymour and T. Cheng, Eds., D. Reidel, Dordrecht, 1986.

Advances in Polyolefins - The World's Most Widely Used Polymers, R. B. Seymour and T. Cheng, Eds., Plenum, New York, 1987.

Transition Metal Catalyzed Polymerizations - Alkenes and Dienes, R. P. Quirk, Ed., MMI Press, Volume 4A and B, Harwood Academic Publishers, New York, 1983.

Transition Metal Catalyzed Polymerizations: Ziegler-Natta and Metathesis Polymerizations, R. P. Quirk, Ed., Cambridge University Press, New York, 1988.

Rubber Technology, 3rd Ed., M. Morton, Ed., Van Nostrand Reinhold, New York, 1987.

J. Skupińska, *Chem. Rev.*, **91**, 613-648 (1991).

"Elastomer Stereospecific Polymerizations," B. L. Johnson and M. Goodman, Eds. *Advances in Chemistry Series 52*, American Chemical Society, Washington, DC, 1952, for an interesting historic perspective.

D. B. Sicilia, "A Most Invented Invention," *American Heritage of Invention and Technology, 6* (Spring/Summer), 45 (1990) presents an interesting perspective on the role of patents and litigation in the area of polyolefins.

Encyclopedia of Polymer Science and Engineering, 2nd ed., H. F. Mark, N. M. Bikales, C. Overberger, and G. Menges, Eds., Wiley Interscience, New York, 1985-89, provides a thorough insight into the economics, process chemistry, and uses of a number of the polymers covered here.

SPECIFIC REFERENCES

1. K. Ziegler, H. Breil, H. Martin, and E. Holzkamp, U.S. Patent 3,257,332 (filed Nov. 17, 1953, issued 1966).
2. J. P. Hogan and R. L. Banks U.S. Patent 2,825,721 (filed Jan. 27, 1953, issued 1958) to Phillips Petroleum.
3. A. Clark, J. P. Hogan, R. L. Banks, and W. C. Lanning, *Ind. Eng. Chem.*, **48**, 1152 (1956).
4. F. J. Karol, G. L. Karapinka, C. Wu, A. W. Dow, R. N. Johnson, and W. L. Carrick, *J. Polym. Sci.*, **10**, 2621 (1972).
5. G. L. Karapinka, U.S. Patent 3,709,853 (1973) to Union Carbide.
6. W. L. Carrick, *Macromol. Synth.*, **2**, 33 (1966) .
7. M. A. Hamilton, D. A. Harbourne, C. G. Russell, V. G. Zboril, and R. M. Mulhaupt, EPO 0,131,420 (1988) and C. G. Russel, U.S. Patent 4,612,382 (1986), both to Du Pont Canada.
8. R. Spitz, A. Revillon, and A. Guyot, *J. Catal.*, **35**, 335, 345 (1974).
9. A. Zecchina, E. Garrone, G. Ghiotti, and S. Coluccia, *J. Phys. Chem.*, **79**, 972 (1975)
10. E. A. Benham, P. D. Smith, E. T. Hsieh, and M. P. McDaniel, *J. Macromol. Sci., Chem.*, **A25**, 259 (1988).

11. J. P. Hogan, D. D. Norwood, and C. A. Ayers, *J. Appl. Polym. Sci. Appl. Polym. Symp.*, **36**, 49-60 (1981).

12. M. P. McDaniel in *Advances in Catalysis*, Vol. 33, D. D. Eley, H. Pines, and P. B. Weisz, Eds., Academic Press, New York , 1985, p. 47.

13. I. J. Levine and F. J. Karol, U.S. Patent 4,011,382 (1977); F. J. Karol, G. L. Goeke, B. E. Wagner, W. A. Fraser, R. J. Jorgensen, and N. Friis, U.S. Patent 4,302,566 (1981); and K. J. Cann, D. L. Miles, and F. J. Karol, U.S. Patent 4,670,526 (1987), all to Union Carbide.

14. F. J. Karol, *Catal. Rev. - Sci. Eng.*, **26**, 557-595 (1984).

15. R. J. Jorgensen, U.S. Patent 4,710,538 (1987) and R. J. Bernier and G. E. Keller, U.S. Patent 4,910,294, both to Union Carbide.

16. R. J. Bernier, J. O. Buhler-Vidal, U.S. Haapala, and B. R. Rozenblat, EP 0,348,907 (1990).

17. A. W. Anderson and G. S. Stamatoff, Canadian Patent 664,211 (1963) and U.S. Patent 4.076,698 (1978, filed in 1957) to Du Pont.

18. C. T. Elston, U.S. Patent 3,645,992 (1972) to Du Pont of Canada.

19. S-B. Samuels and F. J. Karol, U.S. Patent 4,918,038 (1990) to Union Carbide.

20. H. Sinn and W. Kaminsky, *Adv. Organomet. Chem.*, **18**, 99 (1980).

21. H. Sinn, W. Kaminsky, H. J. Vollmer, and R. Woldt, *Angew. Chem.*, **92**, 396 (1980).

22. L. Resconi, S. Bossi, and L. Abis, *Macromolecules*, **23**, 4489-91 (1990).

23. *Chem. Eng. News*, 42, March 19 (1992).

24. W. Kaminsky, K. Kulper, H. H. Brintzinger, and F. R. W. P. Wild, *Angew. Chem., Int. Ed.*, **24**, 507 (1985).

25. J. A. Ewen, *J. Am. Chem. Soc.*, **106**, 6355 (1984).

26. J. A. Ewen, L. Haspeslagh, J. L. Atwood, and H. Zhang, *J. Am. Chem. Soc.*, **109**, 6544 (1987).

27. W. Kaminsky, K. Kulper, H. H. Brintzinger, and F. R. W. P. Wild, *Angew. Chem. Int. Ed. Engl.*, **24**, 507 (1985).

28. P. Pino, P. Cioni, M. Galimberti, and J. Wei, *J. Am. Chem. Soc.*, **109**, 6189 (1987).

29. P. Pino, P. Cioni, M. Galimberti, J. Wei, and N. Picollrovazzi in *Transition Metals and Organometallics as Catalysts for Olefin Polymerization*, W. Kaminsky and H. Sinn, Eds., Springer, Berlin, 1988, p. 269.

30. P. Pino, M. Galimberti, P. Prado, and G. Consiglio, *Makromol. Chem.*, **191**, 1677-88 (1990).

31. J. A. Ewen, R. L. Jones, A. Razavi, and J. D. Ferrara, *J. Am. Chem. Soc.*, **110**, 6255 (1988).

32. R. F. Jordan, R. E. LaPointe, C. S. Bagjur, S. F. Echols, and R. Willet, *J. Am. Chem. Soc.*, **109**, 4111 (1987).

33. G. G. Hlatky, H. W. Turner, and R. R. Eckman, *J. Am. Chem. Soc.*, **111**, 2728-9 (1989); H. W. Turner, European Patent 277,004 (1988).

34. E. J. Vandenberg, *Macromol. Synth.*, **5**, 95 (1974).

35. R. J . Kern, H. Schnecko, W. Lintz, and L. Kollar, *Macromol. Synth., Coll. Vol. I*, 425 (1977).

36. R. J. Kern, *Macromol. Synth., Coll. Vol. I*, 1 (1977).

37. P. J. T. Tait, in *Developments in Polymerization*, Vol. 2, R. N. Howard, Ed., Applied Science Publishers, Barking, Essex, UK, 1979.

38. P. Cossee, *J. Catal.*, **3**, 80 (1964); E. J. Arlman, *ibid.*, **5**, 178 (1966).

39. M. Ystenes, *J. Catal.*, **129**, 383-401 (1991).

40. D. R. Armstrong, P. G. Perkins, and J. J. P. Stewart, *J. Chem. Soc. Dalton*, 1972 (1972).

41. O. Novaro, E. Blaisten-Barojas, E. Clementi, G. Giunchi, and M. E. Ruiz-Vizcaya, *J. Chem. Phys.*, **68**, 2337 (1978).

42. P Galli, P. C. Barbe, and L. Noristi, *Die Angew. Makromol. Chem.*, **120**, 73-90 (1984).

43. U. Giannini, E. Albizzati, and S. Parodi, U.S. Patent 4,149,990 (1979) to Himont.

44. G. Crespi, A. Valvasorri, and U. Flisi, "Olefin Copolymers," in W. M. Saltman, Ed., *The Stereo Rubbers*, Wiley, New York, 1977, pp. 365-427.

45. S. Cesca, *Macromol. Rev.*, **10**, 1 (1975).

46. R. German, G. Vaughan, and R. Hank, *Rubber Chem. Tech.*, **40**, 569 (1967).

47. P. J. T. Tait, "Ziegler-Natta and Related Catalysts," in R. N. Haward, Ed., *Developments in Polymerization*, Vol. 2, Applied Science Publishers, London, 1979, pp. 81-148.

48. G. Henrici-Olive and S. Olive, *Angew. Chem. Int. Ed.*, **10**, 776 (1971).

49. A. G. Evans, J. C. Evans, and E. H. Moon, *J. Chem. Soc. Dalton*, 2390 (1974).

50. D. Apotheker and N. van Gulick, U.S. Patent 3,723,348 (1973) to Du Pont.

51. T. A. Cooper, *Macromol. Prep.* (24th IUPAC) 257 (1975).

52. G. G. Evans, E. M. J. Pijpers, and R. H. M. Seevens, in *Transition Metal Catalyzed Polymerizations: Ziegler-Natta and Metathesis Polymerizations*, R. P. Quirk, Ed., Cambridge University Press, New York, 1988, p. 782.

53. D. L. Beran, K. J. Cann, R. J. Jorgensen, F. J. Karol, N. J. Maraschin, and A. E. Marcinkowsky, U.S. Patent 4,508,842 (1985) to Union Carbide.

54. S. Floyd and E. L. Hoel, European Patent application EP 347129 (1989).

55. J. W. Collette and C. W. Tullock, U.S. Patent 4,335,225 (1982) to Du Pont.

56. J. W. Collette, C. W. Tullock, R. N. MacDonald, W. H. Buck, A. C. L. Su, J. R. Harrell, R. Mulhaupt, and B. C. Anderson, *Macromolecules*, **22**, 3851 (1989).

57. C. W. Tullock, F. N. Tebbe, R. Mulhaupt, D. W. Ovenall, R. A. Setterquist, and S. D. Ittel, *J. Polym. Sci., Polym. Chem. Ed.*, **27**, 3063 (1989).

58. J. W. Collette, D. W. Ovenall, W. Buck, and R. C. Ferguson, *Macromolecules*, **22**, 3858 (1989).

59. R. A. Setterquist, U.S. Patents 3,932,307 (1976), 3,950,269 (1976), 3,971,767 (1976), and 4,017,525 (1977), all to Du Pont.

60. R. A. Setterquist, F. N. Tebbe, and W. G. Peet in *Coordination Polymerization*, C. C. Price and E. J. Vandenberg, Eds., Plenum, New York, 1983.

61. N. Kashiwa, *Nikkei New Material*, 65-69 (May 7, 1990); Japanese Kokai Tokkyo 60-168708 (1985), 2-276842 (1990), and 2-276843 (1990) to Mitsui Petrochemical.

62. *Nikkei New Material*, 82-96 (Sept. 18, 1989).

63. W. Kaminsky A. Bark, R. Spiehl, N. Moller-Lindenhof, and S. Nieboda in *Proceedings, International Symposium on Transition Metal and Organometallic as Catalysts for Olefin Polymerization.*, W. Kaminsky and H. Sinn, Eds., Springer Press, Berlin, 1988, p. 291; W. Kaminsky, A. Bark, M. Joachim, and H. Cherdron, Ger. Offen. DE 3,835,044 (1990) to Hoechst.

64. S. Collins, W. J. Gauthier, W. M. Kelly, and D. G. Ward, *9th International Symposium on Olefin Metathesis and Polymerization*, Collegeville, PA, July 1991.

65. A. Sen, T. -W. Lai, and R. R. Thomas, *J. Organomet. Chem.*, **358**, 567 (1988).

66. N. Seehof, C. Mehler, S. Breunig, and W. Risse, *9th International Symposium on Olefin Metathesis and Polymerization*, Collegeville, PA, July 1991.

67. V. M. Mohring and G. Fink, *Angew. Chem. Int. Ed.*, **24**, 1001 (1985).

68. G. Fink, *NATO ANSI Ser., Ser. C.*, **215**, 515-33 (1987).

69.   G. Fink and V. Mohring, Eur. Pat. Appl. EP194456 (1986).

70.   L. J. Kuzma, "Polybutadiene and Polyisoprene Rubbers," in *Rubber Technology, 3rd. Ed.*, M. Morton, Ed., Van Nostrand Reinhold, New York, 1987.

71.   E. W. Duck and J. M. Locke, "Polydienes by Anionic Catalysts," in W. M. Saltman, Ed., *The Stereo Rubbers*, Wiley, New York, 1977, p. 139.

72.   W. M. Saltman and E. Schoenberg, *Macromol. Syn.*, **2**, 50 (1966).

73.   For instance, see W. M. Saltman and T. H. Link, *Ind. Eng. Chem. Prod. Res. Dev.*, **3**, 199 (1964).

74.   J. S. Lasky, *Macromol. Syn., Coll., Vol. I*, 141 (1977) .

75.   G. Natta, L Porri, and A. Mazzie, *Chem. Ind.*, **41**, 116 (1959).

76.   A. J. Bell, U.S. Patent 4,645,809 (1987) to Goodyear.

77.   Y. Takeuchi, *Chem. Econ. Eng. Rev.*, **4**, 37 (1972).

78.   C. F. Gibbs, V. L. Folt, E. J. Carlson, S. E. Horne, and H. Tucker, British Patent UK 916,383 (1963).

79.   M. Gippin, *Macromol. Synth.*, **2**, 42 (1966)

80.   M. Gippin, *Ind. Eng. Chem. Prod. Res. Dev.*, **1**, 32 (1962).

81.   V. N. Zgonnik, B. A. Dolgoplosk, N. I. Nikolaev, and V. A. Kropachev, *Vyosokomol. Soedin.*, **4**, 1000-4 (1962).

82.   G. Henrici-Olive, S. Olive, and E. Schmidt, *J. Organomet. Chem.*, **39**, 201 (1972).

83.   Ph. Teyssie and F. Dawans, in W. M. Saltman, Ed., *The Stereo Rubbers*, Wiley, New York, 1977, p. 79.

84.   J. P. Durand, F. Dawans, and Ph. Teyssie, *J. Polym. Sci.*, A-l, **8**, 979 (1970).

85.   C. A. Tolman *J. Am. Chem. Soc.*, **92**, 6777-84 and 6785-90 (1970).

86.   Ph. Teyssie, *Proceedings, 1st International Symposium on Homogeneous Catalysis*, Corpus Christi, TX, Nov. 29-Dec. 1, 1978 .

87.   J. P. Durand, F. Dawans, and Ph. Teyssie, *J. Polym. Sci.*, A-l, **8**, 979 (1970).

88.   J. Furukawa, *Acc. Chem Res.*, **13**, 1-6 (1980).

89.   K. L. Lindsay, "Alpha-Olefins" in J. J. McKetta and W. A. Cunningham, *Encyclopedia of Chemical Processing and Design*, Vol. 2, M. Dekker, New York, 1977, p. 482.

90.   A. Lundeen and R. Poe, "Alpha-Alcohols," in J. J. McKetta and W. A. Cunningham, *Encylcopedia of Chemical Processing and Design*, Vol. 2, M. Dekker, New York, 1977, p. 465.

91.   K. Ziegler in *Organometallic Chemistry*, H. Zeiss, Ed., Reinhold, New York, 1960, pp. 194-195, 229-231.

92.   A. J. Lundeen and J. E. Yates, U.S. Patent 3,450,735 (1969) to Conoco.

93.   G. W. Kabalka and R. J. Newton, *J. Organomet. Chem.*, **156**, 65 (1978).

94.   *Chem. Week*, 70 (23 Oct. 1974); *Chem. Market. Reporter* (18 April 1977).

95.   E. R. Freitas and C. R. Gum, *Chem. Eng. Proc.*, 73 (Jan. 1979).

96.   R. S. Bauer, H. Chung, P. W. Glockner, and W. Keim, U.S. Patent 3,644,563 (1972); R. F. Mason, U.S. Patent 3,737,475 (1973) .

97.   F. H. Kowaldt, Ph.D. Thesis, Aachen, 1977.

98.   G. Lefebvre and Y. Chauvin, in R. Ugo, Ed., *Aspects of Homogeneous Catalysis*, Vol. 1, Carlo Manfredi, Milan, 1970, p. 108.

99.   P . W. Jolly and G . Wilke, *The Organic Chemistry of Nickel,* Vol. II, Academic Press, New York, 1975.

100.   G. Henrici-Olive and S. Olive, *Transition Met. Chem.,* **1,** 109 (1976).

101.   R. Cramer, *J. Am. Chem. Soc.,* **87,** 4717 (1965).

102.   B. Bogdanovic, H. Biserka, H. G. Karmann, H. G. Nussel, D. Walter, and G. Wilke, *Ind. Eng. Chem.,* **62,** 34 (Dec. 1970).

103.   R. Cramer, *J. Am. Chem. Soc.,* **87,** 4717 (1965).

104.   J. C. Stevens and W. A. Fordyce, U.S. Patents 4,695,669 (1987) and 4,855,523 (1989), both to Dow.

105.   *Chemistry in Britain,* 400, May (1990).

106.   Y. Chauvin, A. Hennico, G. Leger, and J. L. Nocca, *Erdol, Erdgas, Kohle,* **106,** 309-315 (1990).

107.   J. Leonard and J. F. Gaillard, *Hydrocarbon Proc.,* **99,** March (1981).

108.   S. J. McLain and R. R. Schrock, *J. Am. Chem. Soc.,* **100,** 1315 (1978).

109.   R. H. Grubbs, A. Miyashita, M. Liu, and P. Burk, *J. Am. Chem. Soc.,* **100,** 1300, 2418, 7416 (1978).

110.   J. R. Briggs, *J. Chem. Soc., Chem. Commun.,* 674 (1989); U.S. Patent 4,668,838.

111.   A. C. L. Su, *Adv. Organomet. Chem.,* **17,** 269 (1978).

112.   H. M. J. C. Creemers, U.S. Patent 3,767,717 (1973).

113.   R. Cramer, *J. Am. Chem. Soc.,* **89,** 1633 (1967)

114.   R. G. Miller, T. J. Kealy, and A. L. Barney, *J. Am. Chem. Soc.,* **89,** 3756 (1967).

115.   R. Cramer, U.S. Patents 4,025,570 and 4,028,429 (1977) to Du Pont.

116.   A. C. L. Su and J. W. Collette, *J. Organomet. Chem.,* **36,** 177 (1972).

117.   C. A. Tolman, *J. Am. Chem. Soc.,* **92,** 6777-84 (1970).

118.   C. A. Tolman, *J. Am. Chem. Soc.,* **92,** 6785-90 (1970).

119.   M. Iwamoto and S. Yuguchi, *J. Org. Chem.,* **31,** 4290 (1966).

120.   A. Miyake, G. Hata, M. Iwamoto, and S. Yuguchi, *Proceedings of the 7th World Petroleum Congress,* Mexico City, 1967, p. 37.

121.   W. Griehl and D. Ruestem, *Ind. Eng. Chem.,* **62,** 16 (March 1970).

122.   P. Heimbach, "Nickel Catalyzed Syntheses of Methyl-Substituted Cyclic Olefins," in R. Ugo, Ed., *Aspects of Homogeneous Catalysis,* Vol. 2, D. Reidel, Dordrecht, 1974, p. 79.

123.   A. J. de Jong, Ger. Offen. 2,704,547 (1977).

124.   A. D. Josey, *J. Org. Chem.,* **39,** 139 (1974); U.S. Patent 3,925,497 (1975) to Du Pont.

125.   W. Brenner, P. Heimbach, H. Hey, E. W. Muller, and G. Wilke, *Liebigs Ann. Chem.,* **727,** 161 (1969).

126.   J. Feldman, O. Frampton, B. Saffer, and M. Thomas, *Am. Chem. Soc., Pet. Div. Prepr,* **9,** A-55-64 (1964); U.S. Patents 3,284,529 (1966) and 3,480,685 (1969) to USI Chemicals.

127.   F. J. Weigert and W. C. Drinkard, *J. Org. Chem.,* **38,** 335 (1973).

128.   J . F . Kohnle, L . H . Slaugh, and K. L. Nakamaye, *J. Am. Chem. Soc.,* **91,** 5904 (1969).

129.   Y. Sasaki, Y. Inoue, and H. Hashimoto, *Chem. Commun.,* 605 (1976).

130.   K. E. Atkins, R. M. Manyik, and G. L. O'Connor, U.S. Patent 3,992,456 (1976).

131.   P. W. Jolly, S. Stobbe, G. Wilke, R. Goddard, C. Kruger, J. C. Sekutowski, and Y. H. Tsay, *Angew. Chem. Int. Ed.,* **17,** 124 (1978).

132.   P. W. Jolly, I. Tkatchenko, and G. Wilke, *Angew. Chem. Int. Ed.,* **10,** 329 (1971).

133.   C. R. Graham and L. M. Stephenson, *J. Am. Chem. Soc.,* **99,** 7098 (1977).

134.   H. Breil, P. Heimbach, M. Kroner, H. Muller, and G. Wilke, *Makromol. Chem.,* **69**, 18 (1963).

135.   G. Wilke, *Angew. Chem. Int. Ed.,* **2**, 105 (1963).

136.   W. Ring and J. Gaube, *Chem. Ing. Tech.,* **36**, 1041 (1966).

137.   B . Bogdanovic, P . Heimbach, M. Kroner, and G . Wilke, *Liebigs Ann. Chem.,* **727**, 143 (1969).

138.   P. Heimbach and G. Wilke, *Liebigs Ann. Chem.,* **727**, 183 (1969).

139.   L. I . Zakharkin and V. M. Akhmedov, *Zh. Org. Khim.,* **2**, 998 (1966).

140.   V. M. Akhmedov, M. T. Anthony, M. L. H. Green, and D. Young, *J. Chem. Soc. Dalton*, 1419 (1975).

141.   M. M. Baizer, *Chemtech*, 161-4 (1980); D. E. Danley, *Chemtech*, 302-311 (1980).

142.   F. Beck, *Angew Chem. Int. Ed.,* **11**, 760 (1972).

143.   M. Hidai and A. Misono, "Dimerization of Acrylic Compounds," in R. Ugo, Ed., *Aspects of Homogeneous Catalysis,* Vol. 2, D. Reidel, Dordrecht, 1974, p. 159.

144.   W. Strohmeier and A. Kaiser, *J. Organomet. Chem.,* **114**, 273 (1976).

145.   T. Alderson, E. L. Jenner, and R. V. Lindsey, *J. Am. Chem. Soc.,* **87**, 5638 (1965).

146.   T. Alderson, U.S. Patent 3,013,066 (1961) to Du Pont.

147.   M. G. Barlow, M. J. Bryant, R. N. Haszeldine, and A. G. Mackie, *J. Organomet. Chem.,* **21**, 215 (1970).

148.   W. A. Nugent and F. W. Hobbs, *J. Org. Chem.,* **43**, 5364 (1983).

149.   W. A. Nugent and R. J. McKinney, *J. Mol. Catal.,* **29**, 65-76 (1985).

150.   R. J. McKinney, *Organometallics*, **5**, 1752-3 (1986).

151.   C. Y. Ren, W. C. Cheng, W. C. Chan, C. H. Yeung, and C. P. Lau, *J. Mol. Catal.,* **59**, L1-8, (1990).

152.   D. T. Tsou, J. D. Burrington, E. A. Maher, and R. K. Grasselli, *J. Mol. Catal.,* **22**, 29-45 (1983).

153.   G. F. Schmidt and M. Brookhart, *J. Am. Chem. Soc.,* **107**, 1443-4 (1985).

154.   M. Brookhart and D. M. Lincoln, *J. Am. Chem. Soc.,* **110**, 8719 (1988).

155.   M. Brookhart and S. Sabo-Etienne, *J. Am. Chem. Soc.,* **113**, 2777-9 (1991).

156.   N. Enikolopyan, B. R. Smirnov, G. V. Ponomarev, and I. M. Belgovskii, *J. Polym. Sci., Polym. Chem. Ed.,* **19**, 879 (1981); B. R. Smirnov, A. P. Marchenko, G. V. Korolev, I. M. Bel'govskii, and N. Yenikolopyan, *Polym. Sci. USSR,* **23**, 1158 (1981).

157.   B. R. Smirnov, I. S. Morozova, A. P. Marchenko, M. A. Markevich, L. M. Pushchaeva and N. S. Enikolopyan, *Dolk. Akad. Nauk. SSSR,* **253**, 891-895 (1980); B. R. Smirnov, I. M. Bel'govskii, G. V. Ponomarev, A. P. Marchenko, and N. S. Enikolopyan, *Dolk. Akad. Nauk. SSSR,* **254**, 127-130 (1980).

158.   L. R. Melby, A. H. Janowicz, and S. D. Ittel, Eur. Pat. Appl. EP 86/301444 and 86/301443; A. H. Janowicz and L. R. Melby, U. S. Patent 4,680,352 (1987); A. H. Janowicz, U. S. Patent 4,694,054 (1987), all to Du Pont.

159.   G. M. Carlson and K. J. Abbey, U.S. Patent 4,526,945 to SCM.

160.   R. Amin Sanayei and K. F. O'Driscoll, *J. Macromol. Sci. - Chem.,* **A26**, 1137-1149 (1989).

161.   K. G. Suddaby, R. Amin Sanayei, A. Rudin, and K. F. O'Driscoll, *J. Appl. Polym. Sci.,* **43**, 1565-1575 (1991).

162.   P. Cacioli, G. Moad, E. Rizzardo, A. K. Serelis, and D. H. Solomon, *Polym. Bull.,* **11**, 325 (1984).

5 | REACTIONS OF CARBON MONOXIDE

The organic chemistry of carbon monoxide embraces some of the most important applications of homogeneous catalysis. The hydroformylation of olefins has been practiced commercially since the early 1940s [1], and new carbonylation processes continue to grow in importance. The feedstock fueling these processes is "synthesis gas," a mixture of carbon monoxide and hydrogen which can be made from any carbon source. The "oil shocks" which changed the face of the chemical industry made coal look like the feedstock of the future. As a result, the Japanese Ministry of International Trade and Industry organized a seven-year C_1 effort involving 14 major chemical companies to develop technology for making basic chemicals from CO and H_2. Many radically new CO-based processes went into development around the world. Most of these processes did not proceed to commercialization because they could not anticipate other forces affecting the global petrochemical industry. As an example, prices of methanol have changed so dramatically in the last 20 years that it has been considered both as a primary product from synthesis gas and a feedstock for synthesis gas plants.

The growth of carbonylation chemistry should continue even when conventional organic feedstocks are scarce because carbon monoxide is available from many sources. Reforming of natural gas and naphtha is now the major source of synthesis gas. However, coal and heavy petroleum fractions are also used as feedstocks for production of synthesis gas. It is likely that coal-based synthesis gas will continue to grow as a feedstock for chemical synthesis, displacing ethylene. For instance, in the last 10 years, virtually all new plants for acetic acid synthesis have been based upon synthesis gas, providing economics which shut down processes based upon ethylene or butane [2].

Even though CO is widely used as a feedstock in the manufacture of organic compounds, it is not very reactive by itself. Powerful electrophiles and nucleophiles attack the CO molecule, but a catalyst is generally required to bring about reactions with olefins, alcohols, or hydrogen. In practice, the catalysts are generally low-valent complexes of Group VIII metals; cobalt and rhodium compounds are especially important from an industrial viewpoint. The vital nature of metal catalysts has stimulated the study of the metal carbonyls as catalysts for reactions of carbon monoxide.

5.1   CHEMISTRY OF CARBON MONOXIDE

Carbon monoxide is an effective ligand for stabilization of low-valent metal complexes. Zero-valent carbonyl complexes of each of the first row transition metals from vanadium to nickel are well characterized [3]. The compositions of these compounds are predicted by the "18-electron rule." It predicts monomeric compounds from metals in even-numbered groups of the periodic table, for example, $Cr(CO)_6$, $Fe(CO)_5$, and $Ni(CO)_4$. Odd-numbered groups form paramagnetic $[V(CO)_6]$ or dimeric carbonyl compounds such as $Mn_2(CO)_{10}$ and $Co_2(CO)_8$. The transition metals of the second and third rows form metal-metal bonded cluster compounds; those of the Group VIII metals can be spectacularly complex. In a formal sense, these cluster or multimetallic complexes combine the characteristics of mononuclear carbonyls with those of a metal surface [4].

The catalytic activation of carbon monoxide involves changes in the molecular orbitals induced by coordination to a metal atom. Carbon monoxide is a weak σ-donor ligand, but it can bond quite tenaciously to a metal. The general bonding pattern illustrated in Figure 5.1 is based on a weak ligand-to-metal σ-donor interaction reinforced by back donation from filled metal d-orbitals to vacant ligand π-antibonding orbitals. The two bonding modes are synergistic in transferring electron density between the ligand and the metal [5]. The most stable M-CO bonds seem to result when a strong electron donor ligand such as a phosphine is trans to the carbonyl. In this position, the R_3P and CO ligands share metal orbitals and electron density is transferred from P to CO.

The complex interaction between the CO ligand and the metal ion modifies the reactivity of the carbon monoxide molecule. A CO ligand in a cationic carbonyl complex such as $[Mn(CO)_6]^+$ becomes susceptible to attack by external nucleophiles such as amines, alkoxides, and carbanions. The most important modification from a practical viewpoint, however, is susceptibility to attack by other ligands coordinated to the same metal ion. This insertion or alkyl migration process described below is a key step in all the practical CO-based syntheses discussed in this chapter.

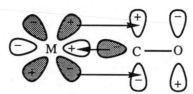

Figure 5.1 Interaction of metal and carbon monoxide orbitals in a metal-carbonyl complex.

Nearly all the reactions discussed in this chapter involve the "insertion" of a carbon monoxide into a C–M σ-bond. The insertion process has been difficult to study in detail, especially with the catalytically active cobalt and rhodium complexes which involve rather labile intermediates. The most detailed studies of insertion have been done with pseudooctahedral $RMn(CO)_5$ compounds. Other studies have used relatively stable $RCo(CO)_3(PR'_3)$ complexes and the kinetically sluggish iridium analogs of the rhodium catalysts. The results indicate that the "insertion" of carbon monoxide into a C–M σ-bond is more correctly viewed as the migration of the alkyl group.

Alkyl Migration		CO Insertion

The CO insertion and decarbonylation reactions with methylmanganese compounds are instructive. The reaction of methylmanganese pentacarbonyl with a ligand, L, such as triphenylphosphine or ^{13}CO occurs entirely within the coordination sphere of the metal in two steps [6]:

$$CH_3Mn(CO)_5 \rightleftharpoons CH_3COMn(CO)_4 \xrightarrow{L} cis\text{-}CH_3COMn(CO)_4L$$

The alkyl group migrates to the carbon of the CO ligand, leaving behind a vacant coordination site. Even in the presence of excess labelled CO, the acetyl group contains no ^{13}C because the methyl migrates to a CO already present on the metal. The first observable product is the cis-substituted acetyl derivative. Evidently the intermediate five-coordinate acetyl complex has enough configurational stability to permit the incoming ligand to seek out the coordination site vacated by the methyl group. The initially observed cis-acetyl complex later isomerizes to the trans isomer by CO dissociation and reassociation without involvement of the acetyl carbonyl group. The motions of the CH_3 and CO ligands during the insertion or migration process have been calculated in detail [7].

The migration reaction is readily reversible unless additional ligand is present to occupy the coordination site vacated by the alkyl group. The reverse reaction, decarbonylation, has some synthetic utility, as discussed in Section 5.5. Chirality of an alkyl group is generally preserved during migration in a carbonylation or decarbonylation process, though it is not clear whether it is with retention or inversion [8,9]. Stereochemistry can also be preserved at the metal center [10].

Several advances in analytical techniques have allowed a closer look at carbonylation reactions under conditions approximating those of industrial processes. High-pressure infrared cells allow spectroscopy under high temperature and pressure [11-14]. While not applicable under as wide a range of conditions, high-pressure nmr provides a wealth of information ranging from identification of side products and reactions to information about gas solubilities in solvents [15-17].

:ETIC ACID SYNTHESIS

Acetic acid and acetic anhydride are major industrial chemicals used in the manufacture of vinyl acetate, cellulose acetate, pharmaceuticals, dyes, and pesticides. Acetic acid is also used as a solvent in the oxidation of *p*-xylene to terephthalic acid, and acetate esters are used extensively as solvents. Acetic acid and its derivatives are made by three major processes [18], each of which employs homogeneous catalysts:

Oxidation of ethylene via acetaldehyde (Section 6.1)
Oxidation of butane or naphtha (Section 10.5)
Carbonylation of methanol

The last of these processes has come to dominate the acetic acid industry, although oldoxidation plants continue to operate.

Acetic acid has been made by carbonylation of methanol for over 30 years [2], but this reaction has become more attractive with the development of a catalyst which operates under low pressure. The original process developed by BASF [1,19] uses a cobalt carbonyl catalyst with an iodine compound as a cocatalyst. Cobaltous iodide is commonly used for in situ generation of $Co_2(CO)_8$ and HI. As is common with cobalt-catalyzed carbonylation, severe conditions (ca. 210°C/700 atmospheres) are required to give commercially acceptable reaction rates. The newer process developed by Monsanto [20,21] and operated by BP [2] is catalyzed by a rhodium salt with an iodide cocatalyst; it can produce acetic acid even at atmospheric pressure, although higher pressures are used in practice. Both catalysts are effective for the conversion of higher alcohols to the homologous carboxylic acids. The cobalt system has been used commercially in Europe for synthesis of propionic acid from ethanol, although carboxylation of ethylene (Section 5.3) may be more economical.

The commercial synthesis of acetic acid with the rhodium catalyst is carried out by reacting methanol with CO at about 180° and 30-40 atmospheres pressure with 10^{-3} molar Rh [20]. The reaction produces acetic acid or its methyl ester with greater than 99% selectivity. The source of rhodium may be almost any soluble rhodium compound; commercial derivatives such as $RhCl_3 \cdot 3H_2O$ are rapidly converted to $[Rh(CO)_2I_2]^-$, the presumed active catalyst, under reaction conditions. Many iodine compounds can be used as cocatalysts, but much of the cocatalyst is present in the plant streams as methyl iodide. Both rhodium and iodine are sufficiently expensive that nearly quantitative recycle is necessary. The rhodium complexes are relatively nonvolatile and reportedly stay in the reaction loop for years. The volatile iodine compounds are more troublesome, especially since iodine contamination of the product is undesirable. As with the BASF catalyst, the use of iodide in an acidic reaction medium requires a highly corrosion-resistant metal reactor.

The role of iodine in the two catalyst systems is the conversion of methanol to the more electrophilic methyl iodide. The iodine is used in a catalytic sense as shown in the right-hand loop in Figure 5.2. This iodine cycle interacts with a cycle involving the metal carbonyl catalyst. The rhodium cycle [22] shown in the figure begins with oxidative addition of methyl iodide to $[Rh(CO)_2I_2]^-$ **1**, a formal transition from Rh(I) to Rh(III). Although oxidative additions to Rh(I) complexes are well known, this one is extraordinarily fast. Evidently the extra electron density in the anionic complex makes it a powerful nucleophile.

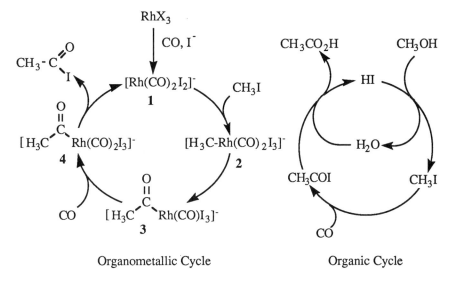

Figure 5.2 Catalytic cycles involved in methanol carbonylation to acetic acid.

The methylrhodium complex **2** formed by the oxidative addition has been identified [23] but is kinetically unstable and rapidly isomerizes to an isolable acetyl monocarbonyl complex **3** [20]. This isomerization is a nice illustration of the stepwise character of the methyl migration ("insertion") reaction. The five-coordinate acetyl derivative reacts with CO to produce a labile six-coordinate acetyl complex **4**. In the absence of water or alcohol, **4** could eliminate acetyl iodide to regenerate the $[Rh(CO)_2I_2]^-$ catalyst and the acetyl iodide would reenter the iodine cycle to form acetic acid and HI by hydrolysis. Under other reaction conditions, acetic anhydride or methyl acetate form by partial hydrolysis or by methanolysis. In the commercial reaction system, the acetylrhodium complex can also react with water yielding acetic acid directly:

$$[CH_3CORh(CO)_2I_3]^- \xrightarrow{\text{H}_2\text{O}} [HRh(CO)_2I_3]^- \xrightarrow{\text{–HI}} [Rh(CO)_2I_2]^-$$
$$+$$
$$CH_3COOH$$

The relative rates of the two acetyl cleavage mechanisms have not been reported, but the water-promoted reaction is probably the more important of the two.

As mentioned, the conditions of the BASF cobalt process are more severe, but the mechanism of the two reactions share some common features. Differences in kinetics and pressure requirements indicate that they do not have common rate-limiting steps. One significant difference in the two mechanisms is that once the acyl is formed on cobalt, it cannot undergo simple reductive elimination, because iodide is not in the coordination sphere of the cobalt. Reaction of RI with $[Co(CO)_4]^-$, which is an 18-electron complex, is much slower than the reaction with $[Rh(CO)_2I_2]^-$, which is a

16-electron complex. Most of the other steps in the reaction will be slower on cobalt, hence the requirement for higher temperatures. In the cobalt system, hydrogen, which is present in the gas stream, is more apt to cause hydrogenation of CH_3-M and CH_3COM intermediates to give methane or acetaldehyde.

Acetic Anhydride

BP Chemicals has modified the original Monsanto process to coproduce acetic acid and acetic anhydride [2]. Tennessee Eastman has developed its own acetic anhydride process [24] modelled after the Halcon process [25]; the highly integrated system is depicted in Figure 5.3. It starts with coal (up to 17 rail cars per day for each of two plants) to make hydrogen-rich synthesis gas. The gas is purified, and a portion of it is separated. Methanol is synthesized from 2/1 H_2/CO, and a portion of the methanol is used to scrub H_2S from the coal-gasification step. The remainder of the methanol is combined with acetic acid to make methyl acetate. The methyl acetate is carbonylated to give acetic anhydride. The acetic anhydride is used to make cellulose acetate in another process, and the resulting acetic acid is recycled to the process. The acetic anhydride step of the process is catalyzed by rhodium. Less expensive but less efficient ruthenium and nickel [26] catalysts have also been developed, but the efficiency of the rhodium retention is such that it has the most favorable economics.

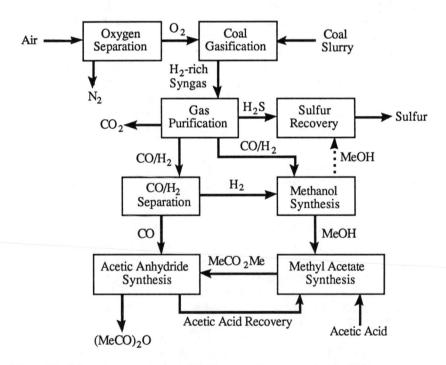

Figure 5.3. Schematic representation of the Tennessee Eastman acetic anhydride process.

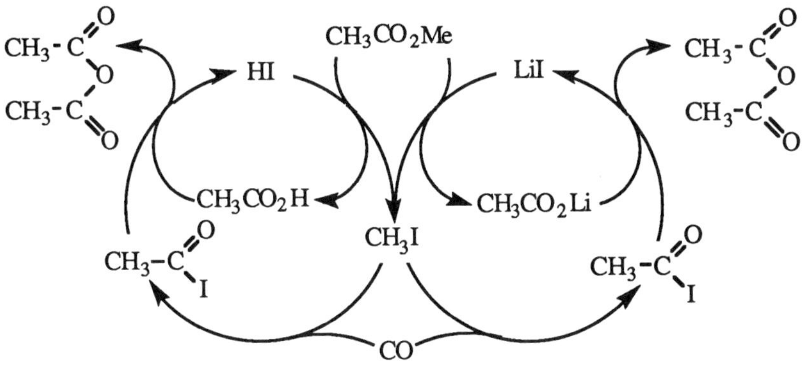

Figure 5.4 Scheme for the organic portion of methyl acetate carbonylation to acetic anhydride.

The organometallic cycle is identical with that shown above for acetic acid synthesis. However, the organic cycle is substantially different and more complex. One portion of the pathway, shown on the left of Figure 5.4, is the same as that for methyl alcohol carbonylation except that methyl alcohol is replaced by methyl acetate; acetate is the leaving group rather than hydroxyl. This, however, is the minor pathway. The acetic anhydride process is promoted by lithium ions [27] which favor the more ionic reaction shown on the right portion of the cycle. Thus LiI rather than HI reacts with methyl acetate to yield the methyl iodide which is carbonylated. Lithium acetate then reacts with the acetyl iodide to give the anhydride. In the absence of water, it is necessary to add hydrogen to the reaction to promote the catalyst; even though it does not appear in the reaction pathway, it is part of the overall mechanism.

Vinyl Acetate Synthesis

Most vinyl acetate is synthesized by the oxidative addition of acetic acid to ethylene catalyzed by palladium catalysts (see Chapter 6). Celanese has a process, operated in Mexico, in which acetaldehyde and acetic anhydride are reacted to form ethylidene diacetate, which is subsequently cracked to yield vinyl acetate [28].

Patents issued to Halcon describe CO chemistry that could lead to a synthesis gas-based process for manufacture of vinyl acetate. In the overall reaction, methyl acetate reacts with synthesis gas to form ethylidene diacetate:

$$CH_3OAc \xrightarrow{\ CO/H_2\ } CH_3CH(OAc)_2 \xrightarrow{\ -HOAc\ } H_2C{=}CHOAc$$

The diacetate can be converted to vinyl acetate and acetic acid in high yield by passage over a metal oxide catalyst at high temperature.

The reaction of methyl acetate with CO is catalyzed by a rhodium or palladium salt in combination with an iodide [29]. For example, a mixture of $RhCl_3 \cdot 3H_2O$,

3-picoline, and methyl iodide, catalyzes the reaction of 1:2 $CO:H_2$ with methyl acetate at 150°C and about 140 atmospheres pressure. The product contains 44% ethylidene diacetate, 6.5% acetic anhydride, and 0.6% acetaldehyde. The catalyst is very much like that used in the Eastman acetic anhydride process. The chemistry seems to combine elements of that process with the Celanese process:

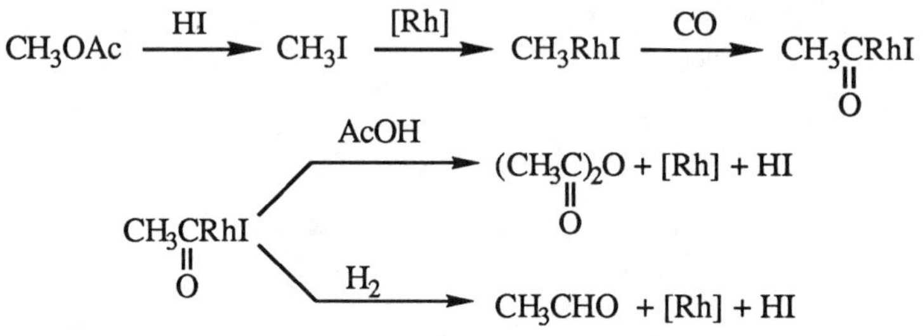

$$CH_3OAc \xrightarrow{HI} CH_3I \xrightarrow{[Rh]} CH_3RhI \xrightarrow{CO} CH_3\overset{\overset{O}{\|}}{C}RhI$$

$$CH_3\overset{\overset{O}{\|}}{C}RhI \begin{cases} \xrightarrow{AcOH} (CH_3\overset{\overset{O}{\|}}{C})_2O + [Rh] + HI \\ \xrightarrow{H_2} CH_3CHO + [Rh] + HI \end{cases}$$

$$CH_3CHO + (CH_3CO)_2O \rightleftharpoons CH_3CH(OCOCH_3)_2$$

As in the Eastman process, it is necessary to convert the CH_3O function of methyl acetate to CH_3I to facilitate reaction with the low-valent metal complex. Oxidative addition produces a CH_3-Rh bond that readily undergoes CO insertion to form an acetyl compound, which is a key intermediate. Reaction with acetic acid forms the anhydride; reaction with hydrogen produces acetaldehyde, just as in the rhodium-catalyzed hydroformylation process. Combination of the anhydride and the aldehyde yields ethylidene diacetate. Although no information about the mechanism of the diacetate synthesis has been published, this scheme accounts for the products observed.

5.3   CARBOXYLATION OF OLEFINS

Carboxylic acids and their derivatives can be synthesized by catalytic reactions of olefins, CO, and hydrogen donors. These syntheses are catalyzed by standard carbonylation catalysts such as $Co_2(CO)_8$ [1]. Such processes complement the acetic acid synthesis from methanol because they provide routes to higher aliphatic acids from α-olefins readily available from oligomerizations of ethylene discussed in Chapter 4. As in the acetic acid synthesis, the products are formed from an acyl metal complex, as shown by the synthesis of propionic acid derivatives from ethylene:

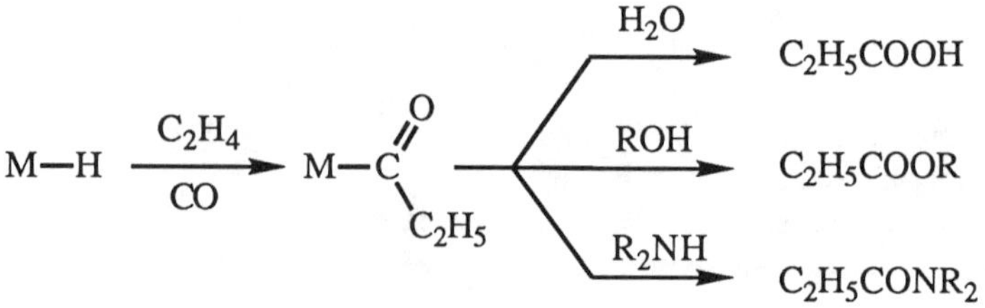

$$M-H \xrightarrow[CO]{C_2H_4} M-\overset{\overset{O}{\|}}{C}\diagdown_{C_2H_5} \begin{cases} \xrightarrow{H_2O} C_2H_5COOH \\ \xrightarrow{ROH} C_2H_5COOR \\ \xrightarrow{R_2NH} C_2H_5CONR_2 \end{cases}$$

Industrial applications of carboxylation have been limited because good alternative processes exist for production of most large-volume acids. For example, the widely used 2-ethylhexanoic acid ("isooctanoic" or simply "octoic" acid) is commonly made by propene hydroformylation, aldol condensation, and oxidation of the corresponding aldehyde (see Section 5.4). However, some propionic acid is made from ethylene as shown above, and fatty esters are made from olefins such as 1-hexene and 1-octene. Interest in the latter processes has been enhanced by the development of long-lived synthetic lubricants based on esters such as pentaerythrityl heptanoate. In a closely related reaction, acrylates have been made commercially by reaction of acetylene, carbon monoxide, and alcohols (Section 8.4).

Propionic acid can be made directly by reaction of ethylene, CO, and water ("hydrocarboxylation"), but the higher acids are produced as methyl or ethyl esters. The hydrocarboxylation reaction requires a solvent or a phase-transfer agent to bring the olefin and water in contact. A minor drawback arising from the presence of water is the formation of hydrogen via the shift reaction (Section 5.7):

$$CO + H_2O \rightleftharpoons CO_2 + H_2$$

As in acetic acid synthesis, hydrogen leads to byproduct formation by hydrogenation or hydroformylation of the olefin, but the hydrogen is necessary to maintain activity with rhodium catalysts. These side reactions can be tolerated to some extent for inexpensive olefins such as ethylene, but not for the more expensive higher olefins.

Most commercial carboxylation processes use cobalt carbonyl catalysts [1], usually promoted with pyridine. Palladium complexes such as $PdCl_2(PPh_3)_2$ have commercial potential because they are effective at lower pressures [30,31]. In a typical carboalkoxylation with a cobalt catalyst, a methanol solution of $Co_2(CO)_8$ and 1-pentene is heated to 140-170°C under 100-200 atmospheres CO pressure for several hours (80-90% conversion of olefin). The resulting product is a solution of 70% methyl hexanoate and 30% branched esters along with minor byproducts [32,33]. Direct synthesis of carboxylic acids is carried out similarly except that the solvent is aqueous dioxane or acetone. With $PdCl_2(PPh_3)_2$ as the catalyst, the carbonylation is carried out at 80-110°C and pressures as low as about 35 atmospheres [34]. The yield of methyl octanoate from 1-heptene under such conditions is less than 50% because the selectivity for formation of the linear product is only about 60%. However, if the reaction is run in the presence of excess $SnCl_2$, the selectivity rises to 87% and the yield of linear ester to 76% [34]. The branched ester is the major product when the $PdCl_2(PPh_3)_2$-catalyzed carbonylation is carried out in the presence of HCl. For example, propylene, CO, and methanol yield predominantly methyl isobutyrate [30], a compound of interest as a possible precursor of methyl methacrylate.

The mechanism of the cobalt-catalyzed carboxylation is sketched in Figure 5.5. Its main features are common to most carbonylation reactions catalyzed by $Co_2(CO)_8$ [35]. The reaction of $Co_2(CO)_8$ **5** with adventitious hydrogen or with an alcohol forms $HCo(CO)_4$ **6**. This "hydride" is actually a strong acid in polar solvents [36]. It is quite unstable and readily decomposes to cobalt metal in a CO-free atmosphere. High pressures are generally required in cobalt carbonyl-catalyzed reactions to stabilize $HCo(CO)_4$ and the intermediates in the catalyst cycle.

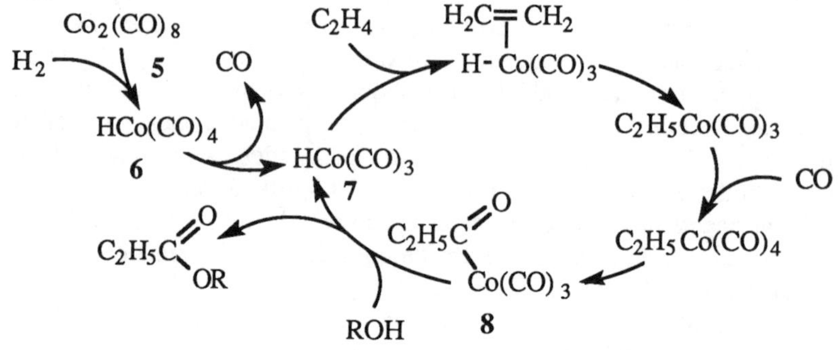

Figure 5.5 Scheme for cobalt-catalyzed carboalkoxylation of ethylene.

It is commonly assumed that the catalyst cycle is entered by dissociation of a CO ligand to give $HCo(CO)_3$ **7** as shown in the figure. However, there is some evidence that the replacement of CO by olefin may occur via an associative substitution. The cycle in the figure indicates coordination of ethylene and insertion of the olefin into the Co–H bond to give the coordinatively unsaturated ethyl complex. The subsequent steps are coordination of CO and insertion of CO into the Co–C_2H_5 bond to give a labile acyl compound **8**. All the steps in the cycle to this point are common to carboxylation and to the hydroformylation reaction discussed in Section 5.4. The two reactions differ mainly in the mode of cleavage of the acyl group from the metal.

Grossly, the acyl cleavage with an alcohol or with water proceeds as shown in Figure 5.5 to give an ester or a carboxylic acid and to regenerate the catalytic $HCo(CO)_3$. The detailed mechanism of cleavage is not established, but studies on the more stable acyl manganese carbonyl derivative **9** are suggestive. Both with water and with alcohols, the initial reaction seems to involve the acid- or base-catalyzed addition of the O-H bond to the acyl carbonyl function [37]:

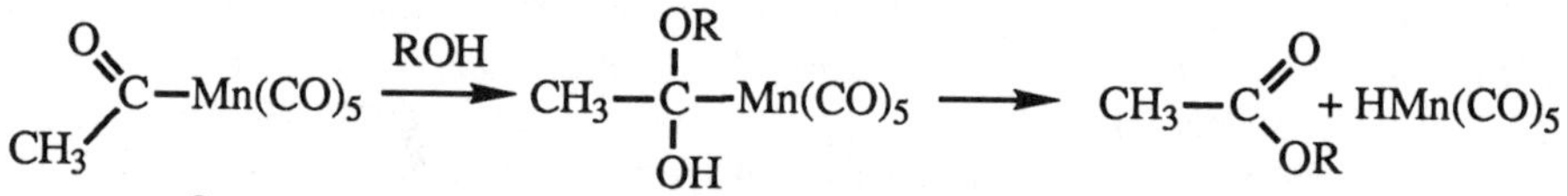

Presumably, a similar carbonyl addition occurs with acyl cobalt carbonyl complexes. The transfer of hydrogen from O-H to metal is analogous to the facile β-hydrogen transfer from C-H to metal that occurs in alkyl derivatives of the transition metals.

The mechanism of carbonylation with the palladium catalysts is less well defined. Two major mechanisms have been proposed. One involves formation of a palladium hydride which acts as a catalyst in very much the same way as $HCo(CO)_4$. The hydride mechanism seems to be generally accepted for the $PdCl_2(PPh_3)_2/SnCl_2$ system [34]. The tin chloride assists in formation of a hydride by creation of labile $SnCl_3^-$ ligands:

$$PdCl_2L_2 \underset{}{\overset{SnCl_2}{\rightleftharpoons}} Pd(SnCl_3)_2L_2 \underset{}{\overset{H_2}{\rightleftharpoons}} HPd(SnCl_3)L_2 + H^+ + SnCl_3^-$$

This assistance to hydride formation is much like that proposed for the $H_2PtCl_6/SnCl_2$ hydrogenation catalyst (Section 3.1). However, the $SnCl_3^-$ ligand probably also has other roles. It affects the direction of addition of the Pd-H bond to the olefin as judged by its effect on the ratio of linear to branched esters.

Another mechanism for the palladium system has been suggested to account for the predominance of branched products in the absence of $SnCl_2$ [31,32]. The version of this mechanism sketched in Figure 5.6 accounts for the role of an acid as a cocatalyst in the reaction. In this scheme, a complex such as $PdCl_2(PPh_3)_2$ **10**, reacts with the alcohol to form a labile alkoxy compound **11**. Coordination of carbon monoxide and insertion of CO into the Pd-O bond forms the carboalkoxy complex **12**. (Palladium compounds that contain the -CO_2CH_3 ligand are well characterized [38]). Insertion of propylene into the Pd-CO_2R bond should yield the Pd-CH_2- derivative **13** when the coordination sphere is crowded with bulky ligands such as triphenylphosphine.

The major role of the acid, HCl, in this scheme is to cleave the Pd-C bond before β-hydrogen elimination can occur. The acid cleavage yields the observed alkyl isobutyrate and regenerates $PdCl_2(PPh_3)_2$. If β-hydrogen abstraction occurs, the organic product is a methacrylate ester, but the catalyst is reduced to metallic palladium or a Pd(0) complex. Recent evidence indicates that the carbomethoxy complex is not an intermediate on the catalytic cycle, but rather a byproduct; the mechanistic pathway involves insertion of olefin into a Pd-H bond followed by carbonylation of the resulting alkyl [39].

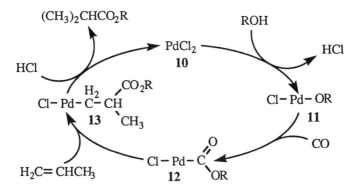

Figure 5.6 Scheme for the palladium-catalyzed carboalkoxylation of propene. PPh₃ ligands are omitted for clarity.

The reactions shown above for olefins can also be carried out on acetylenes using palladium or nickel catalysts. In aqueous systems, acrylic acid is formed; in alcohols, the acrylic esters are obtained. The palladium reaction is of some academic interest, but the nickel catalyst is used commercially in eastern Europe and in some older plants around the world (Section 8.4). In the BASF process, acetylene is reacted with CO and H_2O using a $NiBr_2/CuI$ catalyst system at 180-205°C and 40-55 atmospheres in THF. Yields are 90% on acetylene and 85% on CO. There is no catalyst recovery due to the low concentrations used. Presumably, the copper is present to coordinate and activate the acetylene as an acetylide. Rohm and Haas practices a semibatch reaction which is initiated by stoichiometric reaction of acetylene, water, and $Ni(CO)_4$ at 35-55°C. Once initiated, acetylene, CO, and H_2O are fed to maintain the catalytic reaction.

Adipic Acid from Butadiene

Carbonylation of four-carbon feedstocks is becoming an increasingly attractive route to adipic acid, especially in highly integrated petrochemical complexes where butadiene might be readily available from high-quality C_4 distillation sidestreams. Although there has been considerable interest in this approach, the necessity to carry out two consecutive terminal carboxylations is daunting.

BASF has reported the synthesis of adipic acid by means of direct hydrocarboxylation using rhodium catalysts promoted with methyl iodide [40], but the mole ratios of adipic acid to methylglutaric acid to valeric acid were 2/1/1.5. It is clear that selectivity of the reaction needs to be increased to be commercialized. A step in this direction was the discovery that the same catalyst system could be used to isomerize branched saturated alkylcarboxylic acids to linear alkylcarboxylic acids and vice versa [41]. (This unusual cleavage of a C-C bond is reminiscent of the isomerization required in the hydrocyanation of butadiene to adiponitrile. See Section 3.6). Alternatively, the process can be run in a stepwise manner. With a similar rhodium catalyst in a halocarbon solvent with a limited quantity of water, butadiene can be hydrocarboxylated to 3-pentenoic acid under mild conditions with very high selectivity. In the same solvent at higher temperatures, 3-pentenoic acid can be carbonylated to adipic acid in up to 84% linear selectivity [42]. The hydrocarboxylation reaction is strongly promoted by carboxylic acids [43].

Stepwise carboalkoxylation of butadiene to adipic diesters is another attractive route to adipic acid which has received considerable attention [44]. Butadiene is reacted with an alcohol in the presence of a preformed cobalt carbonyl hydride, a nitrogen base such as pyridine and CO at 130°C and 600 bar to yield the pentenoic ester. Some of the base is removed and pressure is lowered to 150 bar of CO at 170°C before the reaction is continued to give the adipic ester. The cobalt catalyst is then oxidized to cobalt acetate to allow isolation of products. Finally, the cobalt is reduced and put back into the catalytic cycle.

A relatively complete picture of the catalytic cycle of cobalt carboalkoxylation based upon $M-CO_2R$ intermediates has been developed [45]. The cycle shown in

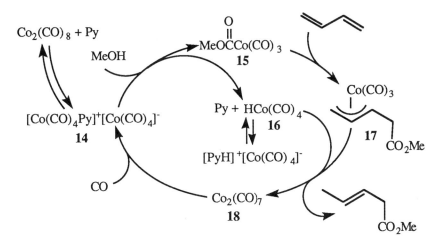

Figure 5.7 Scheme for the cobalt-catalyzed carbomethoxylation of butadiene.

Figure 5.7 is proposed to start with the pyridine-promoted heterolytic cleavage of $Co_2(CO)_8$ [46,47]. Reaction of methanol with the ion pair **14** would yield the two neutral species, $MeOCOCo(CO)_3$ **15** and $HCo(CO)_4$ **16**. Butadiene reacts with the unsaturated **15** much faster than with the saturated **16**, yielding the carbomethoxy-butenyl complex **17** [48]. Subsequent reaction of the substituted butenyl complex with **16** yields free methyl 3-pentenoate and the unsaturated cobalt carbonyl dimer **18** which quickly picks up CO. It is proposed that the high pressures required for this process are not related to the catalytic cycle, but rather retard the facile oxidative decarbonylation of the ion pair which has been demonstrated to be reversible [49]. These studies do not address additional problems such as methylation of the pyridine.

Halide-free palladium catalysts can carboalkoxylate butadiene to pentenoate esters or pentenoic acid [50]. The ligands on these catalysts are generally triphenyl phosphine or bidentate tertiary phosphorus ligands. Selectivities are not adequate for commercialization, but the reaction is mild enough to warrant continued process development. In another interesting variation of palladium catalysis, butadiene is reacted with CO and oxygen in the presence of 1,1-dimethoxycyclohexane which scavenges water as it is formed [51]. The products of the reaction are dimethyl 3-hexene-1,6-dioate and cyclohexanone. This oxidative carbonylation is catalyzed by $PdCl_2$ at 100°C and 1800 psi CO. The palladium would precipitate as the metal, but copper(II) chloride is included in the mixture to reoxidize the palladium to an active state, and the resulting copper(I) is reoxidized with oxygen.

While none of these processes has been commercialized, the BASF process, based upon cobalt-catalyzed carboalkoxylation of butadiene, has been run at a pilot-plant scale [52]. The high investment necessitated by the complexity and high pressure requirements of the process has delayed implementation. Each of these processes also suffers from loss of butadiene to C_9 and C_{10} products derived from concomitant dimerization of the butadiene followed by carboxylation. There have been several attempts to improve selectivity by starting from butadiene derivatives or other C_4 species.

The adipic acid precursor, dimethyl 3-hexene-1,6-dioate is synthesized from 1,4-dimethoxy-2-butene via dicarbonylation [53]. Palladium chloride is employed at temperatures of about 100°C under 200 atmospheres of CO. Dimethyl 3-hexene-1,6-dioate was obtained in 70% yield at 100% conversion. Following the carbonylation, the product is hydrogenated to the saturated diester before hydrolysis to adipic acid. Alternatively, some of the reaction sequences can be entered starting from the appropriate crotyl halides or alcohols or even butanediol [54].

5.4 HYDROFORMYLATION

The oldest and largest homogeneous catalytic reaction of olefins is hydroformylation - the addition of CO and hydrogen to produce an aldehyde. The most prominent example is manufacture of butyraldehyde from propylene and synthesis gas [1]:

$$CH_3CH=CH_2 + H_2 + CO \longrightarrow CH_3CH_2CH_2CHO$$

This aldehyde is a versatile chemical intermediate which is produced on a scale of about four million tons per year worldwide. Some butyraldehyde is hydrogenated to l-butanol, but the greater part is self-condensed to form 2-ethylhexyl derivatives [55]:

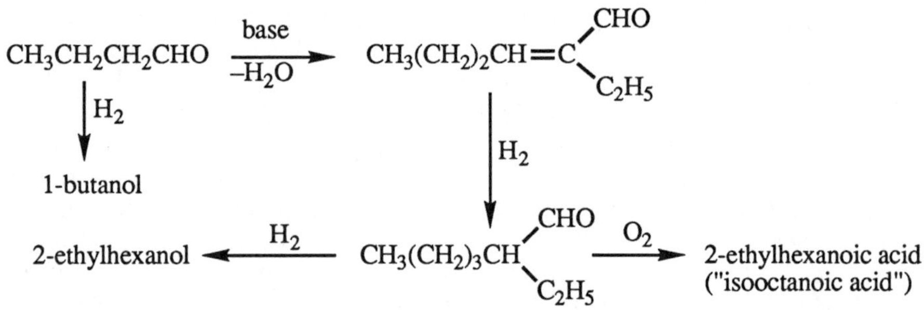

$$CH_3CH_2CH_2CHO \xrightarrow[-H_2O]{\text{base}} CH_3(CH_2)_2CH=C\overset{\displaystyle CHO}{\underset{\displaystyle C_2H_5}{}}$$

$CH_3CH_2CH_2CHO \downarrow H_2$

1-butanol

2-ethylhexanol $\xleftarrow{H_2}$ $CH_3(CH_2)_3CH\overset{\displaystyle CHO}{\underset{\displaystyle C_2H_5}{}}$ $\xrightarrow{O_2}$ 2-ethylhexanoic acid ("isooctanoic acid")

Much of the 2-ethylhexanol is converted to phthalate esters for use as plasticizers. Butanol and other short-chain alcohols made by hydroformylation technology are used extensively as solvents. A second major application of hydroformylation is the synthesis of fatty alcohols from terminal olefins such as l-octene. The process sometimes combines hydroformylation and reduction steps since cobalt carbonyl catalysts are effective for both reactions:

$$RCH=CH_2 \xrightarrow[CO]{H_2} RCH_2CH_2CHO \xrightarrow{H_2} R(CH_2)_3OH$$

The linear alcohols prepared this way, like those made by ethylene telomerization (Section 4.3), are used in the synthesis of biodegradable detergents and for production of adipate esters. The latter serve as high-temperature lubricants and as plasticizers.

In addition to the industrial applications, hydroformylation is a useful laboratory route to aldehydes from both terminal and internal olefins. Such syntheses can be carried out at atmospheric pressure by using $HCo(CO)_4$ as a stoichiometric hydroformylation reagent or by using rhodium-based catalysts such as $HRh(CO)(PPh_3)_3$ [56]. Hydroformylation of internal olefins such as 2-butene generates a chiral center. When the aldehyde synthesis is carried out with a rhodium complex that bears chiral ligands, asymmetric induction occurs and yields optically active aldehydes. The asymmetric hydrogenation catalysts formed from $[Rh(olefin)_2]^+$ salts and chiral ligands are effective for this purpose.

Industrial hydroformylations are carried out with two types of catalysts. Simple cobalt carbonyl catalysts are used for roughly 90% of all production, both of aldehydes and of alcohols. A phosphine-modified cobalt catalyst has been developed by Shell specifically for linear alcohol synthesis. Rhodium-based systems have been commercialized by Union Carbide and Celanese. The cobalt and rhodium systems are discussed separately below.

Cobalt Catalysts

In 1938 Roelen at Ruhrchemie AG discovered that cobalt-containing heterogeneous catalysts effect the addition of carbon monoxide and hydrogen to ethylene to yield aldehydes and ketones [1]. It was soon recognized that the actual catalyst is the soluble cobalt carbonyl formed by reductive carbonylation of cobalt oxide. This chemistry was soon developed into an industrial synthesis of aldehydes and alcohols.

The commercial manufacture of butyraldehyde carried out today does not differ greatly from that developed in the 1940s. Typically, the catalyst is prepared in the hydroformylation reactor by treating finely divided cobalt metal or a cobalt(II) salt with synthesis gas (1:1 H_2:CO).

$$Co^{2+} \xrightarrow[\text{CO}]{H_2} Co^0 \xrightarrow{\text{CO}} Co_2(CO)_8 \xrightarrow{H_2} HCo(CO)_4$$

Addition of propylene to this catalyst/solvent/synthesis gas mixture at 120-140°C and 200 atmospheres pressure [56] forms a mixture of products. The main component is n-butyraldehyde which is formed in 60-70% yield. Isobutyraldehyde is the major impurity, but C_4 alcohols and dipropyl ketones are also present.

The high pressure in the reactor is necessary to stabilize the cobalt carbonyl intermediates, while the requirement for a high temperature reflects the conditions needed to prepare the catalyst. If $HCo(CO)_4$ is prepared in advance in a separate reactor, the hydroformylation proceeds smoothly at 90-120°C. Under the milder conditions, catalyst selectivity is better and butyraldehyde forms in 72-80% yield [57].

The simple cobalt carbonyl catalysts are reasonably satisfactory for industrial hydroformylation processes, but several problems exist. The carbonyls $HCo(CO)_4$ and $Co_2(CO)_8$ are unstable and volatile, making them difficult to separate from aldehyde products during the product purification and catalyst recycle steps. More significantly, the limited selectivity for the desired linear aldehydes and alcohols increases the consumption of starting materials and requires large distillation facilities for product purification. Similarly, the severe reaction conditions impose economic penalties for high-pressure reactor investment and for construction and operation of the gas compressor.

Catalyst systems described below provide better product selectivity and lower pressure reaction conditions [58]. Phosphine-modified rhodium catalysts give high yields of linear aldehydes under mild conditions. Phosphine-modified cobalt carbonyl catalysts are advantageous for direct synthesis of linear alcohols.

Typically, the synthesis of long-chain alcohols from terminal olefins is carried out in two stages [1]. For example, 1-octene is hydroformylated to nonanal at 120-170°C and 200-300 atmospheres pressure. The nonanal is then hydrogenated with a heterogeneous catalyst in a separate step. In principle, the two reactions can be combined by increasing the H_2:CO ratio. However, simple cobalt carbonyls are poor catalysts for aldehyde hydrogenation, and they catalyze several side reactions. The hydroformylation and aldehyde hydrogenation steps can be combined satisfactorily, however, by addition of a trialkylphosphine ligand to the standard cobalt system. In the process developed by Shell [59], the active catalyst appears to be $HCo(CO)_3(PBu_3)$, while $[Co(CO)_3(PBu_3)]_2$ is the resting state [60]. This tributyl-phosphine-modified catalyst is less active for hydroformylation than $HCo(CO)_4$, and yields a lower rate even at 180°C. However, it is much more active for hydrogenation and gives a linear:branched alcohol ratio of about 7:1, compared to 4:1, typical of unmodified cobalt hydroformylation catalysts [1]. The $HCo(CO)_3(PBu_3)$ catalyst is also more stable and is used at about 100 atmospheres pressure vs. 200-300 atmospheres. This stability simplifies catalyst recycle because the alcohols can be distilled from the catalyst. A limitation of this process is that some olefin is lost through hydrogenation to alkane.

Cobalt-catalyzed hydroformylation was one of the first homogeneous catalytic reactions to be characterized from a mechanistic viewpoint [61]. Infrared studies of the reaction have tended to confirm that the intermediates under typical high-pressure conditions resemble those observed earlier at atmospheric pressure [62,63]. Except for the final product-forming step, the chemistry is identical to that in the carboxylation reaction shown in Figure 5.5. The $Co_2(CO)_8$ catalyst is almost completely hydrogenated to $HCo(CO)_4$ at 150°C and 290 atmospheres H_2/CO pressure [61]. The hydride reacts with an olefin to form an olefin complex, an alkyl derivative and an acyl derivative, and an acyl derivative just as shown in the figure. With unsymmetrical olefins such as propylene, the direction of olefin insertion into the Co-H bond determines whether a linear or a branched aldehyde is formed.

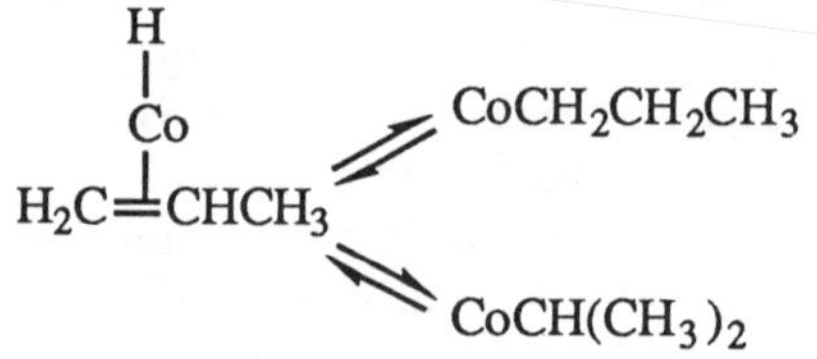

$$
\begin{array}{c}
\text{H} \\
| \\
\text{Co} \\
|\!| \\
\text{H}_2\text{C}\!=\!\text{CHCH}_3
\end{array}
\quad
\begin{array}{l}
\nearrow \text{COCH}_2\text{CH}_2\text{CH}_3 \\
\\
\searrow \text{COCH(CH}_3)_2
\end{array}
$$

A key intermediate in the overall process is the coordinatively unsaturated acylcobalt compound, $PrCOCo(CO)_2L$. It can undergo several reactions with CO and hydrogen:

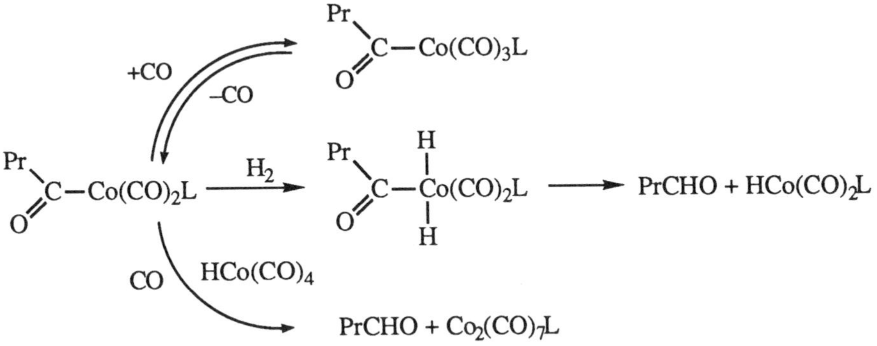

$$L = CO, PR_3 \quad R = \text{Aryl, Alkyl}$$

A reversible reaction favored by high pressures of CO forms the relatively stable complexes, $PrCOCo(CO)_3L$; these complexes constitute a major fraction of the cobalt-containing species in the reactor, but they are not on the catalytic cycle [61]. $PrCOCo(CO)_2L$ reacts with hydrogen to form the six-coordinate dihydride which eliminates butyraldehyde, or it reacts with cobalt hydride to eliminate butyraldehyde and form the cobalt dimer. The relative importance of these two mechanisms is unknown, but this step is often rate-determining in aldehyde synthesis. When L = PBu_3, the hydrido complex is favored in solution; again, the known alkyl and acyl derivatives are not observed. This result suggests that olefin insertion into the Co-H bond may be rate-determining as well as being product-determining. The predominance of the n-propyl derivative presumably results from crowding in the coordination sphere of the cobalt caused by the bulky phosphine ligand. Both steric and electronic effects are probably involved in the stabilization of these complexes by the trialkylphosphine ligand. This stabilization permits operation at high temperatures and low CO pressures, conditions which favor hydrogenation of the aldehyde products to alcohols [64].

Rhodium Catalysts

Rhodium complexes catalyze hydroformylation of olefins under very mild conditions. Despite their high activity, the simple rhodium compounds were not initially attractive because they gave mostly branched aldehydes, for example, isobutyraldehyde from propylene. Addition of phosphorus ligands such as triphenylphosphine or triphenyl phosphite gives active catalysts with excellent selectivity for formation of the desired linear aldehydes [1]. This modified rhodium technology was put in commercial production by Union Carbide in 1976 [65-68]; yields of n-butanal are about 85% [69]. Simultaneously with the commercial development, Wilkinson's investigation of the mechanism of hydroformylation with HRh(CO)(PPh$_3$)$_3$ produced a good understanding of the role of the catalyst [70,71].

The commercial catalyst may be prepared by reaction of high surface-area metallic rhodium (such as rhodium on charcoal) with synthesis gas in the presence of triphenylphosphine or a phosphite. In the presence of this catalyst, propylene reacts with synthesis gas at 10-20 atmospheres pressure and about 100°C to give predominantly n-butyraldehyde. Little or no aldehyde is hydrogenated to C$_4$ alcohols although some olefin is lost through hydrogenation to propane. The selectivity to the desired linear aldehyde is very high, greater than 90% when excess phosphorus ligand is present. The ligand is important to stabilize the catalyst during product recovery as well as to direct the reaction to formation of the desired product. Since both selectivity and stability are favored by a large excess of ligand, molten triphenylphosphine (m.p. 79°C) is an ideal solvent for the reaction, and the catalytic rhodium complex is stable almost indefinitely in this medium [72].

The mechanism of hydroformylation with the modified rhodium catalysts is quite similar to that of the cobalt catalysts. Many of the intermediates isolated in the course of mechanistic studies at atmospheric pressure [70,71] have also been observed by infrared studies at pressures typical of commercial operations [73]. The immediate catalyst precursor is HRh(CO)(PPh$_3$)$_3$. This compound is much more stable than HCo(CO)$_4$, but requires the presence of some excess ligand for stabilization during product distillation and catalyst recycle. As in the cobalt system, ligand dissociation is necessary to create a coordinatively unsaturated complex **19** that can bind the olefin. In this case, one of the bulky phosphine ligands probably dissociates as shown in Figure 5.8.

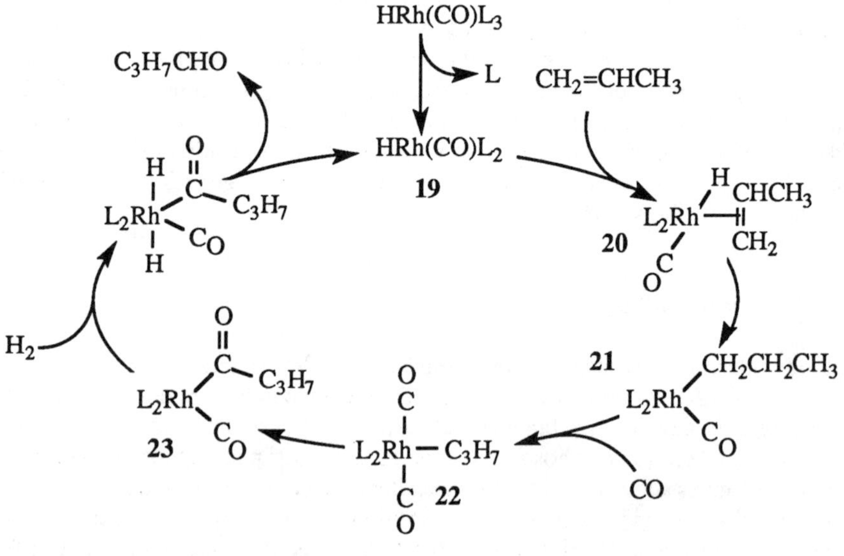

Figure 5.8 Rhodium-catalyzed hydroformylation of propene (L = PAr$_3$).

Propylene enters the vacant coordination site in **19** to form **20** and inserts into the Rh-H bond to form a propyl or isopropyl group in **21**. The effect of the phosphine ligand concentration on the product ratio probably operates at the olefin insertion step. If, as shown in the figure, the olefin π-complex bears two large phosphine ligands, crowding in the coordination sphere of the metal will favor formation of the n-propyl derivative. However, if one of the phosphine ligands has been replaced by a much smaller carbon monoxide ligand, there will be space to form the bulky *iso*-propyl group.

The propyl complex is coordinatively unsaturated and readily binds another CO ligand to give the dicarbonyl complex **22**. Carbon monoxide insertion (or propyl migration) generates a σ-acyl compound **23** which is again coordinatively unsaturated. Oxidative addition of H_2 and reductive elimination of acyl and hydride ligands yield the aldehyde and regenerate the catalytically active rhodium hydride. The major uncertainties in the catalytic cycle concern the number of phosphine and CO ligands in the intermediates.

Despite the advantages of the modified rhodium catalysts (mainly high selectivity and low-pressure operations), the high cost of rhodium inventory and recycle remains a problem [74]. The year 1984 saw the commercialization of a novel application of organometallic chemistry in water by Ruhrchemie [75,76]. Sulfonation of triphenylphosphine followed by neutralization yields the trisodium salt of tri(m-sulfophenyl)phosphine [77]. This water-soluble ligand carries rhodium into the aqueous phase by forming complexes analogous to those described above. While propylene is only slightly soluble in water, the solubility is sufficient to carry out a very efficient hydroformylation. The pH of the system is maintained at about 6.0 to stabilize the butanals. The aldehyde phase is virtually insoluble in the catalyst phase and is simply decanted [78]. This process modification has led to a significant improvement in the economics of rhodium-catalyzed hydroformylation through the rhodium savings and significant savings in propylene and CO/H_2.

Union Carbide has patented rhodium-based hydroformylation catalysts based upon highly hindered phosphite ligands [79]. These ligands are derived from commercially available hindered phenols used in antioxidant and stabilizer packages for polymers. The catalysts provide good selectivity and rate [80] and might find use in specialty applications of hydroformylation chemistry.

Mitsubishi has commercialized an interesting hydroformylation of octene in which they use triphenylphosphine oxide as a ligand for a methanolic rhodium acetate catalyst [81,82]. Triphenylphosphine is added to the mixture before distillation. After distillation, the residues are oxidized giving additional triphenylphosphine oxide before being cycled back to the hydroformylation.

5.5   DECARBONYLATION

The CO insertion step in hydroformylation is reversible and provides an interesting synthesis of alkanes from aldehydes:

$$RCHO \xrightarrow{\text{catalyst}} RH + CO$$

Similarly, many acyl chlorides decarbonylate to give chloroalkanes or chloroarenes when treated with soluble catalysts. Although these reactions are not universal and do not have commercial utility, they have several applications in laboratory syntheses of complex organic molecules. Decarbonylation is a useful way to remove formyl groups in steroid molecules. Synthetic applications have been reviewed [83]. In addition to the aldehyde and acyl chloride reactions, there have been reports of decarbonylation of carboxylic acid anhydrides, ketones, and ketenes.

Decarbonylation is usually effected by heating the aldehyde or acyl chloride with a metal complex in a high-boiling solvent such as benzonitrile. Wilkinson's catalyst, $RhCl(PPh_3)_3$, is one of the most active CO-abstracting reagents. This compound can be used stoichiometrically to remove CO from the organic compound under very mild conditions [84]:

$$C_6H_{13}CHO + RhCl(PPh_3)_3 \xrightarrow[25°C]{CH_2Cl_2} C_6H_{14} + RhCl(CO)(PPh_3)_2 + PPh_3$$

At such a low temperature, the rhodium carbonyl product is stable. However, at high temperatures, CO can be expelled from the complex and catalytic decarbonylation occurs:

$$PhCOCl \xrightarrow[180°C]{RhCl(PPh_3)_3} PhCl + CO$$

For catalytic decarbonylation, the moderately air-stable carbonyl complex, $RhCl(CO)(PPh_3)_2$, seems just as effective as Wilkinson's catalyst and is more convenient to handle. When high reaction temperatures cannot be tolerated, a rhodium complex with chelating phosphine ligands such as the commercially available $Ph_2PCH_2CH_2PPh_2$ can be used catalytically [85].

Zero-valent palladium complexes are effective catalysts for decarbonylation, but equally good results are often obtained with palladium(II) chloride or supported metallic palladium catalysts. For example, an 88% yield of toluene is obtained when p-tolualdehyde is boiled with 5% palladium-on-carbon for 30 minutes [86]. Mitsubishi's purification of terephthalic acid is a major industrial application of catalytic decarbonylation [87]. 4-Formylbenzoic acid, a troublesome impurity in the oxidation of p-xylene to terephthalic acid (see Chapter 10), is decarbonylated cleanly in water to benzoic acid at 250-300°C using Pd/C. The benzoic acid is easily separated from the terephthalic acid.

Although decarbonylation of aromatic aldehydes and acyl chlorides is straightforward and generally proceeds in good yield, several complications are observed with aliphatic derivatives. Alkanoyl chlorides often give olefins rather than alkyl chlorides as products. Isomerization of the alkyl group sometimes occurs. For example, palmitoyl chloride gives mixtures of pentadecenes and isomeric C_{15} chlorides [88]:

$$CH_3(CH_2)_{14}COCl \xrightarrow[-CO]{Pd(acac)_2} CH_3(CH_2)_{12}CH=CH_2 + C_{15}H_{31}Cl + C_{15}H_{31}COCl$$
$$+ \text{isomers} \qquad \text{isomers} \qquad \text{isomers}$$

The acyl chlorides recovered from such reactions can contain grossly altered alkyl groups. Nonanoyl chloride gives 2-methyloctanoyl, 2-ethylheptanoyl, and 2-propyl-hexanoyl chlorides .

The mechanism generally accepted for decarbonylation [83,89] is illustrated in the catalytic cycle for the reaction catalyzed by Wilkinson's catalyst or RhCl (CO)(PPh$_3$)$_2$.

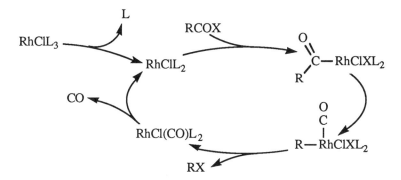

The initial step with either aldehydes or acyl chlorides is oxidative addition to a coordinatively unsaturated rhodium(I) complex. The oxidative addition products of acyl chlorides are five-coordinate acylrhodium(III) compounds which are often isolable [89,90]. The aldehyde addition products are usually not stable and undergo further reaction immediately. Exceptions have been observed for other metals; the reaction of "bis[bis(dimethylphosphino)ethane]iron(0)" with benzaldehyde yields a hydrido benzoyl iron complex which is quite stable, and the reaction with propionaldehyde yields a hydrido acyl complex which decarbonylates yielding the zero-valent iron carbonyl complex as the stable organometallic [91]. The product from 8-quinolinecarboxaldehyde and RhCl(PPh$_3$)$_3$ is stabilized by chelation and has been isolated and structurally characterized [92].

In the rhodium-catalyzed reaction, oxidative addition of either the aldehyde or the acyl halide ultimately gives alkyl or aryl rhodium carbonyl complexes by CO elimination from the acyl-Rh bond. This reaction is the reverse of CO insertion and is most correctly considered as a migration of the alkyl group from CO to Rh. The final organic products arise from coupling of the alkyl and H or Cl ligands of the alkyl complex. As noted previously, the rhodium(I) product of this reductive elimination step is stable at low temperature, but eliminates CO at high temperature to complete the catalytic cycle. The side reactions observed with aliphatic aldehydes and acyl chlorides probably result from rearrangement of the alkylrhodium intermediate.

5.6   ISOCYANATES FROM NITRO COMPOUNDS

Aromatic nitro compounds react with carbon monoxide in the presence of palladium catalysts to give the corresponding isocyanates:

$$ArNO_2 + 3CO \longrightarrow ArN{=}C{=}O + 2CO_2$$

The reductive carbonylation reaction shown above has many advantages over the usual two-step isocyanate synthesis based on phosgene:

$$ArNO_2 \xrightarrow{H_2} ArNH_2 \xrightarrow{COCl_2} ArNCO + 2HCl$$

This unusual reaction is potentially important because diisocyanates such as TDI and MDI are used to prepare polyurethane foams and elastomers (Chapter 11). The conventional process starts from nitroaromatics. TDI is prepared by dinitration of toluene, reduction to the diamine, and reaction with phosgene. MDI is prepared from nitrobenzene by reduction to aniline, coupling with formaldehyde to yield methylene dianiline, and reaction with phosgene to give methylene diisocyanate. In the case of MDI, a variety of isomers and higher oligomers are carried on in the reaction; the polyurethane industry has gotten accustomed to the mixture of products, making "improvements" in the product difficult.

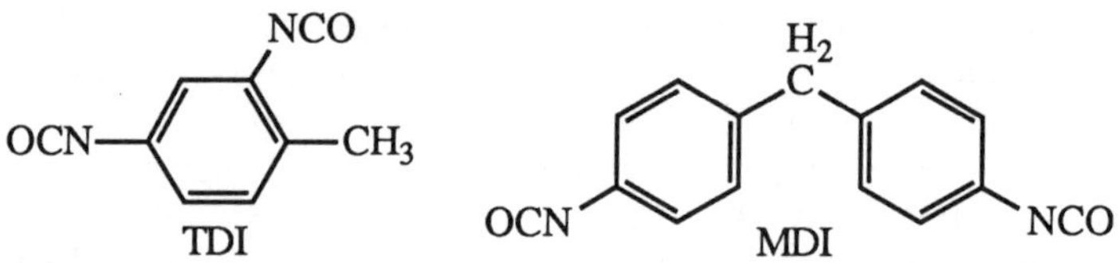

Selenium Catalysis

Arco had announced plans to produce MDI [93] by reductive carbonylation using a selenium catalyst [94] to make the methyl urethane which was then to be cracked to the isocyanate using homogeneous or heterogeneous catalysts [95]. They probably chose to make MDI because it has good market potential and because it has only one NCO group per ring. Attempts to produce TDI and other isocyanates with two NCO groups per ring often give low yields in the reductive carbonylation process. Nitrobenzene was to be converted to the ethyl urethane, condensed to the methylene diurethane, and then cracked to MDI. Despite having started construction on the plant, the program was cancelled, apparently due to difficulty in total removal of the high levels of highly toxic selenium in the waste streams and lack of customer acceptance of even parts per billion levels of selenium in the products. The subtleties of the isocyanate market, which was accustomed to the existing isomer mix and valued product consistency over "higher purity," probably also contributed to the decision.

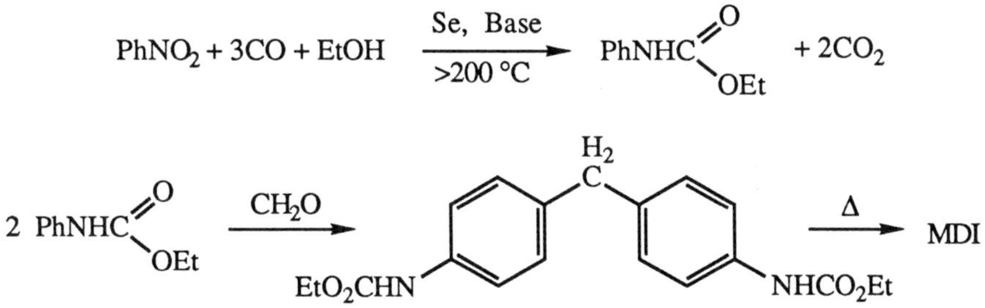

$$PhNO_2 + 3CO + EtOH \xrightarrow[>200\,°C]{Se,\ Base} PhNHC\overset{O}{\underset{OEt}{\diagdown}} + 2CO_2$$

$$2\ PhNHC\overset{O}{\underset{OEt}{\diagdown}} \xrightarrow{CH_2O} EtO_2CHN\!\!-\!\!\langle\bigcirc\rangle\!\!-\!\!\overset{H_2}{\underset{}{C}}\!\!-\!\!\langle\bigcirc\rangle\!\!-\!\!NHCO_2Et \xrightarrow{\Delta} MDI$$

The conditions of the reaction are very reducing and Se^0 is the predominant selenium species. Hydrogen selenide is kept low because it is sufficiently reducing to convert nitrobenzene to aniline; the resulting selenium-black is recycled via the first two reactions [96].

$$H_2O + CO + Se^0 \longrightarrow H_2Se + CO_2$$
$$CO + Se^0 \longrightarrow SeCO$$
$$SeCO + H_2O \longrightarrow H_2Se + CO_2$$
$$3\ H_2Se + ArNO_2 \longrightarrow ArNH_2 + 3\ Se^0 + 2\ H_2O$$

The carbonylation step is initiated by base-promoted nucleophilic addition of carbonyl selenide to the aniline; there is an equilibrium between the protonated and deprotonated species. Base-catalyzed addition of alcohol to the selenocarbamic acid yields hydrogen selenide and the urethane. Additional aniline competes in this reaction and diphenylurea is also observed.

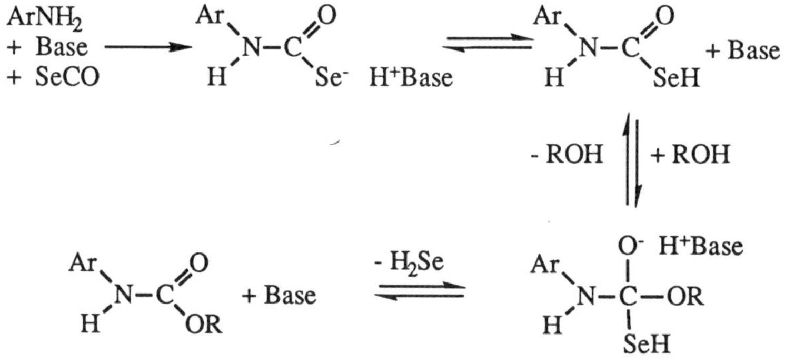

$$\begin{array}{c}ArNH_2\\+\ Base\\+\ SeCO\end{array} \longrightarrow \overset{Ar}{\underset{H}{\diagdown}}N\!-\!C\overset{O}{\underset{Se^-\ H^+Base}{\diagdown}} \rightleftharpoons \overset{Ar}{\underset{H}{\diagdown}}N\!-\!C\overset{O}{\underset{SeH}{\diagdown}} + Base$$

$$- ROH \updownarrow + ROH$$

$$\overset{Ar}{\underset{H}{\diagdown}}N\!-\!C\overset{O}{\underset{OR}{\diagdown}} + Base \underset{-H_2Se}{\rightleftharpoons} \overset{Ar}{\underset{H}{\diagdown}}N\!-\!\overset{O^-\ H^+Base}{\underset{SeH}{\overset{|}{C}}}\!-\!OR$$

The reaction is improved if stronger, bicyclic bases are employed [97].

Transition Metal Reductive Carbonylation

Nitro compounds can be reductively carbonylated by heating under carbon monoxide pressure in the presence of a palladium salt or metallic Pd or Rh [98]. For example, when an acetonitrile solution of nitrobenzene is treated with carbon monoxide at 700 atmospheres pressure and 200°C in the presence of $PdCl_2$, phenyl isocyanate is formed in 55% yield [99]. When the solvent is an alcohol, the products are carbamate esters [100,101], which are useful polyurethane precursors like isocyanates. For example, nitrobenzene in methanol gives a methyl ester:

$$PhNO_2 + 3CO + MeOH \xrightarrow{PdCl_2} PhNHC \overset{O}{\underset{OMe}{}} + 2CO_2$$

It seems likely that the palladium catalyst alternates between Pd(0) and Pd(II) with nitrobenzene as the oxidant and CO as the reductant. Metal salts such as $CoCl_2$ or $CuCl_2$ assist the reaction by promoting the redox cycle [102,103].

The mechanism of this reductive carbonylation of nitro compounds is not well established. The palladium salt may be involved directly in the reduction of the nitro compound. Carbon monoxide reduces $PdCl_2$ to Pd(0) readily in protonic media [104]:

$$Pd^{II} + CO \rightleftharpoons Pd^{II}-CO \xrightarrow{H_2O} Pd^{II}-C \overset{O}{\underset{OH_2}{}} \longrightarrow Pd^0 + CO_2 + 2H^+$$

A palladium(0) complex should function nicely to abstract oxygen from a coordinated aromatic nitro compound [105-107] to form a nitroso compound; this reaction has been demonstrated using Ni^0 and t-$BuNO_2$ [108]. Further oxygen atom abstraction would yield an arylnitrene, ArN, perhaps stabilized as a metal complex [109]. Nitrenes generated by decomposition of aryl azides react rapidly with CO to form isocyanates [110,111]. Interestingly, palladium tetraphenylporphyrin has also been reported to catalyze carbonylation of nitroarenes [112]; no azo products are observed.

Alternatively, metallacyclic species are involved in the oxygen transfer reaction. Zero-valent metals undergo oxidative cyclization reactions with a variety of unsaturated species. Complex **24** has been isolated from the reaction of nitrobenzene and CO with Pd^{II} and o-phenanthroline in ethanol and characterized spectroscopically [113].

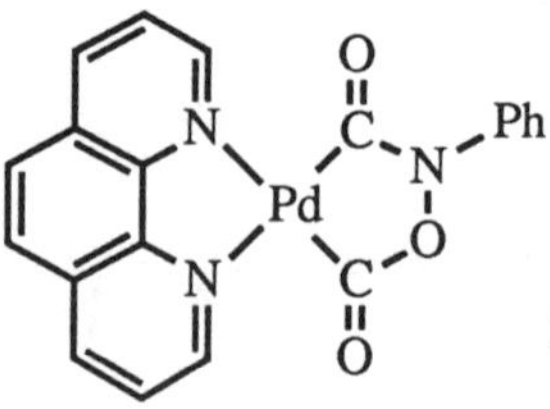

24

This system is very efficient and selective for nitroaromatic carbonylation [114] under conditions more vigorous than those from which the organometallic was isolated. Presumably the product is formed by repeated insertions of CO followed by elimination of CO_2.

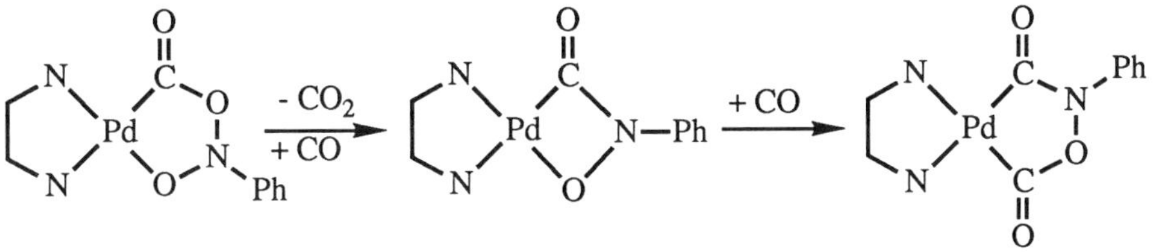

A ruthenium analog of the intermediate shown here, Ru(DPPE)(CO)$_2$[C(O)N-(p-ClC$_6$H$_4$)O], has been characterized crystallographically, though it did not appear to be on the direct catalytic pathway to the carbamate [115]. Rather, under carbonylation conditions in methanol, the aniline was liberated and a bis(carbomethoxy)ruthenium, Ru(DPPE)(CO)$_2$(CO$_2$Me)$_2$ was isolated. Under more vigorous conditions, the liberated aniline reacts with the dicarboxylate to yield the carbamate.

Transition Metal Oxidative Carbonylation

Synthesis gas would be far less expensive than pure CO as a reactant for the reductive carbonylation of nitroaromatics. As a reducing agent, CO is more expensive than H_2, so it would be advantageous to carry out the overall reaction:

$$ArNO_2 + 2 H_2 + CO \longrightarrow ArNCO + 2 H_2O$$

CO would be used only for carbonylation. At present, there is no single system which accomplishes this formidable task. As a result, there has been a focus on carrying out the reaction in a stepwise manner. Nitrobenzene is reduced using conventional supported platinum technology. Aniline is then coupled to methylenedianiline before being subjected to oxidative carbonylation using palladium catalysts in a cooxidation of ArNH$_2$ and CO [116]:

$$\underset{\displaystyle \overset{\displaystyle |}{NH_2Ar}}{Pd^{II}-C\equiv O} \longrightarrow Pd^0 + O=C=NAr + 2H^+$$

The most efficient homogeneous system for aniline oxidative carbonylation involves PdCl$_2$ as the catalyst with FeCl$_3$ or FeOCl as promoter [117]. CuCl$_2$ and VOCl$_3$ have also been found to promote the reaction. The reaction can be described as a Wacker-type transformation, but it is clear that the promoter must not only be redox-active, but must also be Lewis acidic. In the Pd/Fe system, the Pd reduced to Pd^0 is reoxidized to Pd^{II} and the FeCl$_3$ is reduced to FeCl$_2$. Air reoxidation of the FeCl$_2$ is incomplete and iron oxides are formed.

An interesting modification of the process has been presented by Asahi Chemical [118]. In their system, Pd is heterogenized on carbon. The system relies on an alkali metal iodide as promoter. The chemistry resembles that of a homogeneous system, but there is no evidence that the Pd leaves the heterogeneous support.

$$ArNH_2 + \tfrac{1}{2}O_2 + CO + EtOH \xrightarrow[\text{I}^-]{\text{Pd/C}} ArNHCO_2Et + H_2O$$

As a result, isolation of the product is simplified. The reaction is carried out at 150-180°C and pressures of 50-80 kg/cm^2. In 2 hr, the yield is greater than 95% with selectivity of over 97%. Condensation to the methylenediurethane and cracking to MDI follows.

Oxidative carbonylation is being explored as a route to aliphatic isocyanates [119-121]. Macrocyclic complexes of cobalt and other metals have been found to catalyze the reaction [122]. Reductive carbonylation is not practical because nitroalkanes are less readily available than alkylamines which are often derived from hydrogenation of corresponding alkylnitriles (see hydrocyanation in Section 3.6).

5.7 CO OXIDATION AND THE SHIFT REACTION

In most commercial production of hydrogen from hydrocarbons, a mixture of carbon monoxide and hydrogen is produced by steam reforming or partial oxidation. The so-called shift reaction

$$CO + H_2O \rightleftharpoons CO_2 + H_2$$

is then performed in order to maximize hydrogen yield. Ordinarily it is carried out in two stages, first at 350-450°C over a chromium-based catalyst which gives a rapid conversion of most of the CO. A second catalyst based on copper and zinc is then used at 200-300°C, where the equilibrium constant is more favorable for complete conversion [123].

The shift reaction is catalyzed by a variety of soluble complexes such as Ru$_3$(CO)$_{12}$, Pt(PR$_3$)$_3$, [Rh(CO)$_2$I$_2$]$^-$, and [HFe(CO)$_4$] [124-128]. Although the reactions are too slow and too cumbersome for practical use, these studies can provide valuable insights into the mode of operation of the commercial heterogeneous catalysts. The rhodium system [127] is one of the best understood. A solution of [Rh(CO)$_2$I$_2$]$^-$ in aqueous acetic acid containing HCl converts CO to equimolar amounts of hydrogen and carbon dioxide at 80-90°C and 0.5 atmospheres pressure. The rate is low (5-9 cycles per day), but the catalyst seems relatively stable. The reaction becomes rapid at higher temperatures. In fact, it becomes a serious problem in carbonylation reactions in the presence of water because the hydrogen produced consumes olefin by hydrogenation (Section 5.3).

It appears that the rhodium catalyst functions through a two-electron redox cycle. The CO oxidation appears to occur by hydrolysis of a high-valent metal carbonyl complex such as rhodium(III):

$$Rh^{III}-C\equiv O \longrightarrow Rh^{I} + CO_2 + 2H^+$$
$$\overset{\displaystyle\uparrow}{OH_2}$$

In this respect, the shift reaction closely resembles the stoichiometric oxidation of CO by $PdCl_2$ [104]. In both cases, the high-valent metal ion appears to activate the CO by withdrawal of electron density. Partial positive charge on the carbonyl carbon atom renders it susceptible to attack by nucleophiles such as water.

In contrast to $PdCl_2$, the soluble rhodium catalysts are able to complete a catalytic cycle by reoxidation of the low-valent hydrolysis product. A rhodium(I) carbonyl complex is oxidized by protons:

$$Rh^{I}(CO) + 2H^+ \longrightarrow Rh^{III}(CO) + H_2$$

In this way, the protons generated during CO oxidation are converted to the desired product, H_2. It seems quite plausible that similar redox cycles occur on the surface of the solid metal oxides used as commercial shift catalysts.

5.8   CO REDUCTION

As noted at the beginning of this chapter, rising prices and limited supplies of conventional petrochemical feedstocks are a major incentive to use synthesis gas for the production of organic chemicals. This incentive has stimulated much research on selective hydrogenation of carbon monoxide to useful products. Three processes for which soluble catalysts might be used commercially are the syntheses of ethylene glycol, ethanol, and vinyl acetate. In addition to these potential commercial applications, there is much research on homogeneous catalytic equivalents of the Fischer-Tropsch reaction in which CO is hydrogenated to hydrocarbons over a heterogeneous catalyst [129-131]. The three potential industrial processes are discussed individually below.

Union Carbide's work on a synthesis gas-derived route to ethylene glycol provided a major stimulus to transition metal cluster chemistry. Efforts to link homogeneous catalysis of the Fischer-Tropsch reaction with the commercially relevant heterogeneous systems has greatly stimulated both cluster chemistry and surface science [132]. Labelling studies in an active Fischer-Tropsch reaction indicate that the production of hydrocarbons results from the coupling of surface methylene and alkyl groups [133] as originally proposed by Fischer and Tropsch [134]. Stepwise reduction of CO to a bridging methylene has been observed on the face of an osmium cluster [135]. Formation of an ethyl group by migration of a methyl group to a coordinated methylene [136], reduction of a formyl-hydrido to a methyl-hydrido complex [137], and conversion of a formyl-methyl complex to free olefin [138] have all been observed on closely related mononuclear iridium centers. On the other hand, labelling studies on a heterogeneous Rh/Ti on silica catalyst demonstrate that C_2 oxygenates such as ethanol and acetaldehyde are derived from a common intermediate acyl which is formed by CO insertion into surface CH_3 and CH_2 species [139] similar to those observed in homogeneous systems described in the section about acetic acid. A number of metal carbonyl compounds have been reduced to formyl derivatives [140].

Ethylene Glycol

Most ethylene glycol is prepared by the direct oxidation of ethylene to ethylene oxide followed by hydrolysis. Du Pont has practiced a process based upon the carbonylation of formaldehyde to glycolic acid followed by esterification and hydrogenolysis to ethylene glycol [141].

$$C\equiv O + 2\ H_2 \longrightarrow CH_3OH$$
$$CH_3OH + \tfrac{1}{2}\ O_2 \longrightarrow H_2C=O + H_2O$$
$$H_2C=O + C\equiv O + H_2O \longrightarrow HOCH_2CO_2H$$
$$HOCH_2CO_2H + ROH \longrightarrow HOCH_2CO_2R + H_2O$$
$$HOCH_2CO_2R + 2\ H_2 \longrightarrow HOCH_2CH_2OH + ROH$$

Although all of the C and H in this process is ultimately derived from synthesis gas, the multiple steps and sequential over-reduction and over-oxidation render it uneconomical. Use of a Nafion® catalyst apparently improves the process [142].

Union Carbide developed a process in which synthesis gas is converted to ethylene glycol by a soluble catalyst. The process involves high reaction pressures and difficult product separations, but it has major economic potential if these problems can be solved. The process development was carried to a semiworks [143] but was not commercialized.

A variety of rhodium complexes catalyze the transformation of synthesis gas (1:1 H_2:CO) to a mixture of methanol, ethylene glycol, propylene glycol, and glycerol [144,145]. The conditions are severe (210-250°C and 500-3400 atmospheres pressure), but yields of ethylene glycol range up to 70% [143]. Similar results are obtained with mononuclear complexes such as $Rh(CO)_2(acac)$, or with the carbonyl clusters $Rh_4(CO)_{12}$ or $Rh_6(CO)_{16}$. The infrared spectrum of the reaction mixture at 180°C under high pressure shows a pattern of C≡O stretching frequencies assigned to a cluster anion, $[Rh_{12}(CO)_{30}]^{2-}$. However, it has been demonstrated that the larger cluster compounds are not catalytic intermediates [146]. Cobalt compounds such as cobalt(II) acetate also catalyze the hydrogenation of CO to ethylene glycol [147], but they are usually less active than the rhodium catalysts and are less selective for glycol formation. Ruthenium [146,148,149] and mixed rhodium/ruthenium [150] systems have also been reported to be active and selective.

Several early transition metal complexes stoichiometrically reduce CO to a two-carbon ligand. With $Zr(C_5Me_5)_2H_2$ [151], the following sequence of reactions is believed to occur (C_5Me_5 ligands omitted):

$$\underset{\overset{|}{\text{H}}}{\overset{\overset{|}{\text{H}}}{\text{Zr}}} \quad \xrightarrow{\text{CO}} \quad \underset{\overset{|}{\text{H}}}{\overset{\overset{|}{\text{H}}}{\text{Zr}}}-\text{CO} \quad \rightleftharpoons \quad \left[\underset{\overset{|}{\text{H}}}{\overset{\text{H}}{\underset{\text{Zr}}{\diagup}\text{C}=\text{O}}} \right] \quad \longrightarrow \quad \underset{\overset{|}{\text{H}}}{\text{Zr}}\diagdown\text{O}\diagup\overset{\overset{\text{H}}{|}}{\text{C}}=\overset{\overset{\text{H}}{|}}{\text{C}}\diagdown\text{O}\diagup\underset{\overset{|}{\text{H}}}{\text{Zr}}$$

Coordination of CO to the dihydride gives the CO complex which may be in equilibrium with the formyl complex. Dimerization yields the enedioxy compound, a potential glycol precursor. The formyl complex also gives rise to a methoxy compound $(C_5Me_5)_2Zr(H)(OCH_3)$ under other conditions. While there is no direct evidence to link the zirconium and rhodium or ruthenium chemistry, the pathway via a formyl complex seems quite plausible for both methanol and glycol syntheses.

Dialkyl Oxalates and Carbonates

Oxidative carbonylation of alcohols can lead to dialkyl oxalates. While useful in their own right, dialkyl oxalates have also been viewed as intermediates to ethylene glycol [152,153]. The reaction has been demonstrated using palladium catalysts under Wacker-type conditions [38,154,155].

$$2\,CO + 2\,ROH + Pd^{II} \longrightarrow RO_2C\text{-}Pd^{II}\text{-}CO_2R + 2H^+$$
$$RO_2C\text{-}Pd^{II}\text{-}CO_2R \longrightarrow RO_2C\text{-}CO_2R + Pd^0$$
$$Pd^0 + 2\,H^+ + {}^1\!/_2\,O_2 \xrightarrow{\;Cu^I/Cu^{II}\;} Pd^{II} + H_2O$$

Formation of water in the final copper-mediated step is a major limitation for glycol formation. The water hydrolyzes the dialkyl oxalate to the monoester or the diacid. These species are less stable under reaction conditions, decomposing by loss of CO_2 to yield formates. Additionally, the hydrolyzed species are more difficult to reduce cleanly to ethylene glycol. To overcome these problems with water, drying agents such as ethyl orthoformate, zeolites, or boric anhydride may be added to the reaction [156].

Support for the proposed intermediacy of the bis(carboalkoxy)palladium species is provided by platinum model compounds such as $(DPPE)Pt(CO_2Me)_2$ [157] and $Ru(DPPE)(CO)_2(CO_2Me)_2$ mentioned in conjunction with isocyanates [115]. While relatively stable, these complexes can be induced to liberate dialkyl oxalates under appropriate conditions.

Ube Industries have developed both homogeneous and heterogeneous syntheses of oxalate diesters, again based upon oxidative carbonylation of an alcohol substrate [158]. In a marked departure from other oxidative carbonylations, the oxygen carrier is NO which enters the reaction by an initial coupling with the alcohol to yield an alkyl nitrite. This represents a preoxidation of the alcohol; the liberated water is not carried

$$2\,NO + 2\,ROH + {}^1\!/_2\,O_2 \longrightarrow 2\,RONO + H_2O$$

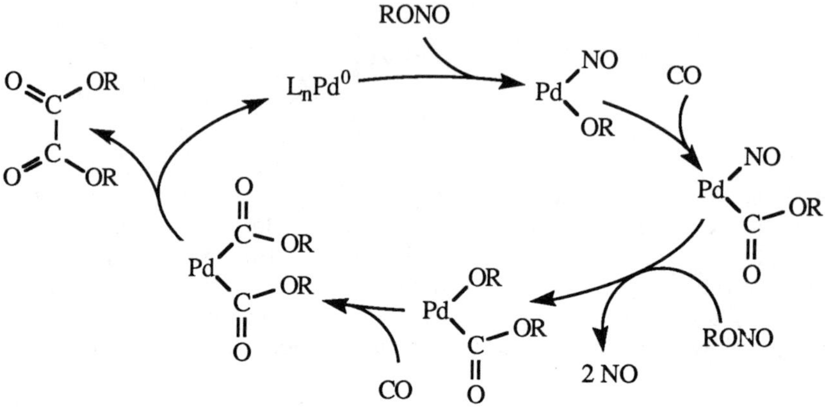

Figure 5.9   Scheme for the nitrite-promoted synthesis of oxalate esters.

into the carbonylation cycle. Thus the limitations imposed by liberation of water in the more conventional Wacker cycle are overcome. The nitrite is thought to undergo oxidative addition to the palladium catalyst [159], initiating the cycle shown in Figure 5.9. While it is likely that the process actually uses a heterogeneous palladium on carbon catalyst, homogeneous catalysts are also effective for the transformation [160]. The homogeneous reaction is carried out using butanol at 110°C and 60 atmospheres CO; yield is about 80% with dibutyl carbonate as the major impurity. The heterogeneous reaction is run at 110°C at one atmosphere of CO to yield 95% of the dialkyl oxalate.

Dialkyl carbonates can be substituted for phosgene in certain applications. As a result, their synthesis by related catalytic cycles has also been of interest [161]. Dialkyl carbonates are known byproducts of the above oxalate syntheses. Clean synthesis of the dialkyl carbonates using palladium catalysis has proved impractical, but Enichem has commercialized a process based upon copper catalysis [162]. A slurry of CuCl in methanol is reacted with oxygen and 2:1 CO/H$_2$ at 120°C and 25 atmospheres. The yields of dimethyl carbonate are 100% based on methanol and 90-95% based on CO when conversions are kept in the range of 30-35%. Some CO is lost to CO$_2$.

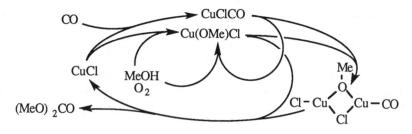

Figure 5.10   Copper-catalyzed synthesis of dimethylcarbonate.

The nature of the intermediates in the catalytic process are ill defined, but a mechanistic scheme has been proposed [163] (Figure 5.10). The cuprous chloride is taken into the cycle by complexation with carbon monoxide. The mixed valence dimer contains the two of the three fragments necessary for formation of the carbonate [164]. Reaction with another copper methoxide liberates the carbonate and regenerates Cu^ICl.

Alcohol Synthesis

As mentioned above, $Co_2(CO)_8$ catalyzes the hydrogenation of CO to methanol and higher alcohols:

$$CO + 2H_2 \longrightarrow CH_3OH \xrightarrow{\text{CO/H}_2} C_2H_5OH + H_2O$$

A 1:1 H_2:CO mixture in dioxane at 182°C and 300 atmospheres yields methanol as the primary product along with some methyl formate [165]. As the reaction proceeds, a second faster process produces ethanol, 1-propanol, and their formate esters. The methanol synthesis is not especially interesting from an industrial viewpoint because copper-based heterogeneous catalysts produce methanol rapidly and cleanly at pressures below 100 atmospheres. However, the homologation of methanol to mixtures of higher alcohols may be useful for the production of synthetic fuels. Methanol homologation to higher alcohols is catalyzed by a variety of metal complexes [166].

It has been proposed [165] that methanol synthesis occurs by a radical process mediated by mononuclear cobalt complexes. The rate-determining step in this scheme is production of a formyl radical from cobalt carbonyl hydride:

$$HCo(CO)_4 + CO \rightleftharpoons H\dot{C}O + \dot{C}o(CO)_4$$

The formyl radical or formaldehyde derived from it would then be hydrogenated by $HCo(CO)_4$ in much the same way that aldehydes are reduced to alcohols in direct alcohol synthesis by hydroformylation.

The methanol homologation reaction has been known for many years [167] and has been reinvestigated recently [168]. When methanol is treated with synthesis gas at 160-180°C and 300 atmospheres pressure in the presence of $Co_2(CO)_8$ or a cobalt(II) salt, ethanol, and higher linear alcohols and formate esters are produced. Typically about half of the methanol is converted to higher alcohols with 35-40% selectivity to ethanol. The reaction is promoted by iodide ion in a manner similar to acetic acid synthesis from methanol. It appears that facilitation of C-O bond cleavage is essential just as in acetic acid synthesis. The proposed mechanism involves acid-promoted nucleophilic attack on methanol. (As mentioned earlier, $HCo(CO)_4$ is a strong acid in polar media [36].)

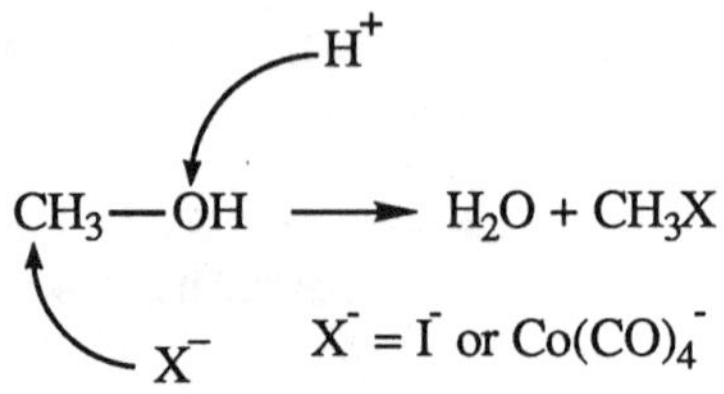

$$CH_3\!-\!OH \xrightarrow{\hspace{1.5cm}} H_2O + CH_3X$$

$$-H^+$$

$$X^- \qquad X^- = I^- \text{ or } Co(CO)_4^-$$

The $CH_3Co(CO)_4$, which is formed directly or via CH_3I, reacts with CO and hydrogen to form acetaldehyde and ethanol, sequentially.

Somewhat similar chemistry seems to be involved in the homologation of dimethyl ether or methyl acetate to ethyl acetate which is catalyzed by ruthenium salts and iodide ion [169]. The reaction may be sequential:

$$CH_3OCH_3 \xrightarrow{CO} CH_3OCCH_3 \xrightarrow[H_2]{CO} CH_3CH_2OCCH_3$$
$$\qquad\qquad\qquad \underset{O}{\|} \qquad\qquad\qquad \underset{O}{\|}$$

In contrast to the cobalt-catalyzed homologation, acetaldehyde is not observed as a byproduct.

5.9 SPECIALTY PRODUCTS

As mentioned in the introduction to this chapter, the area of "C_1 Chemistry" for commodity chemicals has not lived up to the high expectations of a decade ago. We have not seen the introduction of the many new synthesis gas-based processes for the production of major chemicals which had been predicted. Interestingly, there has been activity in the synthesis of specialty, higher value-in-use products using CO-based chemistry. These products are not driven by the cost of feedstocks as much as they are driven by the selectivity of the chemistry involved.

Montedison synthesizes phenylacetic acid by carbonylation of benzyl chloride [170] for use in perfume constituents and the pesticide, Cidial®.

$$PhCH_2Cl + CO + 2OH^- \longrightarrow PhCH_2COO^- + Cl^- + H_2O$$

The reaction is run in a biphasic system employing diphenylether and 40% aqueous sodium hydroxide [171]. A cobalt carbonyl is used as an organometallic catalyst, and a surfactant is employed as a phase-transfer catalyst. Pressures of several atmospheres and temperatures of 50-60°C are employed as benzyl chloride is added continuously. The surfactant presumably transports the $[Co(CO)_4]^-$ into the organic phase where it reacts with the benzyl chloride to form benzylcobalt tetracarbonyl. After rapid CO insertion to form the phenylacetyl complex, the organometallic is hydrolyzed by the small concentration of water present in the organic phase, yielding phenylacetic acid and $HCo(CO)_4$ which continues the cycle.

Aldose sugars are smoothly decarbonylated by $Rh(PPh_3)_3Cl$ in N-methyl-pyrrolidone to give the next lower sugar, alditol. The conversion of commercial 2-deoxyribose to 1-deoxyerythritol requires seven steps by conventional sugar chemistry, but can be carried out in a single step in 30 minutes in 90% yield [172].

Ajinomoto prepared monosodium glutamate via hydroformylation of acrylonitrile from 1963 to 1973 [173-175]. The cobalt catalysts gave good regioselectivity for the linear aldehyde.

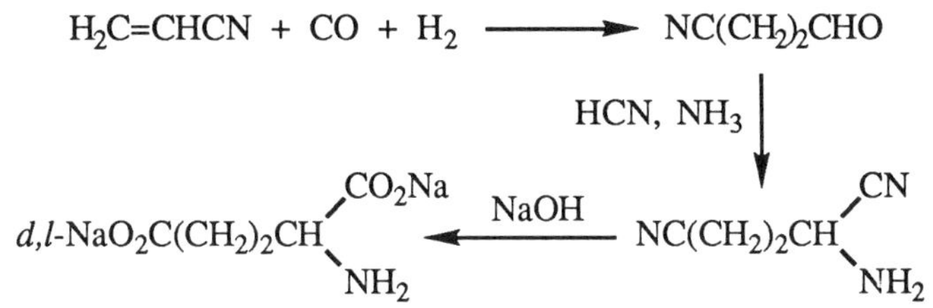

$$H_2C=CHCN + CO + H_2 \longrightarrow NC(CH_2)_2CHO$$

$$d,l\text{-}NaO_2C(CH_2)_2CH \begin{smallmatrix} CO_2Na \\ \\ NH_2 \end{smallmatrix} \xleftarrow{\ \ NaOH\ \ } NC(CH_2)_2CH \begin{smallmatrix} CN \\ \\ NH_2 \end{smallmatrix} \xleftarrow{\ HCN, NH_3\ }$$

The resulting linear aldehyde was reacted with ammonia and HCN to give racemic glutamonitrile. Basic hydrolysis gave the racemic disodium glutamate. Optical resolution gave the desired l-product. The d-product was racemized and recycled to the resolution step. The process has been supplanted by a microbial synthesis, and there was some concern about the use of acrylonitrile in manufacture of a food product.

Vitamin A is produced on a scale which is relatively large for a pharmaceutical. As a result, there have been strong competitive pressures to develop economical syntheses. Both BASF and Hoffmann-La Roche utilize hydroformylation in their commercial production of vitamin A. A key intermediate, 1,4-diacetoxy-2-butene **25**, is synthesized using homogeneous processes described in Chapter 6. From that point, the syntheses diverge. BASF uses a Pt catalyst to isomerize the 1,4-diacetate to the 1,2-diacetate **26** which is lower boiling (Section 2.4).

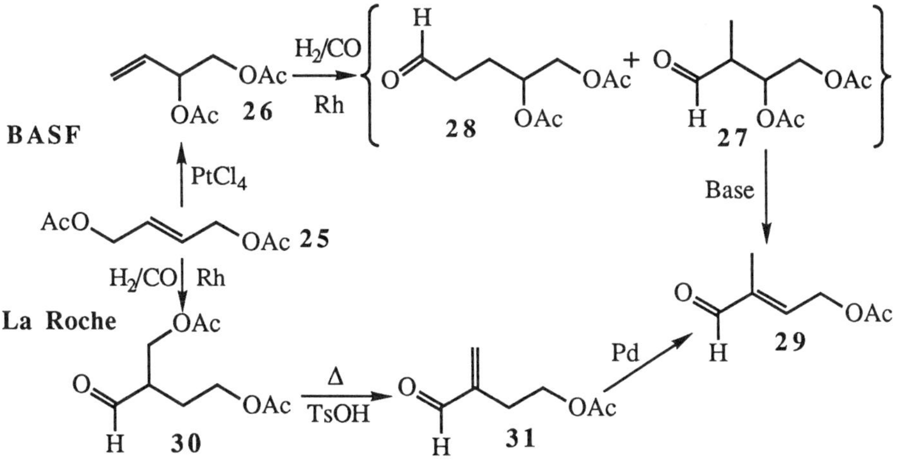

Hydroformylation with a phosphine-free rhodium catalyst is used under relatively high temperatures and pressures. These conditions were chosen to favor synthesis of the branched **27** rather than the usual linear **28** isomer [176]. Treatment of the mixed products with sodium acetate in acetic acid selectively eliminates acetic acid from the branched product yielding the desired 2-methyl-4-acetoxycrotonaldehyde **29**. The linear aldehyde **28** is recovered and converted to 1,2,5-pentanetriol for use in synthetic lubricants [177]. La Roche takes a different approach in their patents. The linear 1,4-diacetoxy-2-butene **25** is hydroformylated using a rhodium catalyst necessarily yielding a branched isomer **30** [178]. Treatment with *p*-toluenesulfonic acid eliminates one equivalent of acetic acid, yielding **31**. Isomerization of **31** with a carbon-supported palladium catalyst yields the desired product **29**.

Application of hydroformylation technology is not limited to simple olefins. Arco has commercialized a route to butanediol based upon hydroformylation of allyl alcohol [179]. Arco's strength in the synthesis of propylene oxide (Section 6.5) makes this route particularly attractive for them.

The isomerization of propylene oxide **32** to allyl alcohol **33** is a curious but well-used [180] reaction catalyzed by lithium phosphate at 250-300°C and 150 psi with >90% selectivity. They then use Kuraray technology for the hydroformylation to 4-hydroxybutyraldehyde **34** [181] which is finally hydrogenated to the diol using Raney nickel.

Shell is offering several unsaturated aldehydes based upon selective hydroformylation of intermediates out of their FEAST specialty metathesis facility (see Chapter 9 for metathesis chemistry). Metathesis of cyclooctene with ethylene yields 1,9-decadiene. Sequential hydroformylation yields 10-undecenal and 1,12-dodecane-dial [182]. The 5- and 6-heptenals and 1,8-octanedial are prepared similarly. Hydroformylation of norbornadiene yields mixed isomers of norbornanedial. These types of products are important in the aroma and perfume industries.

A synthesis of 1,9-nonanediol is based upon the hydrodimerization of butadiene (see Section 4.5) followed by hydroformylation. Kuraray has started a semiworks which hydrodimerizes butadiene using a homogeneous palladium catalyst which employs a monosulfonated triphenylphosphine to solubilize the catalyst in the aqueous phase. The hydrodimer is isomerized using a heterogeneous copper chromite catalyst [183]. Hydroformylation is carried out using a rhodium catalyst followed by hydrocarbon extraction. The ligands may be the same monosulfonated triphenyl-phosphine in a 1:1 sulfolane/water system [184]. Later patents suggest the use of protected enols and a sterically hindered phosphite system using tri(*ortho-t*-butylphenyl)phosphite [185,186].

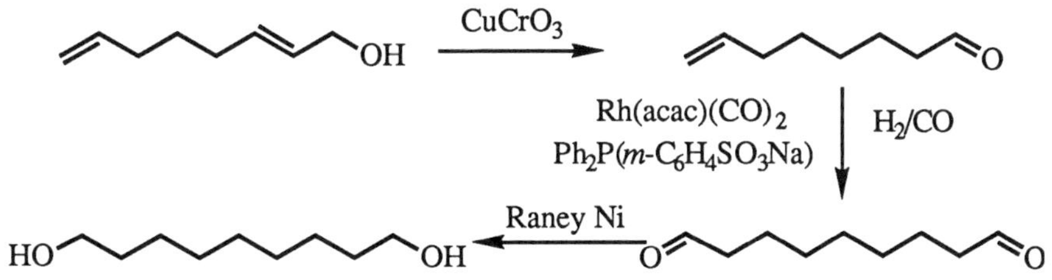

The resulting dialdehyde is extracted from the reaction mixture using an alkane solvent, possibly cyclohexane, and is hydrogenated to the diol over a molybdenum-promoted Raney nickel [187].

There are several carbonylation approaches to cyclic products; 2-hydroxytetrahydropyran can be obtained from the reaction discussed in the previous paragraphs. Kuraray manufactures 3-methyl-1,5-pentanediol from isobutene via a cyclic intermediate. 3-Methyl-3-butenol is available as a side product from their synthesis of isoprene by reaction of isobutene with formaldehyde. It is carbonylated using a rhodium catalyst to give 2-hydroxy-4-methyltetrahydropyran [188].

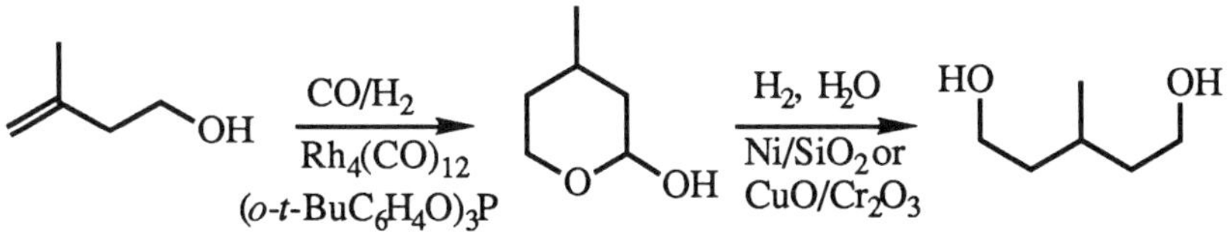

The ether is then hydrogenated in the presence of water using a heterogeneous catalyst to give the desired product [189].

There are relatively few examples of the formation of ketones under hydroformylation conditions. Carbonylation of styrene using weakly solvated cationic palladium complexes yields 1,5-diphenylpentyl-1-ene-3-one with 95% selectivity, the other major product being the saturated 1,5-diphenylpentan-3-one [190]. There are other less interesting examples of carbonylation of ethylene or propylene to give the respective ketones [191]. Thus it is interesting that ethylene or higher α-olefins can be copolymerized with carbon monoxide using a Pd(CN)₂ catalyst to give polyketones [192]. The resulting E/CO polymer is a rigorously alternating, linear, crystalline copolymer. High molecular weight 1:1 E/CO was prepared using HPd(CN)₃ [193] and a 1:1 propylene/CO polymer was reported. These polymerizations could be carried out in a number of polar solvents such as alcohols, acetic acid, ethyl acetate or acetonitrile. Palladium chloride is reported to copolymerize norbornadiene with CO [194]. Many of these copolymerizations can be run under much milder conditions using an improved catalyst derived from triphenylphosphine and "naked palladium" [195,196]. The mildness of these reactions and the dependence on added phosphine implies that the availability of a vacant coordination site on the catalyst is an important feature of this polymerization. This requirement has also been demonstrated in a nickel-catalyzed copolymerization [197]. A cationic palladium bipyridine system displays the first evidence of a living alternating polymerization and should provide valuable mechanistic insight into the polymerization [198]. Shell is proceeding with the commercial-ization of these specialty copolymers as highly photodegradable, environmentally friendly plastics. Their research activities in this area are indicated by a series of patents on E/CO [199,200] and E/P/CO [201] copolymers.

In all of the work discussed to this point, carbonylation has been assumed to lead to the introduction of a single molecule of CO to the reactive group. Palladium-catalyzed double carbonylation of aryl halides can lead to the introduction of two molecules of CO, yielding α-ketocarboxylic acids or their derivatives [202,203]. In special cases, similar products can be derived from aromatic aldehydes through initial oxidative addition of the aldehyde C-H bond and from aromatics through *ortho*-metallation of C-H bonds [204]. Ethyl Corporation has advertised products derived from double carbonylation of haloalkanes and bis(α-ketocarboxylic acids) from bis-double carbonylation of dihaloalkanes [205].

A mechanistic picture has been developed for the formation of α-ketoamides [206]. The reducing conditions of the reaction form zero-valent palladium in situ. Oxidative addition of the aromatic halide yields the aryl palladium halide. Addition of the first CO yields the benzoyl complex. A second CO displaces halide from the complex; the cationic carbonyl is then more susceptible to nucleophilic attack to give the aroyl carbamoyl palladium.

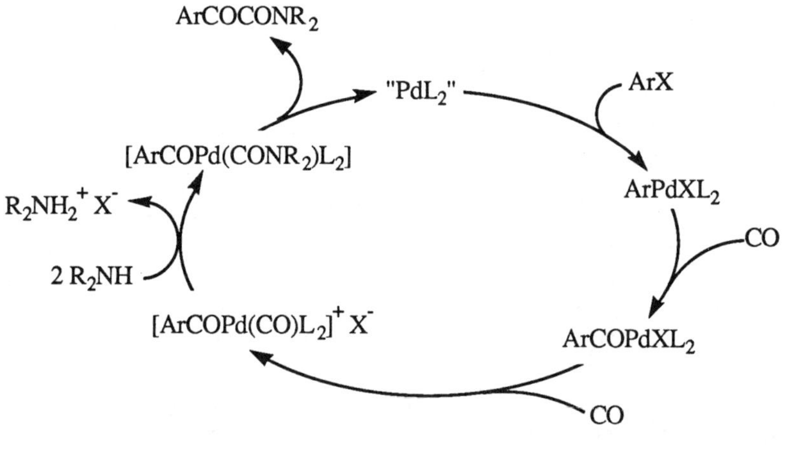

The alternative mechanism is based upon a second CO addition into the benzoyl palladium bond, and has been excluded on the basis of the reactivity of $PhCOCOPdCl(PMePh_2)_2$ [207,208]. Reductive elimination of the benzoyl and carbamoyl groups yields the α-ketoamide and Pd^0 to restart the cycle.

GENERAL REFERENCES

J. Falbe, *Carbon Monoxide in Organic Synthesis*, Springer-Verlag, Berlin, 1970.

I. Wender and P. Pino, Eds., *Organic Syntheses via Metal Carbonyls*, Vols. 1 and 2, Wiley-Interscience, New York, 1968, 1970.

F. G. A. Stone and R. West, Eds., *Advances in Organometallic Chemistry*, Vol. 17, Academic Press, New York, 1979.

W. Keim, Ed., *Catalysis in C1 Chemistry,* Vol. 4, Reidel (Kluwer), Dordrecht, 1983.

R. A. Sheldon, Ed., *Chemicals from Synthesis Gas*, Reidel, Dordrecht, 1983.

D. R. Fahey, Ed., *Industrial Chemicals Via C1 Processes*, ACS Symposium Series 328, Washington, DC, 1987.

Progress in C1 Chemistry, Elsevier, Amsterdam, 1989.

SPECIFIC REFERENCES

1. J. Falbe, *Carbon Monoxide in Organic Synthesis*, Springer-Verlag, Berlin, 1970.
2. *Chemical Week,* 34 (July 26, 1989).
3. F. Calderazzo, R. Ercoli, and G. Natta, "Metal Carbonyls: Preparation, Structure and Properties," in I. Wender, and P. Pino, Eds, *Organic Syntheses via Metal Carbonyls*, Vol. 1, Wiley-Interscience, New York, 1968.
4. E. L. Muetterties, *Science*, **196**, 839 (1977).
5. K. G. Caulton, R. L. DeKock, and R. F. Fenske, *J. Am. Chem. Soc.*, **92**, 515 (1970).
6. K. Noack, M. Ruch, and F. Calderazzo, *Inorg. Chem.*, **7**, 345 (1968); K. Noack and F. Calderazzo, *J. Organomet. Chem.*, **10**, 101 (1967).
7. H. Berke and R. Hoffmann, *J. Am. Chem. Soc.*, **100**, 7224 (1973).
8. F. Calderazzo and K. Noack, *Coord. Chem. Rev.*, **1**, 118 (1966).
9. P. L. Bock, D. J. Boschetto, J. R. Rasmussen, J. P. Demers, and G. M. Whitesides, *J. Am. Chem. Soc.*, **96**, 2814 (1974).
10. T. G. Attig and A. Wojcicki, *J. Organomet. Chem.*, **82**, 397 (1974).
11. R. Whyman, *Chemtech*, 414 (July 1991); K. A. Hunt, R. W. Page, S. Rigby, and R. Whyman, *J. Phys. E: Sci. Instrum.*, **17**, 559 (1984).
12. W. R. Moser, J. E. Cnossen, A. W. Wang, and S. A. Krouse, *J. Catal.*, **95**, 21 (1985).
13. D. E. Morris and H. B. Tinker, *Rev. Sci. Instrum.*, **43**, 1024 (1972).
14. J. M. L. Penninger, *J. Catal.*, **56**, 287 (1979).
15. D. C. Roe, *J. Magn. Res.*, **63**, 388 (1985).
16. B. T. Heaton, L. Strona, J. Jonas, T. Eguchi, and G. A. Hoffman, *J. Chem. Soc., Dalton*, 1159 (1982).

17. J. Jonas, D. L. Hasha, W. J. Lamb, G. A. Hoffman, and T. Eguchi, *J. Magn. Res.*, **42**, 169 (1981).
18. R. P. Lowry and A. Aguilo, *Hydrocarbon Proc.*, 103 (Nov. 1974).
19. P. Ellwood, *Chem. Eng.*, 148 (May 19, 1969).
20. J. F. Roth, J. H. Craddock, A. Hershman, and F. E. Paulik, *Chemtech.*, **1**, 600 (1971); H. D. Grove, *Hydrocarbon Proc.*, 76 (Nov. 1972).
21. F. E. Paulik, U.S. Patent 3,769,329 (1973).
22. D. Forster, *J. Am. Chem. Soc.*, **97**, 951 (1975); D. Forster, *Ann. N. Y. Acad. Sci.*, **295**, 79 (1977); *Adv .Organomet. Chem.*, **17**, 255 (1979).
23. A. Haynes, B. E. Mann, D. J. Gulliver, G. E. Morris, and P. M. Maitlis, *J. Am. Chem. Soc.*, **113**, 8567-8569 (1991).
24. V. H. Agreda, D. M. Pond, and J. R. Zoeller, *Chemtech*, 172-181 (March, 1992).
25. C. Hewlett, German Patent 2,441,502 (1975) to Halcon.; M. Schrod and G. Luft, *Ind. Eng. Prod. Res. Dev.*, **20**, 649 (1981).
26. A. N. Naglieri and N. Rizkalla, U.S. Patent 4,002,678 (1975) to Halcon.
27. T. H. Larkins, S. W. Polichnowski, G. C. Tustin, and D. A. Young, U.S. Patent 4,374,070 (1983) to Eastman Kodak.
28. *Hydrocarbon Proc.*, **44**, 287 (1965); *Hydrocarbon Proc.*, **46**, 146 (1967).
29. N. Rizkalla and C. N. Winnick, Ger. Offenleg. 2,610,035 (1976).
30. K. Bittler, N. von Kutepow, D. Neubauer, and H. Reis, *Angew. Chem. Int. Ed.*, **7**, 329 (1968).
31. D. M. Fenton, *J. Org. Chem.*, **38**, 3192 (1973).
32. P. Pino, F. Piacenti, and M. Bianchi, "Hydrocarboxylation of Olefins" in I. Wender and P. Pino, Eds., *Metal Syntheses via Metal Carbonyls*, Vol. 2, Wiley-Interscience, New York, 1977, p. 233.
33. F. Piacenti, M. Bianchi, and R. Lazzaroni, *Chim. Ind. (Milan)*, **50**, 318 (1968).
34. J. F. Knifton, *J. Org. Chem.*, **41**, 2885 (1976).
35. R. F. Heck and D. S. Breslow, *J. Am. Chem. Soc.*, **85**, 2779 (1963).
36. S. S. Kristjánsdóttir and J. R. Norton, "Acidity of Hydrido Transition Metal Complexes in Solution," in *Transition Metal Hydrides: Recent Advances in Theory and Experiment*, A. Dedieu, Ed., VCH, New York, 1991.
37. R. W. Johnson and R. G. Pearson, *Inorg. Chem.*, **10**, 2091 (1971).
38. F. Rivetti and U. Romano, *J. Organomet. Chem.*, **154**, 323 (1978).
39. G. Cavinato and L. Toniolo, *J. Organomet. Chem.*, **398**, 187-195 (1990).
40. N. Von Kutepow, U.S. Patent 3,876,695 (1975) to BASF.
41. P. M. Burke, U.S. Patent 4,939,298 (1990) to Du Pont.
42. P. M. Burke, U.S. Patent 4,622,423 (1986) and U.S. Patent 4,788,333 (1988) to Du Pont.
43. P. M. Burke, U.S. Patent 4,788,334 (1988) to Du Pont.
44. R. Kummer, H-W. Schneider, and F-J. Weiss, U.S. Patent 4,169,956 (1979); R. Platz, R. Kummer, and H-W. Schneider, U.S. Patent 4,258,203 (1981); R. Kummer, H-W. Schneider, and F-J. Weiss, U.S. Patent 4,259,520 (1981); all to BASF.
45. D. Milstein, *Acc. Chem. Res.* **21**, 428 (1988).

46. L. Wender, H. W. Sternberg, and M. Orchin, *J. Am. Chem. Soc.*, **74**, 1216 (1952).

47. W. Hieber and J. Sedlmeier, *Chem. Ber.*, **87**, 25 (1954).

48. D. Milstein and J. L. Huckaby, *J. Am. Chem. Soc.*, **104**, 6150 (1982).

49. G. Fachinetti, G. Fochi, and T. Funaioli, *J. Organomet. Chem.*, **301**, 91 (1986).

50. J. F. Knifton, U.S. Patent 4,172,087 (1979) to Texaco.

51. *Chem. Eng. News*, 26 (May 13, 1985); H. S. Kesling, *Prepr. Am. Chem. Soc., Div. Pet. Chem.*, **31**, 112 (1986); H. S. Kesling, *ACS Symp. Ser.*, **328**, 77 (1987).

52. *Farbe + Lack*, **90**, 1055 (1984); *Chem. Eng. News*, **65**(21), 14 (1987).

53. *Chem. Eng. News*, **62**(18), 28-9 (1984); *ChemWeek*, **134**, 30 (April 25, 1984); A. S. C. Chan and D. E. Morris, U.S. Patent 4,611,082 (1986); A. S. C. Chan and D. E. Morris, U.S. Patent 4,633,015 (1986) both to Monsanto.

54. British Patent 1,278,353 (1972) to Monsanto.

55. C. E. Loeffler, L. Stautzenberger, and J. D. Unruh, "Butyraldehydes and Butyl Alcohols," in J. J. McKetta and W. A. Cunningham, Eds., *Encyclopedia of Chemical Processing and Design*, Vol. 5, Dekker, New York, 1977, p. 358.

56. P. Pino, F. Piacenti, and M. Bianchi, "Reactions of Carbon Monoxide and Hydrogen with Olefinic Substrates: The Hydroformylation Reaction," in I. Wender and P. Pino, *Organic Syntheses via Metal Carbonyls*, Vol. 2, Wiley-Interscience, New York, 1977.

57. R. Kummer, H. J. Nienburg, H. Hohenschutz, and M. Strohmeyer in D. Forster and J. F. Roth, *Homogeneous Catalysis-II, Advances in Chemistry Series*, Vol. 32, American Chemical Society, Washington, DC, 1974, p. 19.

58. F. E. Paulik, *Catal. Rev.*, **6**, 49 (1972).

59. L. H. Slaugh and R. D. Mullineaux, U.S. Patents 3,239,569 and 3,239,570 (1966) to Shell.

60. J. A. Ibers, *J. Organomet. Chem.*, **14**, 423 (1968).

61. D. S. Breslow and R. F. Heck, *J. Am. Chem. Soc.*, **83**, 4023 (1961).

62. R. Whyman, *J. Organomet. Chem.*, **81**, 97 (1974).

63. N. H. Alemdaroglu, J. L. M. Penninger, and E. Oltay, *Monatsh. Chem.*, **107**, 1153 (1976).

64. G. F. Pregaglia, A. Andreeta, G. F. Ferrari, and R. Ugo, *J. Organomet. Chem.*, **30**, 387 (1971).

65. R. L. Pruett, *Ann. N. Y. Acad. Sci.*, **295**, 239 (1977).

66. R. L. Pruett and J. A. Smith, U.S. Patent 3,527,809 (1970) to Union Carbide.

67. R. Fowler, H. Connor, and R. A. Baehl, *Chemtech.*, 772 (1976); *Hydrocarbon Proc.*, **55**, 247 (Sept. 1976).

68. E. A. V. Brewester, *Chem. Eng.*, **83**, 90 (1976).

69. F. J. Smith, *J. Plat. Met. Rev.*, **19**, 93 (1975).

70. D. Evans, G. Yagupsky, and G. Wilkinson, *J. Chem. Soc. A*, 2660 (1968).

71. G. Yagupsky, C. K. Brown, and G. Wilkinson, *J. Chem. Soc. A*, 1392, 2753 (1970).

72. R. Fowler, British Patent 1,387,657 (1975); G. Wilkinson, U.S. Patent 4,108,905 (1978).

73. D. E. Morris and H. B. Tinker, *Chemtech.*, **2**, 554 (1972).

74. B. Cornils, R. Payer, and K. C. Traenckner, *Hydrocarbon Proc.*, 83 (June 1975).

75. E. G. Kuntz, *Chemtech*, 570 (Sept. 1987).

76.   B. Cornils, J. Hibbel, B. Lieder, J. Much, V. Schmidt, E. Weibus, and W. Konkol, U.S. Patent 4,523,036 (1985); B. Cornils, W. Konkol, H. Bach, G. Dambkes, W. Gick, W. Greb, E. Weibus, and H. Bahrmann, U.S. Patent 4,593,126 (1986); and several intervening patents, all to Ruhrchemie.

77.   J-L. Sabot, U.S. Patent 4,483,801 (1984); J. Jenck and D. Morel, U.S. Patent 4,668,824 (1987), both to Rhone-Poulenc.

78.   H. Bahrmann and H. Bach, *Phosphorus and Sulfur*, **30**, 611 (1987); H. Bach, W. Gick, and W. Konkol, *Proceedings of the 9th International Congress on Catalysis (Calgary)*, Vol. 1, M. J. Phillips and M. Ternan, Eds., Chem. Inst. Canada, Ottawa, 1988, p. 254.

79.   A. G. Abatjoglou, E. Billig, D. R. Bryant, J. M. Maher, and R. E. Murray, U.S. Patent 4,717,775 (1986) and U.S. Patent 4,599,206 (1988), both to Union Carbide.

80.   A. van Rooy, E. N. Orij, P. C. J. Kamer, F. van den Aardweg, and P. W. N. M. van Leeuwen, *J. Chem. Soc., Chem. Commun.*, 1096-7 (1991).

81.   K. Tano, K. Sato, and T. Okoshi, German Patent 3,338,340 (1984) to Mitsubishi Chemical.

82.   C. Miyazawa, H. Mikami, A. Hiroshi, and K. Hamano, European Patent 272,608 (1988) to Mitsubishi Chemical.

83.   J. Tsuji, "Decarbonylation Reactions Using Transition Metal Compounds," in I. Wender and P. Pino, Eds., *Organic Syntheses via Metal Carbonyls*, Vol. 2, Wiley-Interscience, New York, 1977 p. 595; J. Tsuji and K. Ohno, *Synthesis*, 157 (1969).

84.   K. Ohno and J. Tsuji, *J. Am. Chem. Soc.*, **90**, 99 (1968).

85.   D. H. Doughty and L. H. Pignolet, *J. Am. Chem. Soc.*, **100**, 7083 (1978).

86.   J. O. Hawthorn and M. H. Wilt, *J. Org. Chem.*, **25**, 2215 (1960).

87.   *Hydrocarbon Proc.*, **54**, 208 (1975).

88.   T. A. Foglia, P. A. Barr, and M. J. Idacavage, *J. Org. Chem.*, **41**, 3452 (1976).

89.   J. K. Stille and M. T. Regan, *J. Am. Chem. Soc.*, **96**, 1508 (1974); J. K. Stille and R. W. Fries, *J. Am. Chem. Soc.*, **96**, 1514 (1974); J. K. Stille, F. Huang, and M. T. Regan, *J. Am. Chem. Soc.*, **96**, 1518 (1974).

90.   K. S. Y. Lau, Y. Becker, F. Huang, N. Baenziger, and J. K. Stille, *J. Am. Chem. Soc.*, **99**, 5664 (1977).

91.   C. A. Tolman, S. D. Ittel, A. D. English, and J. P. Jesson, *J. Am. Chem. Soc.*, **101**, 1742 (1979).

92.   J. W. Suggs, *J. Am. Chem. Soc.*, **100**, 640 (1978).

93.   *Chem. Eng. News*, 12 (Oct. 10, 1977); *Chem. Week*, 28 (July 26, 1978).

94.   J. G. Zajacek, J. J. McCoy, and K. E. Fuger, U.S. Patent 3,895,054 (1975) and U.S. Patent 3,956,360 (1976) to Arco.

95.   R. Rosenthal and J. G. Zajacek, U.S. Patent 3,919,279 (1975) and U.S. Patent 3.962,302 (1976) to Arco.

96.   T. Miyata, K. Kondo, S. Murai, T. Hirashama, and N. Sonoda, *Angew. Chem., Int. Ed.*, **19**, 1008 (1980).

97.   H. Seiji, H. Yutaka, and M. Katsuhara, U.S. Patent 4,170,708 (1979) to Mitsui Toatsu.

98.   W. B. Hardy and R. P. Bennett, *Tetrahedron Lett.*, 961 (1967); British Patent 1,025,436 (1966).

99.   W. W. Prichard, U.S. Patent 3,576,836 (1971) to Du Pont.

100.   B. A. Mountfield, British Patent 993,704 (1965).

101.   J. G. Zajacek and J. J. McCoy, U.S. Patent 3,993,681 (1976).

102.   G. F. Ottmann, E. H. Kober, and D. F. Gavin, U.S. Patents 3,481,968 (1969) and 3,523,966 (1970).

103.   H. Tietz, K. Unverferth, and K. Schwetlick, Z. *Chem.*, **17**, 368 (1977); **18**, 98 (1978).

104.   V. A. Golodov, Yu. L. Sheludyakov, R. I. Di, and V. K. Fokanov, *Kinet. Katal.*, **18**, 234 (1977).

105.   A. L. Balch and D. Petrides, *Inorg. Chem.*, **8**, 2245 (1969).

106.   R. G. Little and R. J. Doedens, *Inorg. Chem.*, **12**, 536 (1973).

107.   S. Otsuka, Y. Aotani, Y. Tatsuno, and T. Yoshida, *Inorg. Chem.*, **15**, 656 (1976).

108.   R. S. Berman and J. K. Kochi, *Inorg. Chem.*, **19**, 248 (1980).

109.   S. Cenini, M. Pizzotti, and C. Crotti, "Metal Catalyzed Deoxygenation Reactions of Nitroso and Nitro Compounds," in *Aspects of Homogeneous Catalysis, Vol. 6*, R. Ugo, Ed., Reidel (Kluwer), Dordrecht, 1988.

110.   R. P. Bennett and W. B. Hardy, *J. Am. Chem. Soc.*, **90**, 3295 (1968).

111.   F. J. Weigert, *J. Org. Chem.*, **38**, 1316 (1973).

112.   East German Patent 2,960,716 (1981) to Produits Chimiques Ugine Kuhlmann.

113.   P. Leconte, F. Metz, A. Mortreux, J. A. Osborn, F. Paul, and A. Pillot, *J. Chem. Soc., Chem. Commun.*, 1616 (1990).

114.   E. Drent and P. van Leeuwen, European Patent Application 86,281 (1983) to Shell.

115.   J. D. Gargulak, M. D. Noirot, and W. L. Gladfelter, *J. Am. Chem. Soc.*, **113**, 1054 (1991).

116.   E. W. Stern and M. L. Spector, *J. Org. Chem.*, **31**, 596 (1966).

117.   R. Becker, J. Grolig, and C. Rasp, U.S. Patent 4,297,501 (1981) and F. Merger, R. Platz, F. Towae, German Patent 2,910,132 (1980).

118.   S. Fukuoka, M. Chono, and M. Kohno, *J. Chem. Soc., Chem. Commun.*, 399 (1984); S. Fukuoka, M. Chono, and M. Kohno, *Chemtech*, 670 (Nov. 1984).

119.   R. Yamada, H. Tanimoto, and K. Murakami, Japanese Kokai 63,072,666 (1988) to Babcock-Hitachi.

120.   H. Alper and D. J. H. Smith, European Patent 200,556 (1986) to British Petroleum.

121.   F. J. Waller, European Patent 195,515 (1986) to Du Pont.

122.   T. W. Leung and B. D. Dombek, *J. Chem. Soc., Chem. Commun.*, 205-6 (1992).

123.   J. A. Kent, Ed., *Riegel's Handbook of Industrial Chemistry*, 7th ed., Van Nostrand Reinhold, New York, 1974, p. 87.

124.   R. M. Laine and R. B. Wilson, "Recent Developments in the Homogeneous Catalysis of the Water-Gas Shift Reaction," in R. Ugo, Ed., *Aspects of Homogeneous Catalysis, Vol. 5*, Reidel (Kluwer), Dordrecht, 1984.

125.   R. M. Laine, R. G. Rinker, and P. C. Ford, *J. Am. Chem. Soc.*, **99**, 252 (1977).

126.   R. B. King, C. C. Frazier, R. M. Hanes, and A. D. King, *J. Am. Chem. Soc.*, **100**, 2925 (1978).

127.   C. H. Cheng, D. E. Hendriksen, and R. Eisenberg, *J. Am. Chem. Soc.*, **99**, 2791 (1977).

128. T. Yoshida, Y. Ueda, and S. Otsuka, *J. Am. Chem. Soc.*, **100**, 3941 (1978).

129. G. Henrici-Olivé and S. Olivé, *Angew. Chem. Int. Ed.*, **15**, 136 (1976).

130. G. Henrici-Olivé and S. Olivé, *The Chemistry of the Catalyzed Hydrogenation of Carbon Monoxide*, Springer-Verlag, New York, 1984.

131. C. Masters, *Adv. Organomet. Chem.*, **17**, 61 (1979).

132. E. L. Muetterties and J. Stein, *Chem. Rev.*, **79**, 479 (1979).

133. R. C. Brady and R. Pettit, *J. Am. Chem. Soc.*, **103**, 1287 (1981).

134. F. Fischer and H. Tropsch, *Brennst.-Chem.*, **7**, 97 (1926); *Chem. Ber.*, **59**, 830 (1926).

135. G. R. Steinmetz and G. L. Geoffroy, *J. Am. Chem. Soc.*, **103**, 1278 (1981).

136. D. L. Thorn and T. H. Tulip, *J. Am. Chem. Soc.*, **103**, 5984 (1981).

137. D. L. Thorn, *J. Am. Chem. Soc.*, **102**, 7109 (1980).

138. D. L. Thorn, *J. Mol. Catal.*, **17**, 279 (1982).

139. M. Ichikawa and T. Fukushima, *J. Chem. Soc., Chem. Commun.*, 321 (1985).

140. C. P. Casey and S. M. Neumann, *J. Am. Chem. Soc.*, **98**, 5395 (1976).

141. W. F. Gresham, U.S. Patent 2,607,805 (1952); N. F. Cody, U.S. Patent 3,859,349 (1975) both to Du Pont.

142. *Chem. Eng. News*, 41 (April 11, 1983).

143. R. E. Pruett and W. E. Walker, U.S. Patent 3,957,857 (1976); R. L. Pruett, *Ann. N. Y. Acad. Sci.*, **295**, 239 (1977).

144. B. D. Dombeck, *Adv. Catal.*, **32**, 325 (1983).

145. Y. Ohgomori, S-i. Yoshida, and Y. Watanabe, *J. Chem. Soc., Chem. Commun.*, 829 (1987).

146. M. Ishino, M. Tamura, T. Deguchi, and S. Nakamura, *J. Catal.*, **133**, 325-331 and 332-341 (1992).

147. W. F. Gresham, U.S. Patent 2,636,046 (1953).

148. B. D. Dombeck, *J. Am. Chem. Soc.*, **102**, 6855 (1980).

149. J. F. Knifton, *J. Am. Chem. Soc.*, **103**, 3959 (1981).

150. J. F. Knifton, *J. Chem. Soc., Chem. Commun.*, 729 (1983).

151. J. M. Manriquez, D. R. McAlister, R. D. Sanner, and J. E. Bercaw, *J. Am. Chem. Soc.*, **100**, 2716 (1978).

152. H. Sugimoto and D. T. Sawyer, *J. Am. Chem. Soc.*, **107**, 5712 (1985).

153. D. L. King and J. H. Grate, *Chemtech*, 244 (April, 1985).

154. D. M. Fenton and P. J. Steinwald, *J. Org. Chem.*, **39**, 701 (1974).

155. P. L. Burk, D. Van Engen, and K. S. Kampo, *Organometallics*, **3**, 493 (1984).

156. F. Rivetti and U. Romano, *Chim. Ind. Milan*, **62**, 7 (1980).

157. H. E. Bryndza, S. A. Kretchmar, and T. H. Tulip, *J. Chem. Soc., Chem. Commun.*, 977 (1985).

158. K. Fujii, H. Itatani, K.Nishihira, K. Nishimura, S. Uchiumi, and M. Yamashita, U.S. Patent 4,229,589 (1980); K. Fujii, M. Matsuda, K. Mizutare, K. Nishihira, K. Nishimura, and S. Uchiumi, U.S. Patent 4,229,597 (1978); K. Fujii, M. Matsuda, K. Mizutare, K. Nishihira, and S. Tahara, U.S. Patent 4,461,909 (1984); and K. Fujii, M. Matsuda, K. Mizutare, K. Nishihira, and S. Tahara, U.S. Patent 4,467,109 (1984), all to Ube Industries.

159. F. J. Waller, *J. Mol. Catal.*, **31**, 123 (1985).
160. K. Nishimura, K. Fujii, K. Nishihira, M. Matsuda, and S. Uchiumi, U.S. Patent 4,229,591 (1980); K. Nishimura, S. Uchiumi, K. Fujii, K. Nishihira, M. Yamashita, and H. Itatani, U.S. Patent 4,229,589 (1980), both to Ube Industries.
161. A. H. Weiss, S. Trigerman, G. Dunnells, V. A. Likholobov, and E. Biron, *Ind. Eng. Chem. Prod. Res. Dev.*, **18**, 522 (1979).
162. *Chem. Eng.*, **95**, 17 (Mar. 14, 1988).
163. U. Romano, R. Tesel, M. M. Mauri, and P. Rebora, *Ind. Eng. Chem. Prod. Res. Dev.*, **19**, 396 (1980).
164. P. Koch, G. Cipriani, and E. Perrotti, *Gazz. Chim. Ital.*, **104**, 599 (1974).
165. J. W. Rathke and H. M. Feder, *J. Am. Chem. Soc.*, **100**, 3623 (1978).
166. G. Braca and G. Sbrana, "Homologation of Alcohols, Acids, and Their Derivatives by CO + H_2," in R. Ugo, Ed., *Aspects of Homogeneous Catalysis,* Vol. 5, Reidel, Dordrecht, 1984, p. 242.
167. I. Wender, R. Levine, and M. Orchin, *J. Am. Chem. Soc.*, **71**, 4160 (1949).
168. M. Nonotny and L. R. Anderson, Seminar, Catalysis Club of Philadelphia, Chester, PA, May 25, 1978.
169. G. Braca, G. Sbrana, G. Valenti, G. Andrich, and G. Gregorio, *J. Am. Chem. Soc.*, **100**, 6238 (1978).
170. L. Cassar, *Chem. Ind. Milan*, **67**, 256 (1985).
171. L. Cassar and M. Foa, *J. Organomet. Chem.*, **134**, C15 (1977); U.S. Patent 4,128,572 (1978).
172. M. A. Andrews and S. A. Klaeren, *J. Chem. Soc., Chem. Commun.* 1266 (1988).
173. C. Botteghi, R. Ganzerla, M. Lenarda, and G. Moretti, *J. Mol. Catal.*, **40**, 129 (1987).
174. T. Yoshida "L-Monosodium Glutamate" in Kirk-Othmer, Ed. *Encyclopedia of Chemical Technology*, 3rd Ed., Vol. 2, Wiley, New York, 1978, p. 410.
175. T. Kaneko, Y. Izumi, I. Chibata, and T. Itah in *Synthesis, Production and Utilization of Amino Acids*, Wiley, New York, 1974.
176. W. Himmele and W. Aquila, U.S. Patent 3,840,589 (1974); W. Himmele, F. J. Mueller, and W. Aquila, German Patent 2,039,078 (1972) to BASF.
177. H. Pommer and A. Nuerrenbach, *Pure Appl. Chem.*, **43**, 527 (1975).
178. P. Fitton and H. Moffet, U.S. Patent 4,124,619 (1978) to Hoffmann La Roche.
179. *Chemical Week*, 20 (Sept. 13, 1989).
180. Used by FMC for allyl alcohol to glycerol.
181. M. Matsumoto, S. Miura, K. Kikuchi, M. Tamura, H. Kojima, K. Koga, and S. Yamashita, U.S. Patent 4,567,305 (1986) to Kuraray Company and Daicel Chemical Industries; N. Yoshimura and M. Tamura, *Stud. Surf. Sci. Catal.*, **44**, 307-14 (1988).
182. A. J. De Jong, R. van Helden, and R. S. Downing, British Patent 1,555,551 (1979) to Shell.
183. Japanese Kokai 58,118,535 (1983) to Kuraray.
184. Y. Tokito and N. Yoshimura, Japanese Kokai 62,030,734 (1987) to Kuraray.
185. T. Omatsu, Y. Tokito, and N. Yoshimura, European Patent 303,60 (1989) to Kuraray.

186. K. Adachi and N. Yoshimura, Japanese Kokai 02,243,681 (1990) to Kuraray.

187. Japanese Kokai 58,140,030 (1983) to Kuraray.

188. N. Yoshimura and Y. Tokitoh, European Patent 223,103 (1987) to Kuraray.

189. Y. Tokitoh and N. Yoshimura, Japanese Kokai 61,249,940 (1986) to Kuraray.

190. C. Pisano, G. Consiglio, A. Sironi, and M. Moret, *J. Chem. Soc., Chem. Commun.*, 421 (1991).

191. O. Roelen, *Chem Exp. Didakt*, **3**, 119 (1977).

192. D. M. Fenton, U.S. Patent 3,530,109, (1970) to Union Oil.

193. K. Nozaki, U.S. Patent 3,835,123, (1974) to Shell .

194. A. Sen and Ta-Wang Lai, *J. Am. Chem. Soc.*, **104**, 3520 (1982).

195. A. Sen, *Advances in Polymer Science*, Vols. 73/74, Springer-Verlag, New York, 1986, p. 126; Ta-Wang Lai and A. Sen, *Organometallics*, **3**, 866 (1984); A. Sen, *Chemtech*, **16**, 48 (1986).

196. A. Sen and Ta-Wang Lai, *J. Am. Chem. Soc.*, **103**, 4627 (1981).

197. U. Klabunde, T. H. Tulip, D. C. Roe, and S. D. Ittel, *J. Organomet. Chem.*, **334**, 141 (1987).

198. M Brookhart, lecture at the 203rd meeting of the Am. Chem. Soc. (Inorg. 331), San Francisco, CA., April 5-10 (1992); M. Brookhart, F. C. Rix, J. M. DeSimone, and J. C. Barborak, *J. Am. Chem. Soc.*, **114**, 0000 (1992).

199. E. Drent and R. L. Wife, European Patent Application EP 222,454 (1987); J. A. M. Van Broekhoven and E. Drent, EP 239,145, (1987); E. Drent, EP 245,893 (1987); E. Drent, EP 246,683 (1987); E. Drent, P. W. N. M. Van Leeuwen, and R. L. Wife, EP 259,914 (1988), EP 263,564 (1988), EP 277,695 (1988), all to Shell.

200. E. Drent, presented at Euchem Conference, "Perspectives in Organometallic and Coordination Chemistry," Kreuth, Germany, Sept. 17, 1991.

201. J. A. Van Broekhoven, E. Drent, and E. Klei, European Patent 213,671 (1987) to Shell (1987).

202. F. Ozawa, H. Soyama, F. Fujisawa, and A. Yamamoto, *Terahedron Lett.*, **23**, 3383 (1982)

203. T. Kobayashi and M. Tanaka, *J. Organomet. Chem.*, **233**, C64 (1982).

204. F. Ozawa, I. Yamagami, M. Nakano, F. Fujisawa, and A. Yamamoto, *Chem. Lett.*, 125 (1989).

205. J. W. Wolfram, U.S. Patent 4,544,505 (1985) to Ethyl.

206. F. Ozawa, H. Soyama, H. Yanagihara, I. Aoyama, H. Takino, K. Izawa, T. Yamamoto, and A. Yamamoto, *J. Am. Chem. Soc.*, **107**, 3235 (1985).

207. F. Ozawa, T. Sugimoto, T. Yamamoto, and A. Yamamoto, *Organometallics*, **3**, 692 (1984).

208. J. Chen and A. Sen, *J. Am. Chem. Soc.*, **106**, 1506 (1984).

6

OXIDATION OF
OLEFINS AND DIENES

The oxidation of olefins to glycols, aldehydes, ketones, and epoxides is a large and growing part of the industrial application of homogeneous catalysis. Thirty years ago, development of the Wacker process for oxidation of ethylene to acetaldehyde was a milestone in the replacement of acetylene as a major feedstock for production of organic chemicals. Now, as the Wacker process approaches obsolescence, new processes such as the Oxirane propylene oxide synthesis have great potential for growth.

Two families of olefin oxidation catalysts are discernable. One is based on the almost unique ability of palladium(II) salts to oxidize olefins to aldehydes, ketones, enol esters, and other products. The other family, which includes the Oxirane process and the Sharpless oxidations, involve a metal (Mo or Ti) ion in one step of a complex series of organic reactions.

The palladium-catalyzed oxidations are potentially useful for many transformations of olefins. The Wacker processes for acetaldehyde from ethylene and for acetone from propylene are used on a significant scale in the homogeneous catalytic mode and it seems likely that this chemistry will be used to oxidize butenes to methyl ethyl ketone (MEK). Similar chemistry can be used to produce vinyl acetate, allyl acetate, glycol acetates and 1,4-diacetoxy-2-butene, but heterogeneous palladium catalysts are usually used for these reactions. Despite the engineering advantages of the heterogeneous catalysts, studies of the homogeneous palladium-catalyzed acetoxylation reactions contribute much to our understanding of the commercial processes. The palladium-catalyzed oxidations are among the best-studied homogeneous catalytic processes. Intimate details of the mechanisms are controversial, but the scope and character of these reactions have been well explored. Excellent reviews of organopalladium chemistry and its catalytic uses are available [1-4].

In contrast to the widespread study of palladium-catalyzed oxidations, the development of the molybdenum and tellurium chemistry associated with the Oxirane

137

and Halcon processes for olefin oxidation has been empirical. Much of the definition of the scope and practice of these reactions is in the patent literature. Few reviews are available and studies of mechanism are quite limited. The lack of academic attention may reflect the newness and the complexity of these reactions. The organometallic chemist may also find these processes unappealing because the Mo and Te atoms play such a limited part. There seems to be no convincing evidence for organometallic intermediates. Fortunately, the elegance of Sharpless' Ti-catalyzed asymmetric oxidation of allylic alcohols has led to mechanistic understanding that may be transferable to the Mo-catalyzed process.

6.1 WACKER ACETALDEHYDE SYNTHESIS

The synthesis of acetaldehyde by oxidation of ethylene, generally known as the Wacker process, was a major step in displacement of acetylene as a starting material in the manufacture of organic chemicals. Acetylene-based acetaldehyde (Section 8.3) was a major intermediate for production of acetic acid and of butyraldehyde, itself a major intermediate (Chapter 5). The cost was high because a large energy input is required to produce acetylene. When an efficient process for the synthesis of acetaldehyde from ethylene was developed, its adoption was rapid. Ironically this development originated in the laboratories of the Consortium für Elektrochemische Industrie, a branch of Wacker-Chemie [5], which was established to promote acetylene use. Although the Wacker process itself is now being displaced by CO-based technology (Chapter 5), it is still a major application of homogeneous catalysis.

The invention of the Wacker process was a triumph of common sense. It had been known since 1894 [6] that ethylene is oxidized to acetaldehyde by palladium chloride as a stoichiometric reagent (eq. 1). However, it was not until 1956 that this reaction was combined with the known reactions 2 and 3 to yield a catalytic acetaldehyde synthesis (eq. 4).

$$C_2H_4 + PdCl_2 + H_2O \longrightarrow CH_3CHO + Pd^0 + 2\,HCl \qquad (1)$$

$$Pd^0 + 2\,CuCl_2 \longrightarrow PdCl_2 + Cu_2Cl_2 \qquad (2)$$

$$Cu_2Cl_2 + 2\,HCl + 1/2\,O_2 \longrightarrow 2\,CuCl_2 + H_2O \qquad (3)$$

$$C_2H_4 + 1/2\,O_2 \longrightarrow CH_3CHO \qquad (4)$$

This sequence of reactions permits the precious metal salt to be used in a catalytic sense.

In commercial practice, two modes of operation have developed [5,7]. The single-stage Hoechst process involves feeding an oxygen-ethylene mixture to an aqueous solution of $PdCl_2$ and $CuCl_2$ at 120-130°C and 4 atmospheres pressure. The acetaldehyde is fractionated from the gases as they exit from the reactor. In a two-stage process, ethylene is stoichiometrically oxidized to acetaldehyde by a $PdCl_2$-$CuCl_2$ solution in one reactor. The organic-free aqueous solution leaving this reactor contains $PdCl_2$ and Cu_2Cl_2. The copper salt is reoxidized with air in a separate reactor before recycle to the synthesis unit. Both processes give 95% yields of acetaldehyde along with small amounts of 2-chloroethanol, ethyl chloride, acetic acid, chloroacetaldehydes, and acetaldehyde condensation products. The two-stage process requires a higher capital investment than the single-stage process, but permits the use of air rather than oxygen as the oxidant. It also avoids the explosion hazards involved in mixing oxygen and ethylene.

Halide-free catalyst systems for ethylene oxidation have been developed [8]. Such systems typically comprise a palladium salt along with a heteropoly acid salt such as a phosphomolybdovanadate. An obvious advantage is reduction in halide-promoted corrosion of process equipment. This type of process is approaching commercialization in the butene oxidation process being developed by Catalytica, as described below in Section 6.2.

The nature of the products in the palladium-catalyzed oxidation of ethylene is strongly dependent on reaction conditions. The oxidation in acetic acid can be directed to give vinyl acetate (Section 6.3), ethylene glycol acetates, or 2-chloroethyl acetate as major products [4]. Similarly, in methanol, either methyl vinyl ether or $CH_3CH(OCH_3)_2$ can be made to predominate. Even in water [9], 2-chloroethanol becomes a major product at high chloride and copper concentrations.

The major features of the mechanism of ethylene oxidation (Figure 6.1) are well established, but a lively controversy exists about the detailed mechanism. There is general agreement that the initial step is replacement of a chloride ion in $[PdCl_4]^{2-}$ (shown as Pd^{2+} in the figure) by ethylene. Additionally, it is agreed that other chloride ions are replaced by water and hydroxide ligands. As a consequence, the rate of the reaction is sharply decreased by high chloride concentrations.

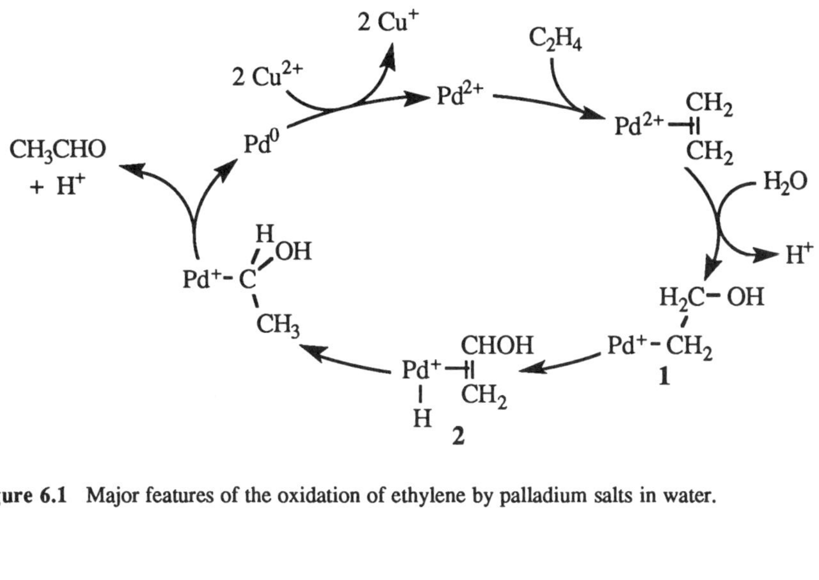

Figure 6.1 Major features of the oxidation of ethylene by palladium salts in water.

The major point of disagreement is the mechanism of the addition of OH to the coordinated ethylene to give a hydroxyethylpalladium intermediate **1**. This intermediate has never been observed but is commonly assumed to be present in the reaction mixture. In one view, the OH addition is considered to be the result of migration of a coordinated OH ligand from palladium to carbon as shown in equation 5.

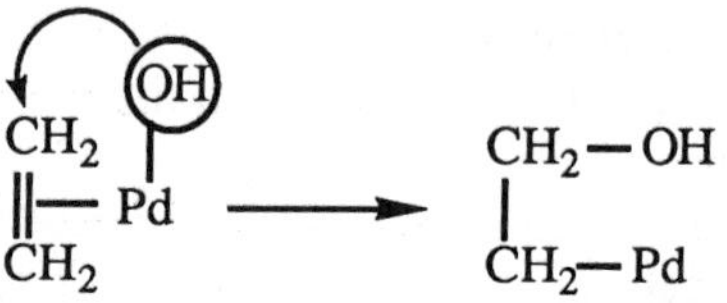

$$
\begin{array}{c}
CH_2 \\
\| - Pd \\
CH_2
\end{array}
\quad \widehat{(OH)}
\longrightarrow
\quad
\begin{array}{c}
CH_2 - OH \\
| \\
CH_2 - Pd
\end{array}
\quad \mathbf{1}
\tag{5}
$$

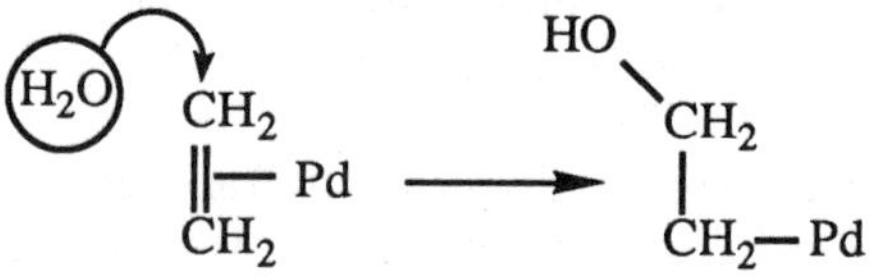

$$
\widehat{(H_2O)}
\quad
\begin{array}{c}
CH_2 \\
\| - Pd \\
CH_2
\end{array}
\longrightarrow
\quad
\begin{array}{c}
HO \\ \diagdown \\ CH_2 \\ | \\ CH_2 - Pd
\end{array}
\quad \mathbf{1}
\tag{6}
$$

This migration (or ethylene insertion into a Pd-OH bond) is consistent with the observed kinetics [10]. The alternative view is that coordinated ethylene is attacked by water (or OH⁻) as indicated in equation 6. This view is supported by stereochemical studies of addition to coordinated *cis*- or *trans*-CHD=CHD which indicate attack by an external nucleophile [11]. It is likely that both mechanisms (cis and trans addition) operate and that reaction conditions determine which predominates.

One remarkable aspect of the ethylene oxidation is that no H/D exchange with solvent occurs. When the reaction is carried out in D_2O, no deuterium appears in the product either in the normal mode of operation or when the Pd catalyst is "heterogenized" by incorporation into the pores of a zeolite [12]. Evidently the hydrogen migrations involved in transformation of the $HOCH_2CH_2Pd$ intermediate **1** to CH_3CHO occur entirely within the coordination sphere of the palladium. β-Hydrogen elimination to give a vinyl alcohol complex **2** as shown in Figure 6.1 seems a likely pathway. Readdition of the hydride ligand to the coordinated vinyl alcohol gives the α-hydroxyethylpalladium complex shown at the 8 o'clock position in the figure. A β-hydrogen elimination from OH completes acetaldehyde formation and results in formation of zero-valent palladium or a palladium hydride. Either species is rapidly reoxidized by copper(II) in a chloride-containing medium.

6.2   KETONES AND ACETALS FROM HIGHER OLEFINS

Chemistry analogous to that of the Wacker acetaldehyde process is used industrially to oxidize higher olefins to ketones. Terminal olefins such as propylene, which could yield either acetone or propionaldehyde, produce the methyl ketones. For example, one of the commercial syntheses of acetone is the palladium-catalyzed oxidation of propylene [13]. This particular reaction is rapid and clean under conditions like those used in the Wacker process. Propylene is oxidized to acetone in 90% yield in 5 minutes at 20°C in aqueous palladium chloride solution [14]. Higher olefins require a polar organic solvent, but also give good yields of the corresponding methyl ketones. 1-Decene, for example, gives 2-decanone in 65-73% yield [15]. Cyclohexene gives cyclohexanone in a similar yield in aqueous palladium chloride solution. The reaction is so straightforward that it is the basis for a student experiment, the oxidation of vinylnaphthalene to acetonaphthone [16].

Much attention has been attracted by the announcement of a process to produce 2-butanone (methyl ethyl ketone, MEK) [17], a widely used solvent. The process involves a palladium-catalyzed oxidation of butenes analogous to the acetone-from-propylene process, but it offers two major advantages: (a) it operates on an inexpensive mixture of 1- and 2-butenes; (b) the catalyst is halide-free and thus avoids the corrosion and halogen byproduct problems of a typical Wacker process.

As described in patents issued to Catalytica [18], the catalyst system contains a halide-free mixture of palladium and copper salts, for example, $Pd(NO_3)_2$ and $Cu(NO_3)_2$ along with a heteropolyanion such as $[PV_{14}O_{42}]^{9-}$ dissolved in aqueous acetonitrile. In typical examples, the olefin (usually 1-hexene for experimental convenience) is oxidized at 75-85°C under approximately 5 atmospheres pressure of oxygen. One remarkable and valuable aspect is that 1-butene and *cis*- and *trans*-2-butenes are oxidized at roughly the same rate to produce 2-butanone in 82-97% yield at 90-100% conversion. This feature should permit the use of a cheap olefin feedstock containing mixed C_4 hydrocarbons from which butadiene and isobutene have been extracted. The nature of the heteropolyanion is defined very broadly, but it seems likely that vanadium-containing species are preferred. Earlier Russian work [8] on halide-free catalyst systems suggests that the V-containing heteropolyanions are especially effective in electron transfer reactions such as the reoxidation of a reduced palladium ion.

The specificity of Wacker-type catalysts for olefin oxidation is illustrated nicely by an *Organic Syntheses* procedure [19] in which an olefin is oxidized to a ketone without attack on an aldehyde function in the same molecule:

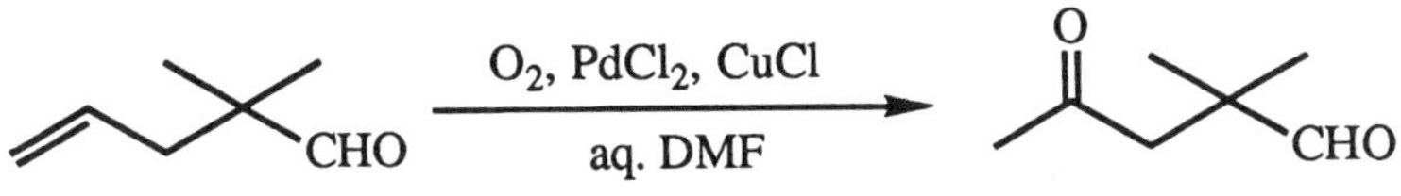

The reaction proceeds at room temperature and atmospheric pressure and produces the keto-aldehyde in 78% yield. There are few other oxidation catalysts that would leave the aldehyde untouched in the presence of molecular oxygen.

In contrast to the preference for methyl ketone formation from simple alkenes, 1-alkenes that bear electron-withdrawing substituents are subject to attack at the terminal carbon. For example, the oxidation of methyl acrylate, acrylonitrile or styrene by a $PdCl_2/CuCl_2$ catalyst system in the presence of 2,4-pentanediol gives the corresponding cyclic acetal **3** in good yield [20,21]:

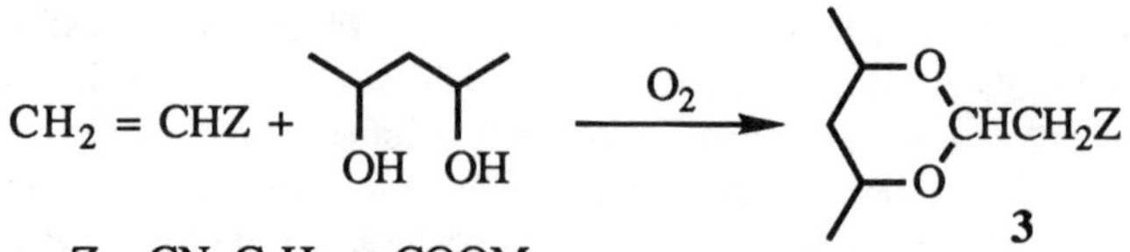

$$CH_2 = CHZ + \underset{OH \quad OH}{\bigvee\bigvee} \xrightarrow{O_2} \text{(3)}$$

Z = CN, C_6H_5 or COOMe

This reaction also proceeds nicely with simple alcohols. The oxidation of ethylene by CH_3ONO in the presence of methanolic $Pd(NO_3)_2$ affords $(CH_3O)_2CHCH_3$ [22]. The corresponding oxidation of acrylonitrile provides the basis for a potentially practical synthesis of vitamin B_1, as described in the final section of this chapter.

The mechanism of acetal formation appears to be generally similar to that proposed for the Wacker acetaldehyde synthesis (Figure 6.1). A key feature in the acetal synthesis is that the presence of an electron-withdrawing substituent **Z** on the olefin directs the regiochemistry of nucleophilic attack on the coordinated olefin:

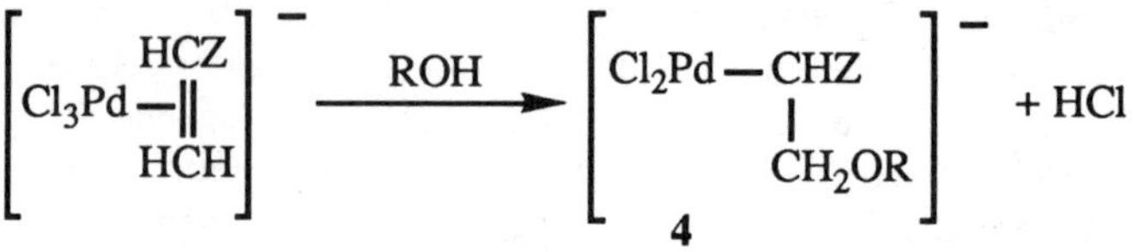

$$\left[\begin{array}{c} HCZ \\ Cl_3Pd - \| \\ HCH \end{array}\right]^{-} \xrightarrow{ROH} \left[\begin{array}{c} Cl_2Pd - CHZ \\ | \\ CH_2OR \end{array}\right]^{-} + HCl$$

Subsequently, β-hydrogen transfer from C to Pd within **4** leads to the vinyl ether complex **5**:

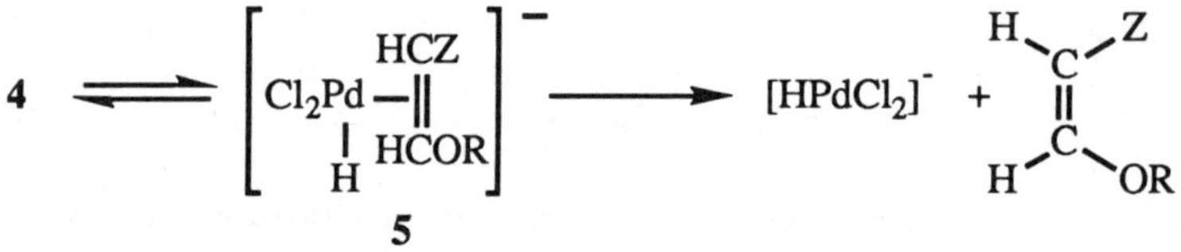

$$\mathbf{4} \rightleftharpoons \left[\begin{array}{c} HCZ \\ Cl_2Pd - \| \\ | \quad HCOR \\ H \end{array}\right]^{-} \longrightarrow [HPdCl_2]^- + \underset{5}{}$$

Dissociation of the vinyl ether product leaves a reduced palladium species which must be reoxidized to complete the catalyst cycle. The vinyl ether can react with more ROH to give the acetal product, which is observed [21].

6.3 ACETOXYLATION OF OLEFINS AND DIENES

The reaction of an olefin with a palladium(II) salt in acetic acid is a versatile synthesis of vinylic and allylic acetates. These reactions are generally used industrially as heterogeneous processes but extensive development effort has been applied to the analogous soluble processes. Some of the most promising reactions are:

$$CH_2=CH_2 \quad \underset{\longrightarrow}{\overset{\longrightarrow}{}} \quad \begin{array}{l} CH_2=CHOAc \\ HOCH_2CH_2OAc \end{array}$$

$$CH_3CH=CH_2 \quad \longrightarrow \quad CH_2=CHCH_2OAc$$

$$CH_2=CH-CH=CH_2 \quad \longrightarrow \quad AcOCH_2CH=CHCH_2OAc$$

The chemistry of these processes, which are discussed in the following sections, closely parallels that of the Wacker process.

Vinyl Acetate and Glycol Acetates from Ethylene

The reaction of ethylene with palladium salts in acetic acid can give several different products under different reaction conditions (Table 6.1). The original version of the reaction was a stoichiometric oxidation of ethylene with $PdCl_2$ and sodium acetate in acetic acid [23]. The reaction proceeded at room temperature to give vinyl acetate as a major product, but yields were low. Roughly 10 g of $PdCl_2$ were needed to make 1 g of vinyl acetate. Such an inefficient use of a precious metal compound obviously had no direct industrial application, but it pointed the way to catalytic processes in which air or oxygen was the ultimate oxidant.

$$CH_2=CH_2 + HOAc + \tfrac{1}{2}O_2 \longrightarrow CH_2=CHOAc + H_2O$$

The two liquid-phase catalytic processes which evolved parallel the one- and two-stage acetaldehyde processes. In the one-stage process [24], a suspension of $PdCl_2$, $Cu(OAc)_2$, KCl, and KOAc in acetic acid is saturated with ethylene at 40-45 atmospheres and 120°C. Small amounts of oxygen are then added as required to produce reaction. This process, in a continuous version, gives high yields of vinyl acetate along with some acetaldehyde. The aldehyde probably forms by reaction of vinyl acetate with water, which is a coproduct in the reaction. In the two-stage process,

Table 6.1 Versatility of Palladium(II)-Catalyzed Reactions of Ethylene

Reactants	Product
H_2O	CH_3CHO
H_2O, HCl, $CuCl_2$	$ClCH_2CH_2OH$
H_2O, HNO_3	$O_2NOCH_2CH_2ONO_2$
HOAc	$CH_2=CHOAc$
HOAc, LiOAc, $CuCl_2$	$ClCH_2CH_2OAc$
HOAc, NaOAc, $CuCl_2$	$AcOCH_2CH_2OAc$
HOAc, $LiNO_3$	$HOCH_2CH_2OAc$

ethylene is stoichiometrically oxidized to vinyl acetate by $PdCl_2$ and $CuCl_2$ (or benzoquinone) in one reactor. The volatile vinyl acetate is separated by distillation. The reduced working solution is sent to a second reactor where it is reoxidized with air before it is recycled to the vinyl acetate synthesis unit.

Neither liquid-phase process is industrially attractive despite attempts to commercialize both versions. The one-stage process is hazardous because oxygen and ethylene are mixed under pressure. Careful control is needed to avoid explosive mixtures. The two-stage process encounters serious mechanical problems in pumping copper(I) chloride/acetate slurries between reactors. In addition, the product distribution is changed by the presence of $CuCl_2$ [10]. Both processes use extremely corrosive working solutions and require exotic materials of construction.

The solution to these problems was development of a heterogeneous catalyst which converted a gaseous mixture of ethylene, oxygen, and acetic acid vapor to vinyl acetate. The composition of the catalysts that were developed for this purpose suggest that the chemistry is similar to that in the liquid phase. Typically, a porous silica is impregnated with Na_2PdCl_4 and $HAuCl_4$ and the catalyst is reduced to give a well-dispersed palladium-gold alloy. This catalyst is impregnated with potassium acetate before use. In use at 140-170°C and approximately 5-10 atmospheres pressure, such a catalyst gives a 96% yield of vinyl acetate (based on ethylene) along with small amounts of carbon dioxide [25]. It is tempting to speculate that the pores of the silica catalyst support are saturated with acetic acid under the reaction conditions and that the vinyl acetate synthesis occurs in solution in an acetic acid film.

Both the liquid- and vapor-phase processes convert ethylene and volatile carboxylic acids to vinyl carboxylates in good yield [4]. For example, isobutyric acid gives vinyl isobutyrate in 96% yield with the heterogeneous catalyst [25]. Nonvolatile acids can be converted to vinyl esters easily by ester interchange with vinyl acetate:

$$C_2H_3OAc + RCOOH \longrightarrow C_2H_3OOCR + HOAc$$

Both palladium(II) and mercury(II) salts are good catalysts for this reaction [26-28].

In contrast to the ease with which acids can be varied in vinyl carboxylate synthesis, the range of operable olefins is very limited. Most olefins other than ethylene give allylic rather than vinylic acetates when they are oxidized in acetic acid under standard vinyl acetate synthesis conditions. Stoichiometric oxidations may, however, give enol acetates. The oxidation of propylene by $Pd(OAc)_2$ in acetic acid without added acetate gives 96% 2-propenyl acetate, whereas oxidation in the presence of sodium acetate gives 94% allyl acetate [29].

The great range of ethylene oxidation products summarized in Table 6.1 probably arise by closely related mechanisms [2]. It is generally agreed that ethylene coordination to palladium sensitizes the olefin to attack by a nucleophile, acetate ion in the case of acetoxylation. As in acetaldehyde synthesis, the reaction with the nucleophile could occur within the coordination sphere of the palladium, but other work suggests attack on coordinated ethylene by an external nucleophile. Stereochemical studies in acetic acid have not resolved this point [30], but, in methanol, coordinated *cis*-CHD=CHD yields a modestly stable 2-methoxyethyl

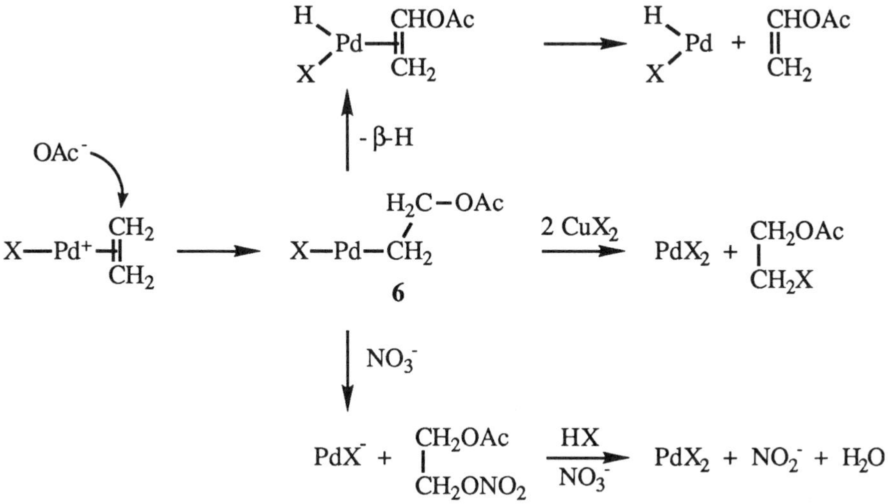

Figure 6.2 Formation of ethylene oxidation products in acetic acid (X = Cl, OAc).

palladium complex [31]. The stereochemistry of this adduct is consistent with trans (or exo) attack by uncoordinated methanol or methoxide. By analogy, the initial reaction in acetic acid is proposed to occur as shown in Figure 6.2. The situation in halide-free acetic acid oxidations is further complicated by the observation that $Na_2Pd(OAc)_4$, $Na_2Pd_2(OAc)_6$, and $Na_2Pd_3(OAc)_8$ oxidize ethylene at different rates [29]. The binuclear complex is most reactive.

The 2-acetoxyethylpalladium intermediate **6** probably yields all the observed ethylene oxidation products as shown in Figure 6.2. In the absence of an added oxidizing agent, β-hydrogen elimination forms a vinyl acetate complex. When the enol acetate dissociates from the metal, the unstable palladium hydride product decomposes to metallic palladium unless an oxidant intercepts it. The β-hydrogen transfer appears to be reversible. Both 1,1- and 1,2-dideuteroethylenes yield the same mixture of six mono- and dideuterated vinyl acetates [30]. This result is consistent with intramolecular H/D scrambling via rapid β-hydrogen elimination and readdition.

The β-acetoxyethyl complex **6** also appears to be the precursor of the 2-substituted ethyl acetates that form when an oxidant such as Cu^{2+} or NO_3^- is added. Evidently the Pd-C bond is cleaved by the oxidant and the alkyl fragment reacts with the predominant nucleophile (OAc⁻, Cl⁻, or NO_3^- ion) in the system. The simplest explanation would be a one-electron oxidation to form a $AcOCH_2CH_2 \cdot$ radical, but the stereochemical specificity of the reaction suggests a more complex process [32]. Oxidation of cis-1-decene-d₁ with $CuCl_2$ gives mostly erythro-$C_8H_{17}CH(OAc)CHDCl$, consistent with "backside" attack of Cl⁻, or $CuCl_2$ on the Pd-C bond [33]. Similarly, the oxidation by nitrate could be viewed as a nucleophilic displacement of XPd by NO_3^- as shown in Figure 6.2. This explanation is undoubtedly oversimplified. It has been observed that an oxygen atom of the oxidant is transferred to the acetate function in the 2-acetoxyethyl nitrate product [34]. In the nitrate-promoted oxidation, the relative yields of mono- and diacetates may reflect the relative rates of nitrate and acetate attack. This ratio varies extensively with reaction conditions [35]. With LiNO$_3$ and PdCl$_2$ in acetic acid, 98% HOCH$_2$CH$_2$OAc is obtained.

Allylic Oxidation

π–Allyl complexes of palladium are remarkably stable and form under many different conditions. Isobutylene gives a π-methallylpalladium complex **8** along with a simple olefin complex **7** [36,37]:

$$Me_2C=CH_2 \xrightarrow[\text{HOAc}]{PdCl_2} \left[\begin{array}{c} Me_2C \\ \| \\ H_2C \end{array} \!\!-PdCl_2 \right]_2 \xrightarrow{\text{-HCl}} \left[Me\!\!-\!\!\left\langle\!\!\left\langle PdCl \right. \right]_2$$

$$\quad\quad\quad\quad\quad\quad\quad\quad\quad\quad\quad\quad \mathbf{7} \quad\quad\quad\quad\quad\quad\quad\quad\quad\quad \mathbf{8}$$

As a result, allylic oxidation competes with vinylic oxidation in the reaction of propylene and higher olefins. Stoichiometric oxidation of propylene in unbuffered acetic acid gives 96% 2-propenyl acetate, but allyl acetate forms in 94% yield in the presence of sodium acetate [29]. The allylic oxidation can be made catalytic in palladium by provision of a secondary oxidizing agent such as MnO$_2$ or benzoquinone. The oxidation of cycloheptene to 2-cyclohepten-1-yl acetate proceeds in good yield on a laboratory scale [38]. On an industrial scale, however, allylic oxidation in the liquid phase has found little use. Vapor-phase oxidation over heterogeneous catalysts has been preferred.

Cooxidation of propylene and acetic acid vapor over a standard heterogeneous catalyst for vinyl acetate synthesis gives allyl acetate in high yield [39]. This finding has been used in two potentially attractive processes for synthesis of 1,4-butanediol and tetrahydrofuran (Figure 6.3). One approach has been direct hydroformylation of allyl acetate [40] with a conventional cobalt-based catalyst (Chapter 5). A second route involves hydrolysis of allyl acetate to allyl alcohol [41] and hydroformylation of the alcohol with HRh(CO)(PPh$_3$)$_3$ as the catalyst. The 1,4-butanediol produced is a valuable polymer intermediate (for polybutylene terephthalate) and can be cyclized to tetrahydrofuran by treatment with acid. In addition to these allyl acetate-based routes, two other oxidation-based routes have been commercialized recently. Arco has announced production of allyl alcohol and butanediol by the scheme at the bottom of Figure 6.3 [42]. Propylene is oxidized to propylene oxide (Section 6.5) which, in turn, is hydrolyzed to provide allyl alcohol for hydroformylation. Another new process is that of Mitsubishi Kasei [43] in which butadiene is oxidized to butenediol diacetate as described below under the heading of Diene Oxidation. These processes have attracted interest because starting material costs in the conventional acetylene-based synthesis (Section 8.2) have escalated dramatically.

The simplest mechanism for allylic oxidation begins with coordination of the olefin to a palladium(II) ion to form a simple π-complex. The actual oxidation step is elimination of a proton from an allyl position to give a π-allyl complex as shown for isobutylene above. Allylic oxidation products form by reactions of the π-allyl ligand.

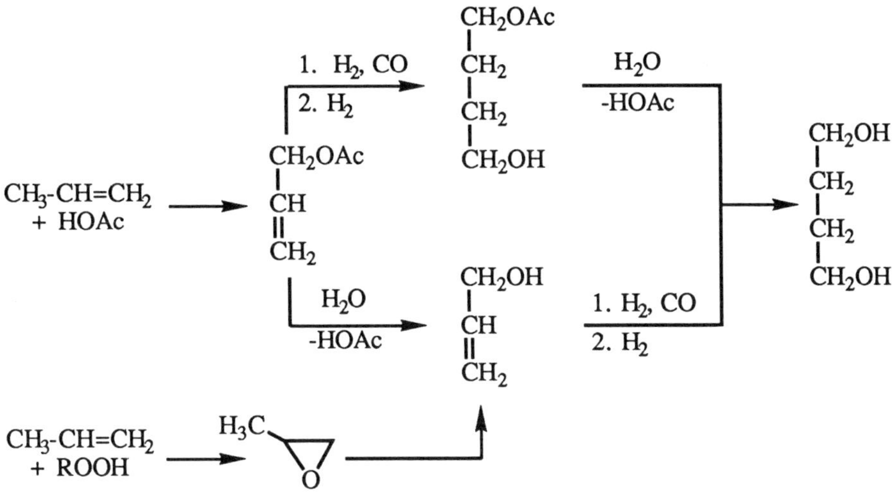

CH3-CH=CH2
+ HOAc
CH2OAc
CH
CH2
1. H2, CO
2. H2
CH2OAc
CH2
CH2
CH2OH
H2O
-HOAc
CH2OH
CH
CH2
1. H2, CO
2. H2
H2O
-HOAc
CH2OH
CH2
CH2
CH2OH
CH3-CH=CH2
+ ROOH
H3C
O

Figure 6.3 Propylene-based processes for synthesis of 1,4-butanediol.

Solvolysis of π-allylpalladium chloride in NaOAc-buffered acetic acid forms allyl chloride and allyl acetate [4]. The latter product predominates in halide-free systems like the surface of the catalyst used in the vapor-phase synthesis. Hydrolysis of π-methallylpalladium chloride gives a mixture of products which include isobutyraldehyde and methacrolein [36]. The unsaturated aldehyde predominates when an oxidizing agent is present.

In all the palladium-catalyzed oxidations, the kinetics of various reaction paths are very sensitive to reaction conditions. Olefin π complexes can either react with a nucleophile directly to form vinylic oxidation products or they may lose allylic protons to form allylic products. Acidity and ionic strength of the reaction medium seem to be major factors in deciding the course of the reaction.

Diene Oxidation

The oxidation of butadiene by palladium salts gives a variety of products (Figure 6.4) under different reaction conditions [44], just as in ethylene oxidation. The stoichiometric oxidation of butadiene by $PdCl_2$ in water gives a 34% yield of crotonaldehyde [14]. Similarly, oxidation by $PdCl_2$ in buffered acetic acid solution gives CH_2=CHCH=CHOAc [45]. These simple stoichiometric oxidations correspond nicely to the acetaldehyde and vinyl acetate syntheses from ethylene. In contrast, when $PdCl_2$ and $CuCl_2$ are used together in acetic acid, a mixture of allylic acetates forms [46] as shown in the figure. These reactions are simple 1,4- and 1,2-additions of acetic acid to butadiene. The 1,4 addition is especially interesting because it opens the way to an allylic oxidation to give the economically desirable 1,4-diacetoxy-2-butene.

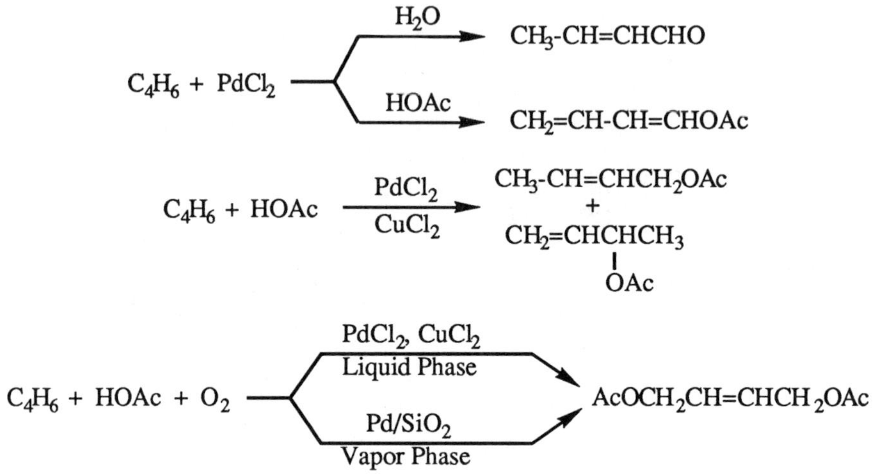

$$C_4H_6 + PdCl_2 \diagup \overset{H_2O}{\longrightarrow} CH_3\text{-}CH\text{=}CHCHO$$
$$\diagdown \overset{HOAc}{\longrightarrow} CH_2\text{=}CH\text{-}CH\text{=}CHOAc$$

$$C_4H_6 + HOAc \xrightarrow[\text{CuCl}_2]{\text{PdCl}_2} \begin{array}{c} CH_3\text{-}CH\text{=}CHCH_2OAc \\ + \\ CH_2\text{=}CHCHCH_3 \\ | \\ OAc \end{array}$$

$$C_4H_6 + HOAc + O_2 \diagup \overset{\substack{PdCl_2, CuCl_2 \\ \text{Liquid Phase}}}{} \diagdown \overset{}{\underset{\substack{\text{Pd/SiO}_2 \\ \text{Vapor Phase}}}{}} \diagup AcOCH_2CH\text{=}CHCH_2OAc$$

Figure 6.4 Oxidation of butadiene by palladium salts and catalysts.

Diacetoxylation of butadiene provides an attractive alternative to the conventional acetylene-based synthesis of 1,4-butanediol (Section 8.2) [47]. 1,4-Diacetoxy-2-butene is readily converted to the diol by successive hydrogenation and hydrolysis steps:

$$AcOCH_2CH\text{=}CHCH_2OAc \xrightarrow[\text{Raney Ni}]{H_2} AcO(CH_2)_4OAc$$

$$AcO(CH_2)_4OAc \xrightarrow{H^+, H_2O} HO(CH_2)_4OH + 2 HOAc$$

The reaction of butadiene with acetic acid in the liquid phase yields a mixture of 1,4-diacetoxy-2-butene and 3,4-diacetoxy-1-butene [44]. As in glycol acetate synthesis, the catalyst system comprises $PdCl_2$, $CuCl_2$, and lithium acetate. Most industrial attention has centered on heterogeneous catalysts, just as in synthesis of vinyl and allyl acetates. When used in a vapor-phase process with butadiene, oxygen, and acetic acid vapor flowing over a tellurium-promoted, silica-supported palladium catalyst, good yields of 1,4-diacetoxy-2-butene are obtained. Addition of CO to the reaction mixture is said to be advantageous [48].

Interestingly, the Mitsubishi process commercialized in 1982 uses a similar tellurium-promoted catalyst in a liquid-phase system [47,49]. The reaction of butadiene, acetic acid, and air with a slurry of Te-promoted palladium-on-carbon gives high yields of the desired 1,4-diacetoxy-2-butene along with small amounts of the 3,4-diacetoxy-1-butene. A major virtue to the liquid-phase process is that high boiling byproducts do not accumulate on the catalyst and reduce its activity.

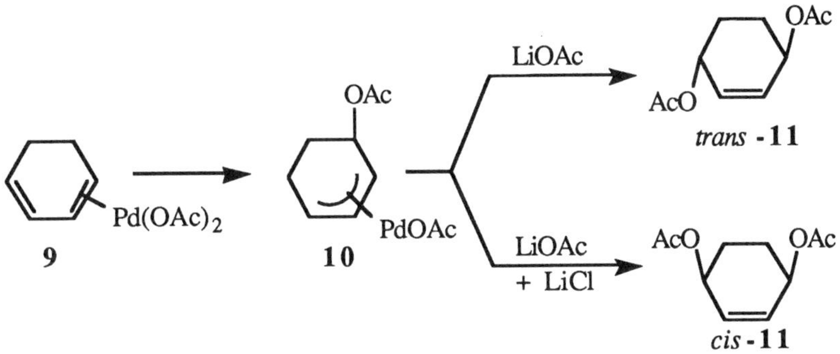

Figure 6.5 Oxidation of 1,3-cyclohexadiene by Pd(OAc)₂/benzoquinone in acetic acid.

The limited studies on the mechanism of palladium-catalyzed diene oxidation suggest a mechanism analogous to the oxidations of simple olefins [1]. As shown in Figure 6.5, the acetoxylation of 1,3-cyclohexadiene by palladium(II) acetate provides evidence for an allylic intermediate of the sort common to palladium chemistry [50]. Addition of acetate to the ring in an initial diene complex **9** yields the π-allylic complex **10**, probably by attack of free acetate from solution rather than by ligand migration from the metal ion. In a chloride-free system, the second step is transfer of acetate from Pd to C to form *trans*-**11**. In the presence of chloride, the dominant process is attack by external nucleophile to form *cis*-**11**.

The oxidative addition of acetate to dienes is a valuable tool for organic synthesis. The palladium-catalyzed reaction has been used to make a series of 4-substituted butenyl acetates [51].

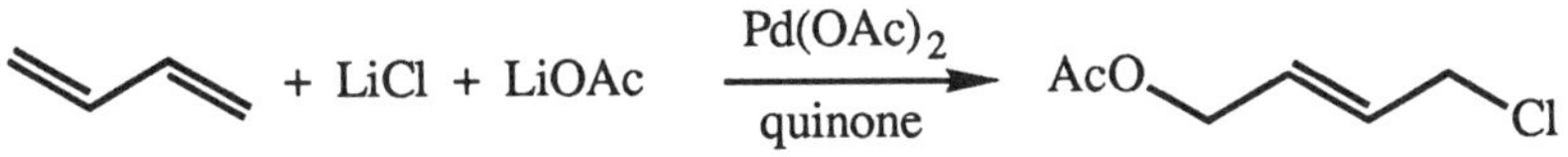

The allylic chloride in the product is easily displaced by amines in a reaction catalyzed by Pd(PPh₃)₄.

6.4 GLYCOL ACETATE SYNTHESIS

Ethylene glycol is a major intermediate for synthesis of polyesters and polyurethanes (Chapter 11) in addition to its well-known use as an antifreeze component. It is made by oxidation of ethylene as shown in Figure 6.6. The conventional process involves reaction of an ethylene/oxygen mixture over a metallic silver catalyst to form ethylene oxide in approximately 75% yield. The glycol is obtained by hydrolysis of the epoxide. In the late 1970s, Oxirane Corporation undertook the development of an alternative process in which ethylene was oxidized in the liquid phase to produce ethylene glycol acetates [52]. These acetates can be hydrolyzed cleanly to give the free glycol (lower equation in Figure 6.6). The Oxirane glycol process [53], like a

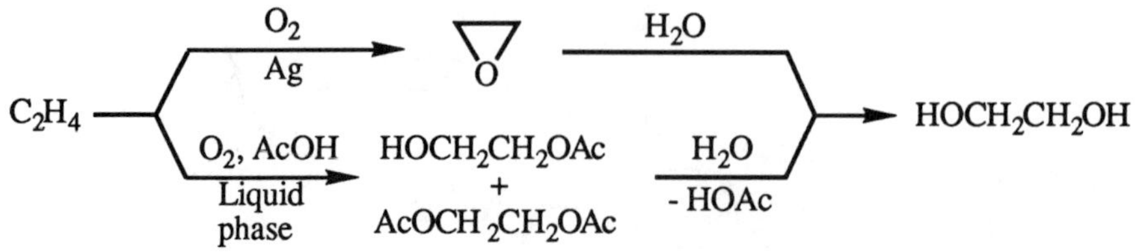

$$C_2H_4 \overset{\begin{array}{c}O_2 \\ \text{Ag}\end{array}}{\underset{\begin{array}{c}O_2,\ \text{AcOH} \\ \text{Liquid} \\ \text{phase}\end{array}}{\Bigg<}}\ \ \overset{H_2O}{\Big<}\ HOCH_2CH_2OH$$

$$HOCH_2CH_2OAc + AcOCH_2CH_2OAc$$

Figure 6.6 Alternative routes for synthesis of ethylene glycol.

closely related technology developed by Teijin [54], is based on nonprecious metals. Although glycol acetates can be made in high yields under mild conditions by the oxypalladation chemistry of Section 6.2, economic factors favor the use of semicatalytic processes based on cheap metals.

The Oxirane glycol acetate process was based on a sequence of fairly conventional reactions that together amount to a catalytic oxidative acetoxylation of ethylene [55]:

$$C_2H_4 + Br_2 \longrightarrow BrCH_2CH_2Br$$

$$BrCH_2CH_2Br \overset{H_2O/HOAc}{\longrightarrow} HOCH_2CH_2OAc + AcOCH_2CH_2OAc + 2\ HBr$$

$$2\ HBr + Te^{6+} \longrightarrow Br_2 + Te^{4+} + 2\ H^+$$

$$Te^{4+} + 2\ H^+ + {}^1\!/_2\ O_2 \longrightarrow Te^{6+} + H_2O$$

$$C_2H_4 + {}^1\!/_2\ O_2 \overset{H_2O/HOAc}{\longrightarrow} HOCH_2CH_2OAc + AcOCH_2CH_2OAc$$

Philosophically, this development was analogous to the Wacker acetaldehyde synthesis in its creative use of known chemistry. However, much less was known about the relevant Te/Br chemistry. This lack of fundamental knowledge may have contributed to the abandonment [56] of the commercial development. Severe corrosion problems in the process equipment and bromine contamination of the organic products made the process economically unattractive.

As the chemistry was practiced [53], a reactor was fed with ethylene, oxygen, and an acetic acid solution of TeO_2, 48% HBr, and 2-bromoethyl acetate. Reaction occurs in about one hour at 160°C and 20-30 atmospheres pressure to produce a liquid effluent which contained 28% $AcOCH_2CH_2OAc$, 16% monoacetate, and 2% ethylene glycol in acetic acid solution. The ratios of mono- and diacetates were controlled by variation of reaction conditions, but extremely careful distillation was necessary to separate them from brominated intermediates such as $BrCH_2CH_2OAc$.

The fundamental chemistry of the Te/Br catalysis of ethylene oxidation is much more complex than that shown in the equation above. Studies of the reaction of TeO_2 with olefins in acetic acid containing halide ion point to formation of a haloalkyltellurium(IV) intermediate [57]:

$$RHC=CH_2 + TeO_2 \xrightarrow[\text{HOAc}]{\text{LiX}} \begin{array}{c} X \\ \backslash \\ CH\text{-}CH_2\text{-}Te(OAc)_{3\text{-}Y}X_Y \\ / \\ R \end{array}$$

This sort of intermediate is unstable and spontaneously decomposes to elemental tellurium plus a mixture of glycol mono- and diacetates. In the Oxirane chemistry, it is likely that bromide ion plays a role in reoxidation of low-valent Te species by oxygen.

6.5   OLEFIN EPOXIDATION BY HYDROPEROXIDES

A large and growing application of homogeneous catalysis is the manufacture of propylene oxide by oxygen transfer from an alkyl hydroperoxide [58]:

$$CH_3CH=CH_2 + ROOH \longrightarrow \begin{array}{c} H \\ | \\ CH_3C\!-\!CH_2 \\ \backslash / \\ O \end{array} + ROH$$

This process is displacing older technology based on dehydrochlorination of propylene chlorohydrin [59]. The impact is substantial because propylene oxide (methyloxirane) is a major intermediate for propylene glycol, glycerine, and polyethers. As practiced by Arco Chemical, propylene is reacted with the hydroperoxides obtained by oxidation of ethylbenzene or isobutane (Chapter 10). The 1-phenylethanol obtained as a coproduct from the ethylbenzene reaction is dehydrated to produce styrene. t-Butyl alcohol, coproduced from t-butyl hydroperoxide, is blended with gasoline to inhibit engine knock [60] or dehydrated to make isobutylene for conversion to methyl t-butyl ether, also a gasoline blending component.

The economic superiority of the Arco process may be challenged by new Olin technology in which propylene is oxidized by O_2-enriched air. The reactants (plus some recycled byproducts) are passed through a molten mixture of alkali metal nitrates, which function as both reaction medium and catalyst [61]. At 200°C part of the propylene is oxidized to acetaldehyde and CO_2. The acetaldehyde, which is carefully recycled, appears to be converted to peracetic acid which converts about 65% of the propylene to propylene oxide. The conversion of the aldehyde to peracetic acid is well known (Chapter 10) as is the oxidation of the olefin by AcOOH [58]. Although the yield is inferior to that of the Arco process (below), the process engineering may be simpler.

In a laboratory experiment [62], which may simulate the conditions of the Arco process [63], excess liquid propylene was reacted with a t-butyl hydroperoxide solution at 105°C for 1 hr in an autoclave under autogenous pressure. A small amount of $Mo(CO)_6$ was used as the catalyst. Under these conditions, 92% of the hydroperoxide was converted and propylene oxide was formed in 86% yield. Similar results can be attained with other alkyl hydroperoxides, but the t-butyl derivative is especially desirable for laboratory use because it is moderately stable. An excess of the olefin is commonly used to ensure complete conversion of the hydroperoxide.

The epoxidation can be carried out with a great variety of olefins [62]. Under conditions similar to those described for propylene, l-octene gives a 92% yield of 1,2-epoxyoctane. Internal olefins such as 2-butene, cyclohexene, and 2-methyl-2-pentene react faster than terminal olefins and give nearly quantitative yields of epoxide. Even for laboratory-scale preparations of epoxides, the metal-catalyzed oxidation with t-BuOOH is often more convenient than the conventional uncatalyzed reaction of a peroxy acid with the olefin [64,65]. The catalytic epoxidations are especially easy and selective for oxidation of allylic alcohols as described later in this section.

The most active catalysts for catalytic epoxidation are molybdenum salts and complexes [62,66-69]. Vanadium compounds are also quite effective and many other metals have some activity. Titanium complexes are less active than those of Mo or V, but they have other advantages that make them useful. Attachment of a Ti complex to a silica support produces a heterogeneous catalyst that is advantageous for both laboratory and industrial-scale epoxidations because of the ease of catalyst separation [65]. Titanium appears to be almost unique amongst the catalytically active metals in the tenacity with which it is attached to the silica support. The same coordination stability is also an important aspect in the choice of Ti for enantioselective catalysts for epoxidation of allylic alcohols. With the less effective epoxidation catalysts such as Ti, both free-radical and catalytic decompositions of the alkyl hydroperoxide become competitive reaction pathways. The formation of acetone in the reactions of t-butyl or cumene hydroperoxides is evidence of these undesired side reactions.

Most molybdenum compounds have some activity as catalysts for epoxidation. Curiously, zero-valent and hexavalent compounds, for example, $Mo(CO)_6$ and MoO_3, can have equivalent activity [62]. Even molybdenum metal dissolves with the help of t-BuOOH and oxalic acid to give an active catalyst [67]. The most likely explanation is that the compounds added to the mixture are catalyst precursors rather than catalysts per se. Indeed there is an induction period before the reaction begins. This time may be needed to form a complex with the olefin and the alkylhydroperoxide. Formation of such a complex fits a commonly observed kinetic pattern in which the rate is directly proportional to the concentrations of molybdenum, olefin, and hydroperoxide.

Two attractive mechanisms for catalytic epoxidation are shown in Figure 6.7. In the upper cycle, an alkyl molybdate **12** undergoes ester exchange with alkyl hydroperoxide to form an alkylperoxomolybdate **13**. Reaction of this complex with an olefin may occur by displacement of a neutral ligand such as ROH. At any event, an intermediate or transition state **14** develops in which the Mo-bound peroxy oxygen transfers to olefin. This transfer forms the epoxide which is still complexed to the metal in **15**. Displacement by alcohol or hydroperoxide completes the catalytic cycle [70].

An alternative mechanism, which is based on stoichiometric reactions of vanadium alkylperoxide complexes, invokes a dioxametallacyclopentane intermediate [71]. The initial step is coordination of an olefin to the peroxide complex to form **16**, as shown in the lower equation in Figure 6.7. A virtual insertion of the olefin into a V-O bond of **16** yields the metallacycle **17** in the rate-determining step of the process. The metallacycle decomposes rapidly to form epoxide and a vanadium alkoxide, which regenerates the peroxide complex through an ester exchange analogous to **12→13**.

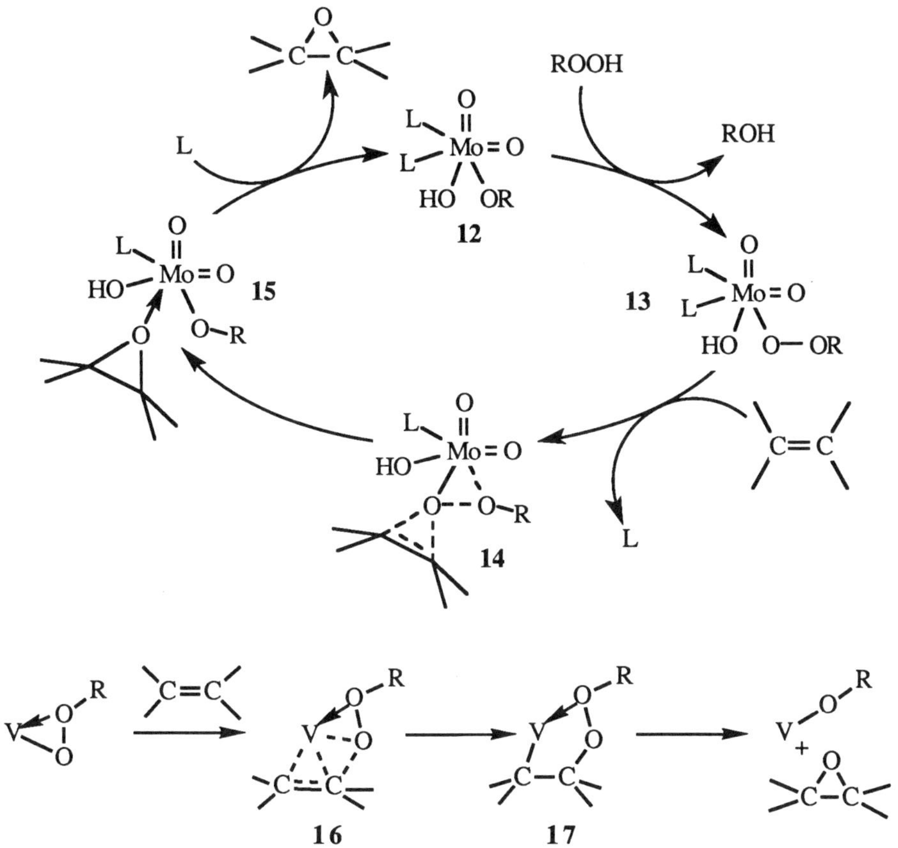

Figure 6.7   Alternative mechanisms for metal-catalyzed epoxidation of olefins.

Recent studies with a well-defined organometallic catalyst, $(C_5Me_5)MoO_2Cl$, could not differentiate between the two mechanisms [68]. Both experimental [72] and computational [73] results on the much-studied titanium system suggest a mechanism like that depicted in the upper part of Figure 6.7 [69].

Asymmetric Epoxidation

One of the most significant recent developments in homogeneous catalysis is the discovery that titanium complexes bearing asymmetric ligands can catalyze the enantioselective oxidation of allylic alcohols to the corresponding epoxy alcohols [74,75]. It has been known previously that chiral complexes of vanadium and molybdenum catalyze asymmetric oxidation of allylic alcohols. The significance of the titanium catalysts is that they are highly selective and convenient to use. Because of these virtues, they are now used commercially in the synthesis of several specialty products, in which biological activity is confined to a particular optical isomer. These applications are described in Section 6.7. In addition to industrial uses, the so-called Sharpless epoxidation finds wide use in organic synthesis. Enantioselective epoxidation of an olefin has the merit that it creates two chiral centers simultaneously. It is also useful for kinetic resolution of racemic mixtures of chiral olefinic compounds [75].

The Sharpless epoxidation employs a catalyst prepared by treating a titanium (+4) alkoxide with a tartrate ester such as the natural L(+) isomer of diethyl tartrate. Although titanium complexes are relatively sluggish catalysts compared with the molybdenum compounds used in propylene oxide manufacture, they give an exceptionally clean, selective reaction. In a typical epoxidation reaction [75], the catalyst is prepared in situ by adding equimolar amounts of L(+)-diethyl tartrate and titanium tetraisopropoxide to cold (-20°C) dichloromethane containing 4-Å molecular sieves. The molecular sieves are essential for a catalytic reaction because they sequester water introduced adventitiously or formed in side reactions. *tert*-Butyl hydroperoxide and the allylic alcohol are added and the reaction is conducted at about -20°C. Chemical yields and enantioselectivities can be very high. With allyl alcohol itself, the yield to glycidol is 65% with 90% enantiomeric excess [75]:

$$\text{/\!\!\!\diagup\!\!\!\diagdown}\text{OH} + Me_3COOH \longrightarrow \text{(epoxide)}\text{OH} + Me_3COH$$

In the absence of a dehydrating agent such as molecular sieves, the process requires a nearly stoichiometric amount of the titanium tartrate complex to get high enantioselectivity. Operation in this mode is illustrated by the *Organic Syntheses* procedure [76] for oxidation of *trans*-2-hexen-1-ol:

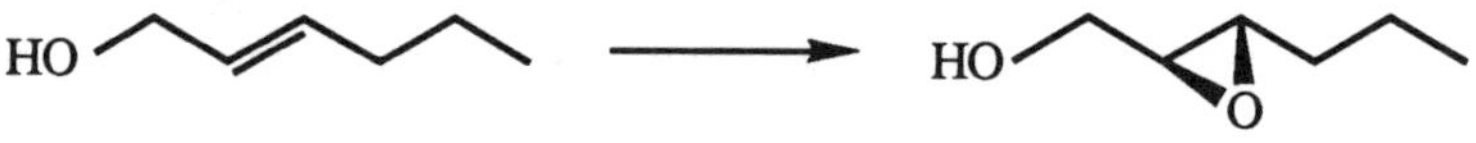

$$HO\text{/\!\!\!\diagup\!\!\!\diagdown\!\!\!\diagup} \longrightarrow HO\text{/\!\!\!\diagup}\text{(epoxide)}$$

The epoxide is produced in 63-69% chemical yield with an ee > 95%.

The mechanism of the epoxidation is assumed to be that of the upper equation in Figure 6.7. The key point is that the chiral tartrate ligand creates an asymmetric environment about the titanium center. When the allylic alcohol and the *t*-butyl hydroperoxide bind through displacement of alkoxy groups from the metal, they are disposed in such a way as to direct oxygen transfer to a specific face of the C=C bond. This situation is sketched in **18**.

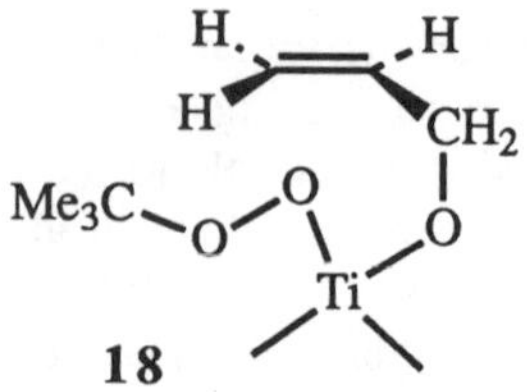

18

This sort of asymmetric ligand distribution is evident in the crystal structure of a catalytically active tartramide complex [77]. The solution structure of the tartrate complex is fluxional and less rigid than that observed in the solid state, but probably similar [78]. The intimate mechanism for asymmetric induction is not fully defined, but the kinetic parameters are well established [72].

A major limitation to the Sharpless epoxidation is that its utility is largely confined to oxidation of allylic alcohols. Homoallylic alcohols are oxidized less cleanly and the oxidation of simple olefins shows little enantioselectivity. Evidently the stereochemical control depends on "anchoring" the substrate to a particular site on the metal by means of the alcohol function. This need for an auxilliary coordinating function on the substrate parallels the requirement noted earlier for the asymmetric hydrogenation of olefins (Chapter 3).

An exception to this need for an auxilliary function on the olefinic substrate has been discovered recently. A manganese complex **19** bearing a chiral tetradentate ligand catalyzes enantioselective oxidation of simple olefins by iodosomesitylene [79]. The reaction seems to function best with styrene derivatives. 2-Vinylnaphthalene, for example, gives a 72% yield of the corresponding epoxide having an ee of 67%. The improved selectivity with aryl-substituted olefins may indicate some kind of "phenyl effect," an affinity between the arene rings of the substrate and those of the ligand that steer the olefin into a particularly favorable interaction with the metal center. Sodium hypochlorite is also a suitable oxidizing agent for this catalyst system [80]. With this oxidant, cis-β-methylstyrene gives its epoxide in 68% yield and 86% ee.

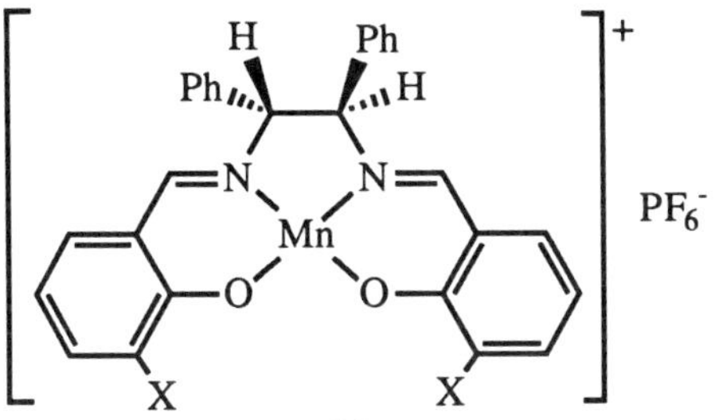

19

6.6   VICINAL HYDROXYLATION OF OLEFINS

Another approach to the synthesis of epoxides and diols has received much attention in recent years. It has been known for decades that osmium tetroxide catalyzes the H_2O_2 oxidation of olefins to cis-1,2-diols, but the cost, toxicity, and volatility of OsO_4 have limited its use to the organic synthesis laboratory. Potential industrial interest has been aroused by three new developments: (a) the invention of a system for vicinal hydroxylation using air as the ultimate oxidant, (b) the discovery of conditions for asymmetric dihydroxylation of olefins, and (c) modification of the reaction to produce 1,2-aminoalcohols.

The classical dihydroxylation is illustrated by the *Organic Syntheses* procedure [81] for synthesis of *cis*-1,2-cyclohexanediol:

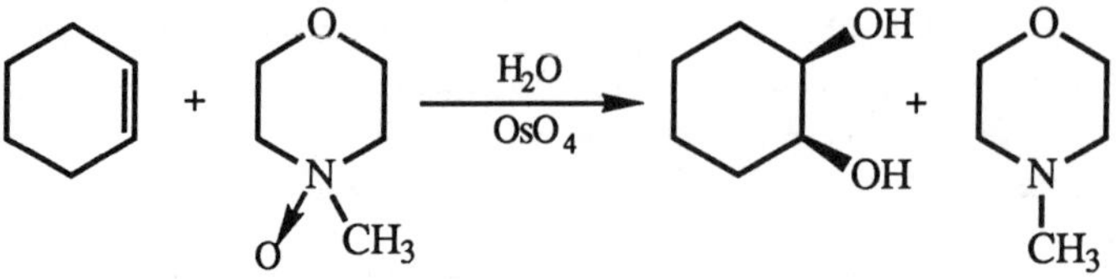

The oxidation is carried out in a two-phase aqueous/organic system using N-methylmorpholine-N-oxide as the oxidant and 1 mole % of OsO_4 as a catalyst. This system gives approximately 90% yield of the desired cis-diol in contrast to much lower yields obtained with oxidants such as H_2O_2, $NaClO_3$, or t-BuOOH.

The clean reactions obtainable in osmium-catalyzed oxidations of olefins would be highly desirable for commercial production of ethylene or propylene glycols if air or oxygen could be used as the ultimate oxidant. In general, however, systems based on O_2/OsO_4 have given inferior yields and low rates and have operated only in limited pH ranges. A bimetallic catalyst system has now been found to couple oxygen and OsO_4 efficiently for oxidation of propylene to propylene glycol [82]:

$$CH_3CH=CH_2 \xrightarrow{\quad O_2, H_2O \quad} \begin{array}{c} CH_3CH-CH_2 \\ | \quad\quad | \\ OH \quad OH \end{array}$$

The reaction of propylene with oxygen in the presence of a 1:15 molar ratio of OsO_4 and $CuCl_2$ at 100°C/28 atmospheres gives propylene glycol in nearly quantitative yield albeit at low conversion [83]. The catalytic species that reacts with the olefin is believed to be a trioxoosmium halide **20** as shown in Figure 6.8. The olefin reaction

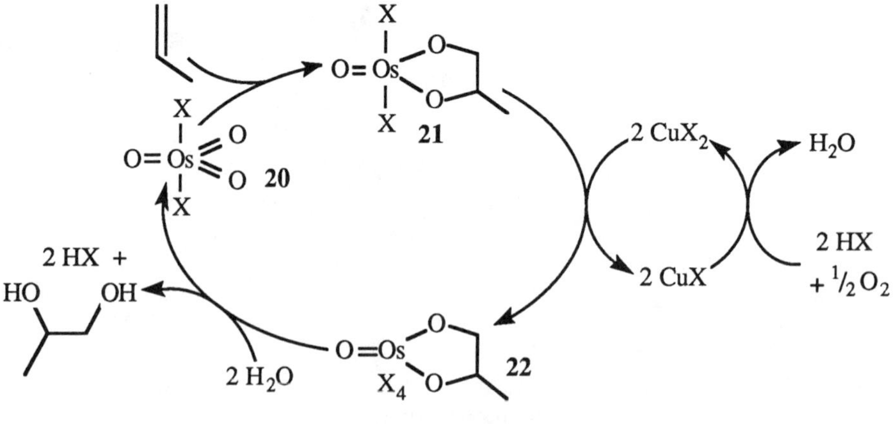

Figure 6.8 The role of $CuCl_2$ in air oxidation of propylene via an osmate ester.

is considered to be, in effect, a cycloaddition to form an osmate ester **21**, as proposed in classical OsO_4 reactions. The role of the copper chloride is to reoxidize the osmium from a formal oxidation state of +6 in **21** to the +8 state in **22**. This oxidation does not proceed well with oxygen alone, but is catalyzed by the copper(+2) which is reoxidized readily by O_2. This reaction has been studied extensively for possible manufacture of ethylene and propylene glycols, but has not been used commercially yet.

Asymmetric Dihydroxylation

Another major development in OsO_4-catalyzed oxidation is the discovery of conditions that lead to enantioselectivity in the dihydroxylation of a prochiral olefin. As in the Ti-catalyzed epoxidations described in Section 6.5, an appropriate chiral ligand in the coordination sphere of the metal directs a coordinated oxygen electrophile to a particular face of the olefin. In contrast to the epoxidation chemistry, it is not necessary to have an auxilliary functional group (such as the OH in an allylic alcohol) in order to attain enantioselectivity. The asymmetric dihydroxylation seems reasonably general, a characteristic that should be useful in preparation of complex, biologically active organics.

The general methodology is illustrated by the *Organic Syntheses* procedure [84] for oxidation of *trans*-stilbene to the corresponding R,R-diol:

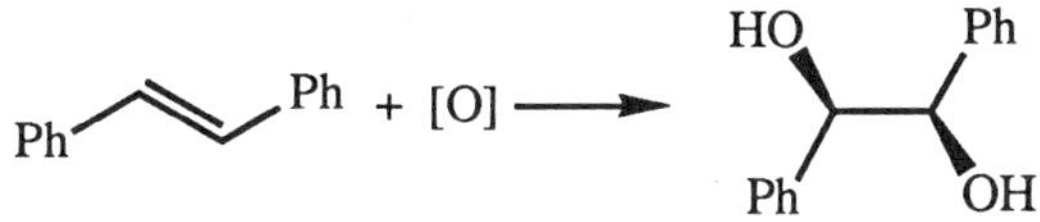

In this example, stilbene is oxidized in aqueous acetone with morpholine-N-oxide as the oxidant, dihydroquinidine 4-chlorobenzoate as the directing ligand, and 0.4 mole % OsO_4 as the catalyst. The reaction produces a quantitative yield of the diol in 90% optical purity. Recrystallization gives a 72-75% yield of enantiomerically pure R,R-diol.

As illustrated by this example, the chemical yields of cis diols are high. The enantioselectivity is critically dependent on olefin structure, ligand choice, and reaction conditions. The procedure using N-methylmorpholine-N-oxide [85] gives 80-95% yields with 50-89% ee. More recent results with $K_3Fe(CN)_6$ as the oxidant and aqueous *t*-butanol as the solvent [86] give 70-99% ee [87]. Selectivity for a particular optical isomer depends on the choice of alkaloid ligand. Dihydroquinidine derivatives direct attack to the *re* face of the olefin whereas dihydroquinines lead to oxygen introduction at the *si* face [85].

Mechanistic understanding of the reaction has been important in improving the enantioselectivity [87]. As shown in Figure 6.9, there are two competing catalytic cycles under the standard reaction conditions. The cycle on the left which retains the alkaloid ligand throughout the reaction leads to high selectivity and high rates. The cycle on the right in which ligand is displaced by a second molecule of olefin has low selectivity. Fortunately, the latter reaction is slow and is largely absent in the $K_3Fe(CN)_6/t$-butanol system. The presumed initiation reaction in both cycles is cycloaddition of cis oxo functions of an $OsO_4 \cdot L$ complex to the C=C bond (illustrated with *trans*-stilbene in the figure). The resulting osmate ester **23** is formally Os(+6). It reacts with the oxidant such as $K_3Fe(CN)_6$ to form an Os(+8) ester **24**. In the more selective cycle, **24** is hydrolyzed to produce the chiral diol and the $OsO_4 \cdot L$ complex which can repeat the reaction cycle. In the cycle on the right, the chiral ligand is lost and a second mole of olefin reacts with the osmium complex to form the *bis*(alkylene) ester **25**. In the absence of the alkaloid, there is little selectivity for attack at a particular face of the olefin. Fortunately, ligand can reassociate with the hydrolysis product of **25** to reenter the preferred catalyst cycle. Crystal structures of complexes like **23** have been determined [88]. Alternative origins for the enantioselectivity have been suggested [89].

Vicinal Oxyamination

An interesting offshoot of the work on osmium-catalyzed oxidations is a new method to introduce alcohol and amine functions in a cis relationship. This reaction is exemplified in a laboratory procedure for oxyamination of cyclohexene [90]:

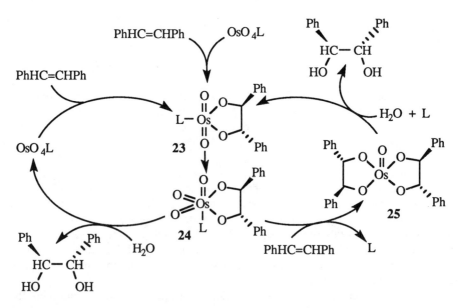

Figure 6.9 Competing catalytic cycles in the asymmetric dihydroxylation of stilbene (L = alkaloid ligand).

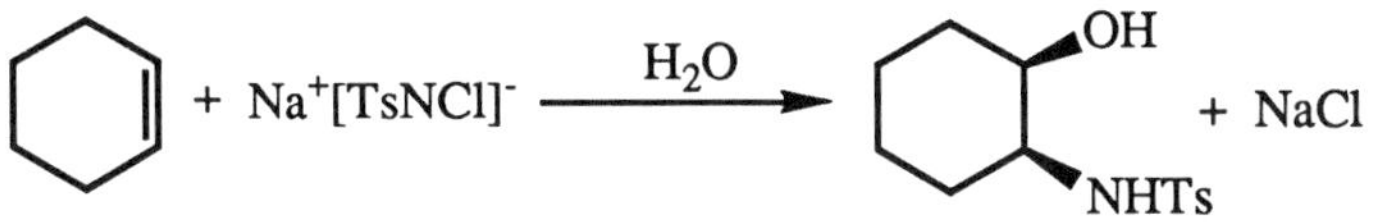

The reaction may be carried out either in a polar solvent such as *t*-BuOH or under phase transfer conditions with partition of the reactants between water and chloroform. Osmium tetraoxide is used in catalytic amounts. Presumably it forms a reagent such as $O_3Os=NTs$, which cycloadds to the olefin much as does OsO_4. This reaction is not used industrially, but may be a useful tool for organic synthesis.

6.7   SPECIALTY CHEMICALS

As noted in Chapter 1, the past decade has seen many new applications of homogeneous catalysis, especially for the production of specialty chemicals. This trend is evident in the commercial applications of olefin oxidation. The traditional uses have been for production of large volume chemical intermediates such as acetaldehyde and propylene oxide. The same basic oxidation reactions have now been applied to making intermediates for small-volume, high-value products such as vitamins and insect attractants.

Vitamin B₁ Precursors

The palladium-catalyzed oxidation of a terminal olefin to an acetal, a close analog of the Wacker acetaldehyde synthesis, has been applied to the practical synthesis of vitamin B_1 [20]. As mentioned in Section 6.2, the oxidation of terminal olefins bearing electron-withdrawing substituents leads to acetals when conducted in the presence of an alcohol. When acrylonitrile is oxidized with methyl nitrite in the presence of $PdCl_2$, the acetal **26** is formed in high yield:

$$H_2C=CHCN + CH_3ONO \longrightarrow (CH_3O)_2CHCH_2CN$$

26

In a large laboratory experiment, which could simulate commercial practice [91], methyl nitrite is generated continuously by passage of O_2 and NO through liquid methanol. The gases are circulated through a hot (63°C) methanol solution containing acrylonitrile and a catalytic amount of $PdCl_2$. The off-gas from the reaction is recirculated to the methyl nitrite generator. The desired acetal **26** accumulates in the reaction mixture. The reaction can also be carried out as a vapor phase process with a mixture of $PdCl_2$ and $FeCl_3$ (or $CuCl_2$) supported on an inorganic oxide such as alumina [92]. The acetal **26** is used as a building block to construct the pyrimidine ring of synthetic vitamin B_1 [93] as shown in Figure 6.10.

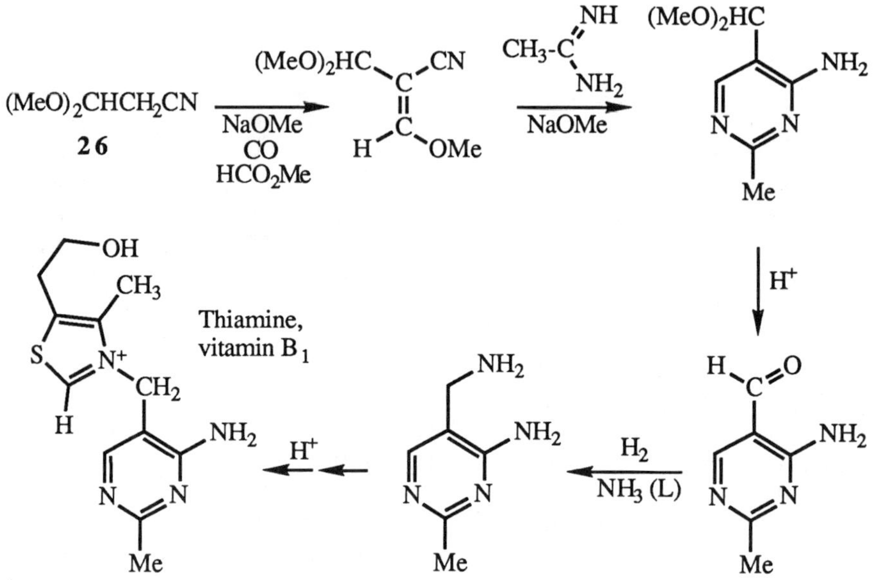

Figure 6.10   Synthesis of vitamin B₁ from an acrylonitrile-derived acetal.

Chiral Epoxides

The Sharpless epoxidation of allylic alcohols to produce specific enantiomers of epoxides (Section 6.5) has found several applications in production of chiral epoxides as intermediates to specialty chemicals. Arco plans to make several such products [94,95] including the two enantiomers of glycidol **27**. These are produced by oxidizing allyl alcohol, of which Arco is a basic producer [42], with titanium complexes containing the (+) or (-) isomers of diethyl tartrate:

The best known of the commercial applications of the Sharpless chemistry is the synthesis of a chiral epoxide as an intermediate to disparlure, a sex pheromone for the gypsy moth. This leaf-feeding insect causes much damage to trees in the eastern United States, but attempts to control its spread through conventional techniques using insecticides have had only limited success. The new strategy is to use disparlure, the sex attractant emitted by the female gypsy moth, to lure male moths to a trap. Alternatively, it can be used to confuse the males and prevent them from finding females. This latter tactic has been quite successful in reducing mating, even in heavily infested areas.

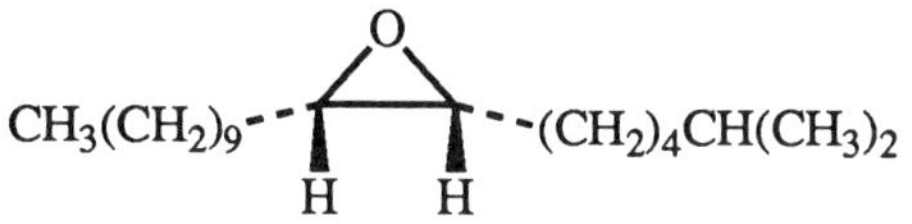

$$CH_3(CH_2)_9 \cdots \overset{O}{\underset{H \quad H}{\triangle}} \cdots (CH_2)_4CH(CH_3)_2$$

(+)-Disparlure

Success depends on use of the (+) isomer of disparlure; the (-) isomer appears to interfere with the male moth's sensor. The synthesis of (+)-disparlure employed by Upjohn [95] employs the enantioselective oxidation of Z-2-tridecenol **28**:

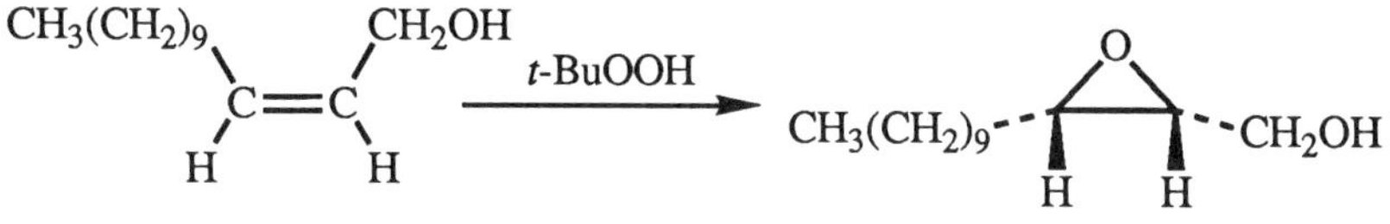

$$\begin{array}{ccc}
CH_3(CH_2)_9 \quad \underset{|}{\overset{|}{C}} = \underset{|}{\overset{|}{C}} \quad CH_2OH & \xrightarrow{\text{t-BuOOH}} & CH_3(CH_2)_9 \cdots \overset{O}{\underset{H \quad H}{\triangle}} \cdots CH_2OH \\
H \qquad\qquad H & & \\
\mathbf{28} & & \mathbf{29}
\end{array}$$

The oxidation is carried out as described in the examples in Section 6.5 using a complex of $Ti(OR)_4$ and D(-)-diethyl tartrate as the catalyst. The epoxy alcohol **29** is produced in 90-95% enantiomeric purity before recrystallization. Conversion of **29** to (+)-disparlure requires three conventional organic steps.

It seems likely that the epoxidation will find extensive use in the production of pharmaceuticals, given the regulatory trend toward treatment of enantiomers of the same compound as distinct therapeutic agents. This trend may also create commercial applications for the enantioselective osmium-catalyzed vicinal hydroxylation reaction.

GENERAL REFERENCES

R. A. Sheldon and J. K. Kochi, *Metal-Catalyzed Oxidations of Organic Compounds,* Academic Press, 1981.

SPECIFIC REFERENCES

1. S. F. Davison and P. M. Maitlis, "Oxidation Using Palladium Compounds," in W. J. Mijs and C. R. H. I. de Jonghe, Eds., *Organic Syntheses by Oxidation with Metal Compounds,* Plenum Press, New York, 1986, pp. 469-480.

2. P. M. Henry, *Palladium-Catalyzed Oxidation of Hydrocarbons,* Reidel, Dordrecht, 1980, pp. 41-223.

3. R. F. Heck, *Palladium Reagents in Organic Syntheses,* Academic Press, New York, 1985, pp. 59-100.

4. R. Jira and W. Freiesleben, "Olefin Oxidation and Related Reactions with Group VIII Noble Metal Compounds," *Organomet. React.,* **3,** 1-83 (1972).

5. R. Jira, W. Blau, and D. Grimm, *Hydrocarbon Proc.,* 97-100 (Mar. 1976).

6. F. C. Phillips, *Am. Chem. J.,* **16,** 255-77(1894).

7. R. Jira, "Manufacture of Acetaldehyde directly from Ethylene," in S. A. Miller, Ed., *Ethylene and its Industrial Derivatives*, Ernest Benn Ltd., London, 1969, p. 650-9.

8. K. I. Matveev, *Kinet. Katal.*, **18**, 862-77 (1977); British Patent 1,508,331 (1978).

9. H. Stangl and R. Jira, *Tetrahedron Lett.*, 3589-92 (1970).

10. N. Gregor, K. Zaw, and P. M. Henry, *Organometallics*, **3**, 1251-6 (1984); W. K. Wan, K. Zaw, and P. M. Henry, *ibid.*, **7**, 1677-83 (1988).

11. J-E. Backvall, B. Akermark, and S. O. Ljunggren, *J. Am. Chem. Soc.*, **101**, 2411-6 (1979); J. K. Stille and R. Divakaruni, *J. Organomet. Chem.*, **169**, 239-48 (1979).

12. P. H. Espeel, M. C. Tielen, and P. A. Jacobs, *J. Chem. Soc., Chem. Commun.*, 669-71 (1991).

13. K. Weissermel and H-J. Arpe, *Industrial Organic Chemistry*, Verlag Chemie, Weinheim, 1978, pp. 244-5.

14. J. Smidt, W. Hafner, R. Jira, J. Sedlmeier, R. Sieber, R. Ruttinger, and H. Kojer, *Angew. Chem.*, **71**, 176-82 (1959).

15. J. Tsuji, H. Nagashima, and H. Nemoto, *Org. Syn.*, **62**, 9-13 (1984).

16. J. H. Byers, A. Ashfaq, and W. R. Morse, *J. Chem. Educ.*, **67**, 340-1 (1990).

17. *Chem. Eng. News*, **64**, 20 (4 Aug. 1986); *Eur. Chem. News*, 34 (27 Jan., 1992).

18. J. Vasilevskis, J. C. DeDeken, R. J. Saxton, P. R. Wentrcek, J. D. Fellmann, and L. S. Kipnis, U.S. Patents 4,720,474; 4,723,041; 4,738,943 (1988).

19. D. Pauley, F. Anderson, and T. Hudlicky, *Org. Syn.*, **67**, 121-4 (1989).

20. T. Hosokawa and S-I. Murahashi, *Acc. Chem. Res.*, **23**, 49-54 (1990).

21. T. Hosokawa, T. Ohta, S. Kanayama, and S-I. Murahashi, *J. Org. Chem.*, **52**, 1758-64 (1987).

22. Ube Industries, Jap. Kokai 81-05429 (1981); *Chem. Abstr.*, **95**, 24272 (1981).

23. I. I. Moiseev, M. N. Vargaftik, and Y. K. Syrkin, *Dokl. Akad. Nauk SSSR*, **133**, 377-80 (1960).

24. R. Jira and W. Freisleben, *Organomet. React.*, **3**, 154 (1972).

25. W. Kronig and G. Scharfe, U.S. Patent 3,822,308 (1974).

26. A. Sabel, J . Smidt, R. Jira, and H. Prigge, *Chem. Ber.*, **102**, 2939-50 (1969).

27. R . N. Pandey and P . M. Henry, *Can. J. Chem.*, **53**, 2223-31 (1975).

28. R. L. Adelman, *J. Org. Chem.*, **14**, 1057-77 (1949).

29. S . Winstein, J.McCoskie, H. B. Lee, and P. M. Henry, *J. Am. Chem. Soc.*, **98**, 6913-8 (1976).

30. M. Kosaki, M. Isemura, Y. Kitaura, S. Shinoda, and Y. Saito, *J. Mol. Catal.*, **2**, 351-9 (1977).

31. T. Majima and H. Kurosawa, *J. Chem. Soc., Chem. Commun.*, 610-1 (1977).

32. P. M. Henry, *J. Org. Chem.*, **39**, 3871-4 (1974).

33. J. E. Backvall, *Tetrahedron Lett.*, 467-8 (1977).

34. N. I. Kuznetsova, V. A. Likholobov, M. A. Fedotov, and Yu. I. Yermakov, *J. Chem. Soc., Chem. Commun.*, 973-4 (1982).

35. M. G. Volkhonskii, V. A. Likholobov, and Yu. I. Ermakov, *Kinet. Katal.*, **18**, 790-1 (1977); E. V. Gusevskaya, I. E. Beck, A. V. Karandin, A. G. Stepanov, and V. A. Likholobov, *Proceedings 7th International Symposium Homogeneous Catalysis*, Lyon, France, 5-9 Sept. 1990, pp. 283-4.

36. R. Huttel, J. Kratzer, and M. Bechter, *Chem. Ber.*, **94**, 766-80 (1961).

37. H. C. Volger, *Rec. Trav. Chim.*, **87**, 225 (1968).

38. A. Heumann, B. Akermark, S. Hansson, and T. Rein, *Org. Syn.*, **68**, 109-13 (1989).
39. H. J. Schmidt and G. Roscher, *Compend. Deut. Ges. Mineraloelwiss. Kohlechem.*, **1975**, pp. 75-76, 318-326; W. Kronig and G. Scharfe, British Patent 1,247,595 (1971).
40. W. E. Smith, British Patent 1,461,831 (1977).
41. P. Hayden, W. Featherstone, and J. E. Lloyd, Canadian Patent 900,510 (1972).

43. *Chem. Week*, 20 (13 Sept. 1989).
44. J. Tsuji, *Acc. Chem. Res.*, **6**, 8-13 (1973).
45. E. W. Stern and M. L. Spector, *Proc. Chem. Soc.*, 370 (1961).
46. T. Inagaki, Y. Takahashi, S. Sakai, and Y. Ishii, *Bull. Jpn. Petrol Inst.*, **13**, 73-7 (1971).
47. Mitsubishi Kasei Corp., *Chemtech*, 759-63 (Dec. 1988); Y. Tanabe, *Hydrocarbon Proc.*, 187-90 (Sept. 1981).
48. H. M. Weitz and J. Hartig, U.S. Patent 4,038,307 (1977).
49. T. Onoda and J. Haji, U.S. Patent 3,755,423 (1973).
50. J. E. Backvall and R. E. Nordberg, *J. Am. Chem. Soc.*, **103**, 4959-60 (1981).
51. J. E. Nystrom, T. Rein, and J. E. Backvall, *Org. Syn.*, **67**, 105-13 (1989).
52. G. E. Weismantel, *Chem. Eng.*, 67-70 (15 Jan. 1979).
53. J. Kollar and R. Hoch, U.S. Patent 3,985,795 (1976); British Patent 1,351,243 (1974); British Patent 1,351,242 (1974).
54. I. Hirose and H. Okitsu, U.S. Patent 4,008,286 (1977).
55. A. M. Brownstein, *Hydrocarbon Proc.*, **53**, 129-32 (June 1974).
56. *Chem. Eng. News*, 6-7 (3 Dec. 1979).
57. J. Bergman and L. Engman, *J. Am. Chem. Soc.*, **103**, 5196-5200 (1981).
58. R. Landau, G. A. Sullivan, and D. Brown, *Chemtech*, 602-7 (1979).
59. R. B. Stobaugh, V. A. Calarco, R. A. Morris, and L. W. Stroud, *Hydrocarbon Proc.*, **52**, 99-108 (Jan. 1973) .
60. D. A. O'Sullivan, *Chem. Eng. News*, 10-11 (18 Mar. 1985).
61. J. L. Meyer and B. T. Pennington, U.S. Patent 4,992,567 (1991).
62. M. N. Sheng and J . G . Zajacek, "Hydroperoxide Oxidations Catalyzed by Metals," in F. R. Mayo, Ed., *Oxidation of Organic Compounds*, Vol. 2, *Advances in Chemistry Series*, **76**, American Chemical Society, Washington, DC, 1968. pp. 418-431.
63. J. Kollar, U.S. Patent 3,351,635 (1967).
64. K. B. Sharpless and T. R. Verhoeven, *Aldrichim. Acta*, **12**, 63-74 (1979).
65. H. P. Wulff, U.S. Patent 3,923,843 (1975).
66. R. A. Sheldon, *Aspects Homog. Catal.*, **4**, 3-70 (1981).
67. B. H. Isaacs, U.S. Patent 4,590,172 (1986).
68. M. K. Trost and R. G. Bergman, *Organometallics*, **10**, 1172-8 (1991).
69. K. A. Jorgensen, *Chem. Rev.*, **89**, 431-58 (1989).
70. A. O. Chong and K. B. Sharpless, *J. Org. Chem.*, **42**, 1587-90 (1977).
71. H. Mimoun, M. Mignard, P. Brechot, and L. Saussine, *J. Am. Chem. Soc.*, **108**, 3711-8 (1986).
72. S. S. Woodard, M. G. Finn, and K. B. Sharpless, *J. Am. Chem. Soc.*, **113**, 106-13 (1991).
73. K. A. Jorgensen, R. A. Wheeler, and R. Hoffmann, *J. Am. Chem. Soc.*, **109**, 3240-6 (1987).
74. K. B. Sharpless, *Chemtech*, 692-700 (1985).

75. Y. Gao, R. M. Hanson, J. M. Klunder, S. Y. Ko, H. Masamune, and K. B. Sharpless, *J. Am. Chem. Soc.*, **109**, 5765-80 (1987).

76. J. G. Hill, K. B. Sharpless, C. M. Exon, and R. Regenye, *Org. Syn.*, **63**, 66-78 (1985).

77. I. D. Williams, S. F. Pedersen, K. B. Sharpless, and S. J. Lippard, *J. Am. Chem. Soc.*, **106**, 6430-1 (1984).

78. M. G. Finn and K. B. Sharpless, *J. Am. Chem. Soc.*, **113**, 113-26 (1991).

79. W. Zhang, J. L. Loebach, S. R. Wilson, and E. N. Jacobsen, *J. Am. Chem. Soc.*, **112**, 2801-2 (1990).

80. W. Zhang and E. N. Jacobsen, *J. Org. Chem.*, **56**, 2296-8 (1991).

81. V. Van Rheenen, D. Y. Cha, and W. M. Hartley, *Org. Syn., Coll. Vol. 6*, 342-50 (1988).

82. R. G. Austin, R. Ç. Michaelson, and R. S. Myers in R. L. Augustine, Ed., *Catalysis of Organic Reactions*, Marcel Dekker, New York, 1985, pp. 269-80.

83. R. G. Austin and R. C. Michaelson, U.S. Patent 4,390,739 (1983).

84. B. H. McKee, D. G. Gilheany, and K. B. Sharpless, *Org. Syn.*, **70**, 47-50 (1991).

85. B. B. Lohray, T. H. Kalantar, B. M. Kim, C. Y. Park, T. Shibata, J. S. M. Wai, and K. B. Sharpless, *Tetrahedron Lett.*, **30**, 2041-4 (1989).

86. M. Minato, K. Yamamoto, and J. Tsuji, *J. Org. Chem.*, **55**, 766-8 (1990).

87. H. L. Kwong, C. Sorato, V. Ogino, H. Chen, and K. B. Sharpless, *Tetrahedron Lett.*, **31**, 2999-3002 (1990).

88. R. M. Pearlstein, B. K. Blackburn, W. M. Davis, and K. B. Sharpless, *Angew. Chem. Int. Ed.*, **29**, 639-41 (1990).

89. E. J. Corey and G. I. Lotto, *Tetrahedron Lett.*, **31**, 2665-8 (1990).

90. E. Harranz and K. B. Sharpless, *Org. Syn.*, **61**, 85-93 (1983).

91. K. Matsui, S. Uchiumi, A. Iwayama, and T. Umezu, U.S. Patent 4,504,422 (1985).

92. K. Matsui, S. Uchiumi, A. Iwayama, and T. Umezu, U.S. Patents 4,501,705 and 4,504,421 (1985).

93. K. Fujii, K. Nishihira, H. Sawada, S. Tanaka, M. Nakai, H. Yoshida, and I. Inoue, U.S. Patents 4,492,792 and 4,539,403 (1985).

94. *Chem. Week*, 29 (28 Feb. 1990); H. Mazurek, *Chem. Eng. News*, 2 (18 Feb. 1991).

95. *Chem. Eng. News*, 24 (2 June 1986).

7 | ARENE REACTIONS

Benzene and its derivatives form a broad range of π-complexes and σ-aryl derivatives with transition metal ions. These arene and aryl complexes are intermediates in many catalytic reactions that have potential application in industry and in laboratory syntheses. Some palladium(II)-catalyzed reactions accomplish substitutions or couplings of arenes that are difficult to effect by conventional methods, but these have found little industrial use. On the other hand, copper complexes catalyze some unique reactions that are used on a moderate scale. "Ziegler catalysts" related to those used in olefin polymerization have become important in the hydrogenation of benzene. A variety of soluble catalysts are used in synthesis of specialty chemicals, especially in the venerable dyes industry.

7.1 ELECTROPHILIC AROMATIC SUBSTITUTION

One major limitation on use of the metal-catalyzed reactions is that classical organic methods for electrophilic aromatic substitution work so well. Many large-scale arene reactions (Figure 7.1) are catalyzed by simple Lewis or Bronsted acids. Although these acids are soluble catalysts in a broad sense, they fall outside the general context of this book. Almost universally, the primary interaction of the acid is with the attacking reagent rather than with aromatic substrate. In chlorination, for example, $FeCl_3$ reacts with chlorine to form a polarized complex which behaves as though free Cl^+ were formed:

$$Cl_2 + FeCl_3 \rightleftharpoons [Cl_2 \cdot FeCl_3] \rightleftharpoons Cl^+ FeCl_4^-$$

This activated reagent attacks benzene to replace H with Cl. Similar mechanisms generate incipient NO_2^+ from nitric acid or $C_2H_5^+$ from ethylene. These reactions are

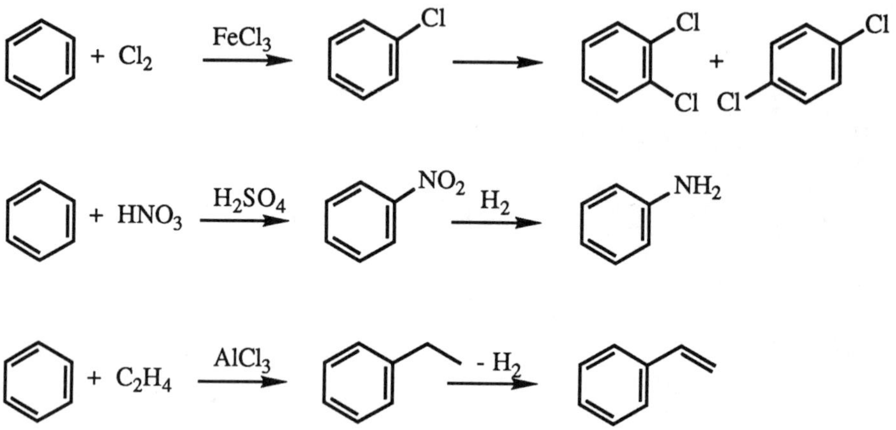

Figure 7.1   Some important industrial processes based on electrophilic substitution reactions of benzene.

discussed thoroughly in standard organic texts and in reviews [1]. The Monsanto process for $AlCl_3$-catalyzed ethylation of benzene as a route to styrene has been described [2].

The palladium-catalyzed reactions discussed in Section 7.2 seem to involve electrophilic attack of Pd^{2+} on the aromatic ring. However, the arylpalladium compounds which result react by conventional organometallic mechanisms like those observed for olefins. In many catalytic reactions, the π-complex involving one double bond is believed to be transformed to a σ-aryl complex via electrophilic metallation or oxidative addition processes [3]:

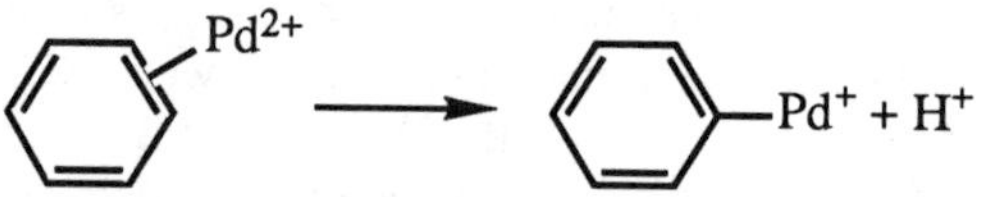

The preliminary π-complex formation facilitates attack on the aryl C-H bond. Alkanes react with metals less readily than do arenes because the alkanes do not undergo π-complexation, but some alkane functionalization chemistry has been reported [4].

The chemistry of the aryl-palladium σ-bonds formed by these processes is very similar to that of the alkyl-metal bonds generated in olefin reactions. Insertion of CO and olefins into Ar-Pd bonds is well known, and reactions with cleavage reagents such as HX and X_2 are similar to those of R-M bonds [5]. Interactions between the arene π-bonds and the metal d-orbitals in σ-aryl complexes are detectable by sensitive methods such as ^{13}C nmr [6], but do not substantially affect the chemistry of the Ar-Pd bond.

7.2   PALLADIUM-CATALYZED REACTIONS

Palladium(II) salts, especially the acetate, catalyze many oxidative substitution reactions of benzene and other aromatic hydrocarbons [3,7-9]. These reactions are not used commercially, but have been studied as potential processes for manufacture of styrene, phenol, and substituted biphenyls. The reaction types closely parallel the Pd-catalyzed oxidations of olefins discussed in Chapter 6. For example, reaction of benzene with palladium acetate can give phenyl acetate, a reaction closely analogous to liquid-phase vinyl acetate synthesis. Like the olefin reactions, the arene reactions are very sensitive to reaction conditions. A change in acetate concentration in the benzene-$Pd(OAc)_2$ reaction makes biphenyl the major product.

These arene oxidations also resemble the olefin oxidations mechanistically as illustrated in the oxidative coupling of olefins and arenes. This cross coupling may be regarded as a hybrid between olefin-olefin coupling (Section 6.3) and the arene-arene coupling discussed later in the present section.

Arene-Olefin Coupling

When styrene and $Pd(OAc)_2$ are heated in benzene that contains acetic acid, *trans*-stilbene is a major product [10]:

$$PhCH=CH_2 + Pd(OAc)_2 \xrightarrow{\;C_6H_6\;} PhCH=CHPh + Pd^0 + 2\,HOAc$$

Labelling studies show that one of the phenyl groups in the stilbene comes from benzene. Although this example uses the palladium salt as a stoichiometric oxidant, the reaction becomes catalytic when it is conducted under oxygen pressure [11]. Styrene, benzene, and palladium(II) acetate react at 80-100°C under 10-20 atmospheres oxygen pressure to form stilbene in amounts corresponding to 2-11 catalyst cycles. Similarly ethylene can be phenylated catalytically to form styrene, stilbene, and higher derivatives [12]:

$$C_2H_4 \longrightarrow PhCH=CH_2 \longrightarrow PhCH=CHPh \longrightarrow Ph_2C=CHPh \longrightarrow Ph_2C=CPh_2$$

The arene-olefin coupling reactions, whether stoichiometric or catalytic, generally do not give high yields of a single product. Arene-arene coupling is not observed in the presence of olefin [11], but addition of acetic acid to the C=C bond occurs. The styrene-$PdCl_2$ complex in benzene/acetic acid that contains sodium acetate gives α-phenylethyl acetate in addition to stilbene and a little β-acetoxystyrene [12]. This lack of selectivity has reduced the appeal of the benzene-ethylene reaction as a potential industrial process for styrene production.

The mechanism of arene-olefin coupling is unclear. Two very plausible proposals appear in the literature [8,9,13]. Both involve metallation of the benzene ring by a

Pd^{2+} electrophile as a key step. It seems likely that the electrophile coordinates to the face of the ring (perhaps off-center) and displaces a proton as illustrated for toluene:

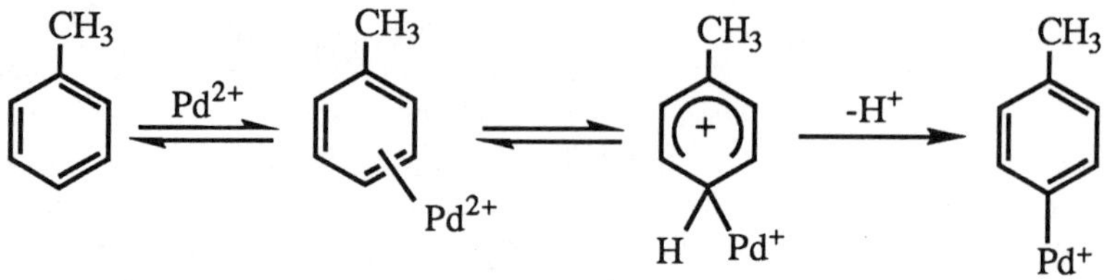

The formation of the σ-aryl derivative appears to be a common step in arene-olefin coupling, arene-arene coupling, and arene acetoxylation. The para specificity observed with toluene in some of these reactions can be interpreted as evidence for attack by a bulky electrophile.

The coupling proceeds through coordination of the olefin to the Ar-Pd compound to give a complex **1**. Figure 7.2 shows two conventional mechanisms for formation of styrene from an ethylene complex. In the upper pathway [14], ethylene inserts into the Ar-Pd bond to form a σ-phenethyl complex **2**. Hydrogen elimination by transfer to the metal gives styrene and an unstable Pd-H species which decomposes to palladium metal. Alternatively (lower equation), the ethylene is metallated to form a σ-vinyl derivative **3**. This compound then undergoes reductive elimination of the two organic ligands to form the styrene directly [12]. There is good precedent for both pathways. Preformed σ-vinyl complexes react with benzene to form styrenes as in the lower pathway [15]. On the other hand, preformed σ-phenylpalladium compounds react with olefins to form similar products.

In a related reaction that may be useful in organic synthesis, aryl halides and olefins react to form styrenes. For example, o-bromotoluene and ethylene react at 125°C and 9 atmospheres pressure to form o-methylstyrene in 86% yield [16]:

$$ArBr + C_2H_4 + Et_3N \xrightarrow{Pd(OAc)_2} ArCH=CH_2 + Et_3NHBr$$

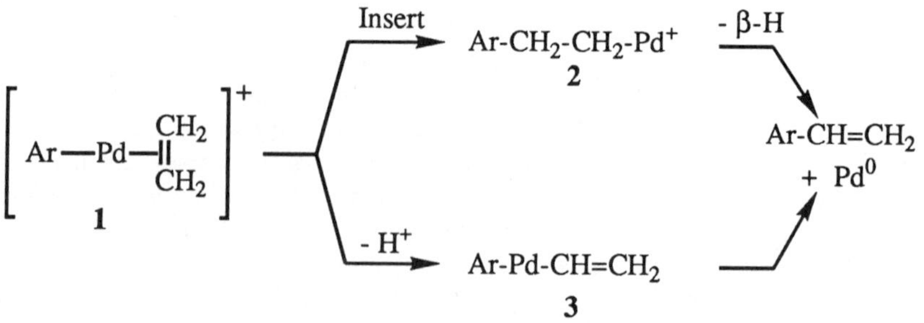

Figure 7.2 Two pathways for styrene synthesis from an ethylene complex of an arylpalladium compound (other ligands omitted).

A palladium complex formed in situ from palladium acetate and triphenylphosphine is used as a catalyst rather than a reagent. It appears that the reaction proceeds via an arylpalladium complex formed by oxidative addition of aryl bromide to a palladium(0) complex.

$$Pd^{2+} + n\ Ph_3P \xrightarrow{\text{[H]}} Pd^0(PPh_3)_n \xrightarrow{\text{ArBr}} ArPdBr(PPh_3)_2$$

Once the aryl-Pd bond is formed, coordination, insertion, and coupling of the olefin proceed as shown in Figure 7.2.

Arene-Arene Coupling

Benzene, palladium chloride, and sodium acetate in acetic acid react to form biphenyl in high yield [17]:

$$2\ C_6H_6 + PdCl_2 \longrightarrow C_6H_5\text{-}C_6H_5 + Pd^0 + 2\ HCl$$

This reaction uses the palladium salt as a stoichiometric oxidant, but the reaction becomes catalytic in palladium when it is carried out under oxygen pressure [18]. The stoichiometric coupling to form biphenyl is ordinarily accompanied by phenyl acetate formation, but acetoxylation is almost completely suppressed by 50 atmospheres oxygen pressure. Heteropolymolybdate ions serve as catalysts to couple the palladium and O_2 redox systems [19,20]. With a $Pd(OAc)_2/Hg(OAc)_2/H_5(Mo_{10}V_2PO_{40})$ catalyst, toluene is converted to dimethylbiphenyls with good rates and yields at 1.5 atmospheres pressure and 50-90°C.

The oxidative coupling occurs with a variety of aromatic compounds. Much of the industrial interest in this reaction lay in its potential use to couple toluene to 4,4'-dimethylbiphenyl and o-xylene to 3,4,3',4'-tetramethylbiphenyl [18]. These compounds are possible precursors of biphenyldi- and tetracarboxylic acids, which yield polyamides and polyimides with interesting physical properties. Ube Industries makes Upirex®, a high-performance polyimide, from the tetracarboxylic acid obtained by oxidative coupling of dimethyl phthalate to the tetraester:

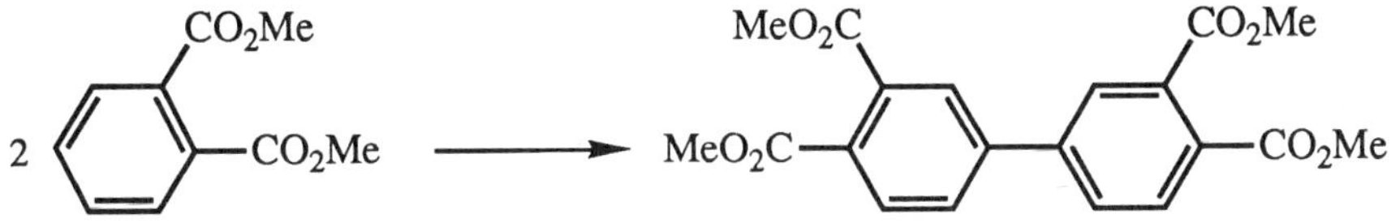

Neat dimethyl phthalate is treated with a palladium(II) chelate, copper(II) acetate and oxygen. The reaction may be conducted at atmospheric pressure [21]. With the 1,10-phenanthroline complex of $Pd(OAc)_2$, 93-94% selectivity to the desired isomer is obtained at approximately 10% conversion. When the reaction is carried out under more severe conditions (160-180°C, 10 atmospheres pressure) to get commercially acceptable reaction rates, the selectivity is about 82% at 9% conversion [22].

As in the palladium-catalyzed arene-olefin coupling, it seems likely that a π-arene-palladium compound **4** forms in the initial reaction of benzene with a palladium (2+) salt [7-9], but the subsequent C-C bond formation mechanism is unclear. Kinetic studies [13] suggest formation of intermediate phenylpalladium **5** and diphenylpalladium **6** intermediates which lead to coupling by reductive elimination of two C-Pd bonds (upper sequence in Figure 7.3). On the other hand, product studies in the toluene and xylene couplings are more consistent with a simple radical-ion mechanism [13] initiated by electron transfer from arene to metal.

Oxidative Substitution

Arenes react with palladium (2+) salts in the presence of anionic nucleophiles to form substitution products. The most studied reaction is acetoxylation [7-9,13].

$$C_6H_6 + Pd^{2+} + HOAc \longrightarrow C_6H_5OAc + Pd^0 + H^+$$

The acetoxylation was of interest as a potential phenol synthesis because phenyl acetate is easily hydrolyzed to phenol and acetic acid. An analogous oxidation of phenyl acetate has been studied as a route to the acetates of catechol, resorcinol, and hydroquinone [23]. The acetoxylation of benzene can be made catalytic in palladium by addition of inorganic oxidants such as $K_2Cr_2O_7$ [13,24], but it is repressed by oxygen. The best results in a catalytic sense result from use of a heterogeneous catalyst, just as in the closely analogous vinyl acetate synthesis (Section 6.3). Nearly quantitative yields of phenyl acetate and phenol are obtained by passing benzene and acetic acid vapors in a dilute oxygen stream over a supported palladium metal catalyst at 130-190°C [25]. Recently, it has been reported that $Pd(OAc)_2$ as a phenanthroline complex catalyzes the direct oxidation of benzene to phenol under an atmosphere of CO and O_2 (15 atmospheres of each) [26].

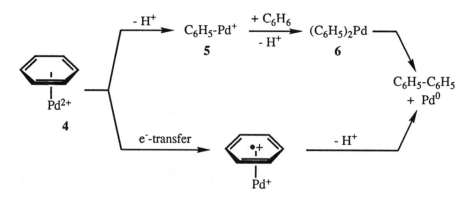

Figure 7.3 Alternative pathways for arene-arene coupling.

Palladium(II) trifluoroacetate reacts readily with electron-rich arenes such as anisole and toluene to form trifluoroacetoxy derivatives with preference for ortho and para substitution [4]. The reaction can be made catalytic in palladium by use of $K_2S_2O_8$ as a cooxidant, just as was done earlier in acetoxylations with $Pd(OAc)_2$ in acetic acid [27]. The reaction is moderately fast and clean with a range of arene substituents from CH_3 to CO_2CH_3. Curiously, in the acetoxylation [27], meta products predominate even with toluene, for which electrophilic attack by Pd^{2+}, might be expected to give ortho and para isomers. A possible explanation for this phenomenon and for the delicate balance between substitution and coupling involves competitive reactions of an arene π-complex **4** as in arene-arene coupling [28]. An alternative explanation is that the palladation of arenes to form σ-arylpalladium species like **5** in Figure 7.3 is simply a nonselective aromatic substitution [13].

Oxidative Carbonylation

The reaction of palladium(II) salts with CO and an arene is a potentially interesting synthesis of arylcarboxylic acid derivatives. Most attention has been given to carbonylation of arylmercury [29,30] and arylthallium [31] compounds in the presence of palladium acetate. Presumably, metal-metal exchange forms an arylpalladium complex which carbonylates readily [7,9]:

$$Ar\text{-}M + Pd^{2+} \longrightarrow \begin{matrix} Ar\text{–}Pd^+ \\ + M^+ \end{matrix} \xrightarrow{CO} \begin{matrix} Ar\text{—}\underset{\displaystyle \underset{O}{\|}}{C}\text{—}Pd^+ \end{matrix}$$

$$M = HgX \text{ or } TlX_2$$

The acylpalladium species thus formed reacts with alcohols to give esters. Particular attention has been given to carboxylation of toluene by this route because the metallation is highly para-specific. Both the mercury and thallium reactions are said to yield over 90% methyl para-toluate, an intermediate in terephthalic acid synthesis for polyester manufacture (Chapters 10 and 11). Difficulties in reoxidizing the reduced metal salts have inhibited industrial use of this chemistry.

For laboratory syntheses of benzoic acid derivatives, a closely related carbonylation of aryl halides may be useful [32]. Palladium complexes such as $PdBr_2(PPh_3)_2$ catalyze the reaction of bromobenzene with CO and butanol under mild conditions (100°C, 1 atmosphere):

$$C_6H_5Br + CO + BuOH + Bu_3N \longrightarrow C_6H_5CO_2Bu + Bu_3NH\,Br$$

With bromo- and iodoarenes, yields are high and many kinds of functional groups are tolerated by the catalyst. The mechanism of the reaction is not clearly established. It seems likely that the palladium(II) complex is reduced by CO or the alcohol. The resulting zero-valent complex reacts with the aryl halide to form an arylpalladium complex:

$$PdBr_2(PPh_3)_2 \xrightarrow[CO]{[H]} Pd(CO)(PPh_3)_2 \xrightarrow[-Ph_3P]{ArX} Ar-\underset{\underset{PPh_3}{|}}{\overset{\overset{CO}{|}}{Pd}}-X$$

Carbonylation of the arene can then occur by mechanisms like those described in Chapter 5.

7.3   COPPER-CATALYZED OXIDATIONS

Copper(II) salts catalyze several synthetically useful oxidations of aromatic compounds [33]. Two have been used industrially on a moderate scale. An oxidative decarboxylation of benzoic acid yields phenol. Oxidative coupling of 2,6-disubstituted phenols produces polymers or quinoid dimers. These reactions differ in mechanism from the analogous oxidative substitution and coupling reactions with palladium catalysts discussed above. The palladium-catalyzed oxidations seem to involve organometallic intermediates like the olefin oxidations of Chapter 6. The copper-catalyzed reactions are commonly described as free-radical processes like those of Chapter 10, although organocopper intermediates may be present.

Decarboxylation

Copper salts catalyze both decarboxylation and oxidative decarboxylation of benzoic acid and its derivatives:

$$C_6H_5CO_2H \begin{cases} \xrightarrow{Cu^I} C_6H_6 + CO_2 \\ \xrightarrow{O_2, Cu^{II}} C_6H_5OH + CO_2 \end{cases}$$

The simple decarboxylation is often used in organic synthesis [34] and the oxidative decarboxylation has been used commercially for manufacture of phenol [35]. In both reactions, a copper(I) salt catalyzes CO_2 elimination but, in the oxidative process, copper(II) also plays an important part.

The simple decarboxylation is very clean with arene carboxylic acids. When cuprous benzoate is heated above 200°C in a high-boiling solvent such as quinoline, benzene is formed in 99% yield [34]. The salt need not be preformed, but can be prepared in situ by reaction with CuO_2CCH_3, CuO_2CCF_3, or an arylcopper(I) complex. The copper(I) compound may be used in catalytic quantities and is quite tolerant of other functional groups. For example, 0.1 equivalent of $[CuC_6F_5]_4$ catalyzes the decarboxylation of an indolecarboxylic acid in high yield:

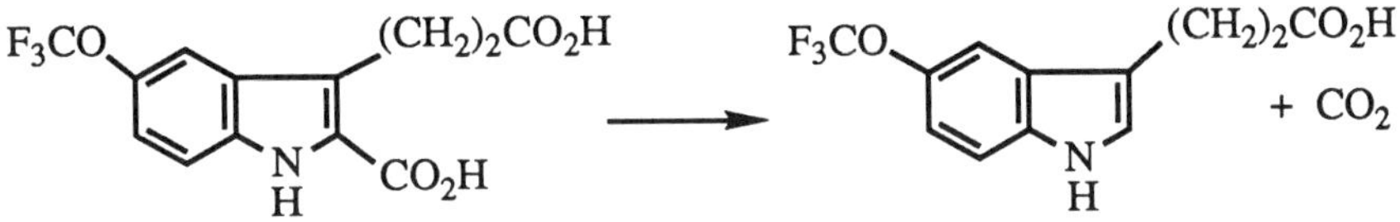

Copper(II) benzoate, in contrast to the Cu(I) salt, undergoes oxidative decarboxylation. This reaction is the basis for phenol syntheses developed by Dow [35] and by Lummus [36]. Steam and air are blown through a solution of copper(II) and magnesium salts in molten benzoic acid at 230-240°C [37]. Carbon dioxide evolves rapidly and phenol distills from the mixture in about 80% yield.

One particularly interesting feature of the oxidative decarboxylation is that the phenolic hydroxyl group occupies a position ortho to that of the original carboxyl group. For example, p-toluic acid yields m-cresol. Similarly, 1-^{14}C-benzoic acid gives 2-^{14}C-phenol [38]. A clue to the origin of the ortho placement of the entering substituent comes from study of the stoichiometric pyrolysis of copper(II) benzoate [39]. Heating this salt in mineral oil at 250°C produces a mixture of copper(I) salts and some free benzoic acid:

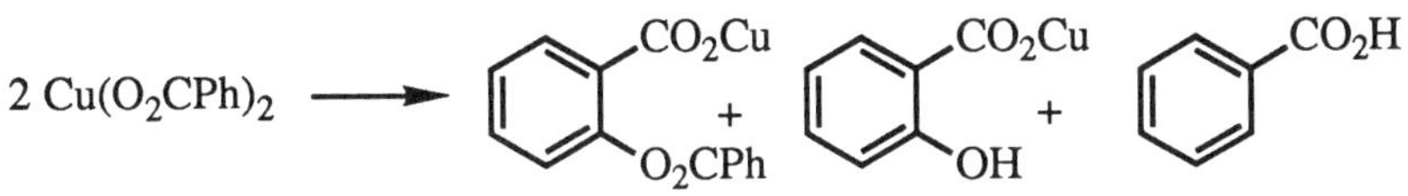

The *ortho*-benzoatobenzoate and the salicylate salts almost certainly result from intramolecular attack on an *ortho*-C-H of the benzene ring. The attack is usually described as the result of two one-electron transfers:

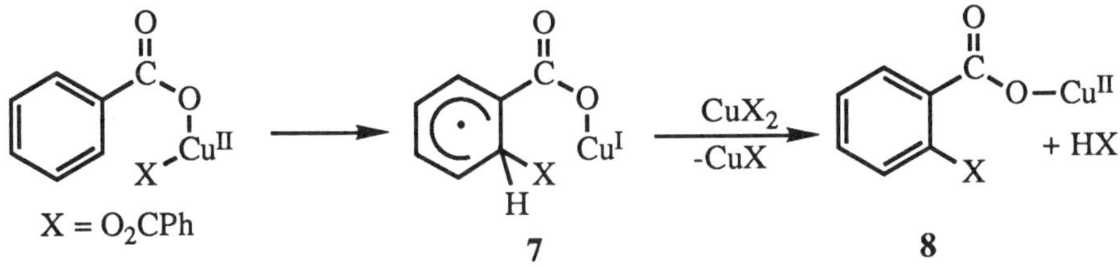

This formulation of the mechanism is based on attack of the ortho position by an incipient benzoate radical. The arenium radical **7** thus formed is oxidized by a second copper(II) ion to produce the observed copper(I) o-benzoatobenzoate **8** (X = PhCO$_2$). No organocopper intermediates are involved in this description of the reaction. The catalytic phenol synthesis is a combination of several reactions as shown in Figure 7.4. The first is the pyrolysis of copper(II) benzoate discussed above. The copper(I) o-benzoatobenzoate decarboxylates rapidly at 230-250°C to form phenyl benzoate. This ester is hydrolyzed under the phenol synthesis conditions to give phenol and benzoic acid. The copper(I) salts are reoxidized by air to regenerate copper(II) benzoate.

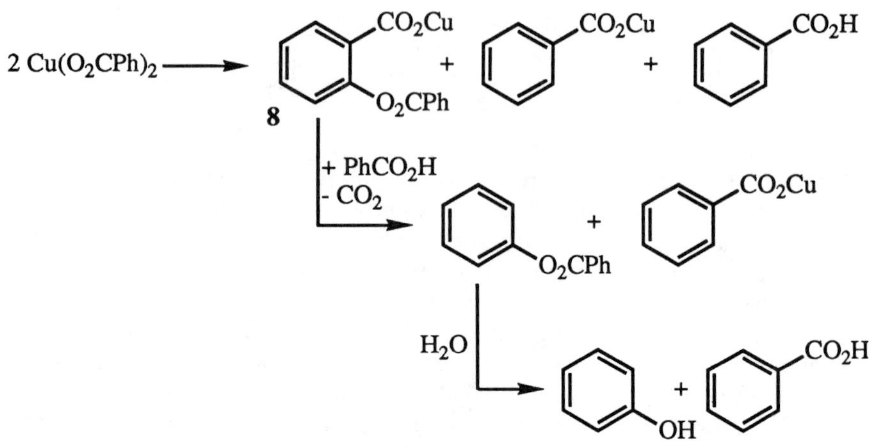

$$2\ CuO_2CPh + 2\ HO_2CPh \xrightarrow{\frac{1}{2} O_2} 2\ Cu(O_2CPh)_2 + H_2O$$

Figure 7.4 Reactions involved in phenol synthesis by oxidative decarboxylation of benzoic acid.

The Dow process is not used extensively for phenol production. In contrast to the major route, based on oxidation of cumene, however, it has the virtue of being based on toluene, which is often much cheaper than benzene. Hence, oxidation of toluene to benzoic acid (Chapter 10) and oxidative decarboxylation can sometimes compete economically with benzene-based phenol syntheses.

Phenol Coupling

The oxidation of phenols by air in the presence of copper(I) salts can take several different pathways as shown in Figure 7.5 [33]. Phenol itself is oxidized by oxygen to *p*-benzoquinone in 80% yield in a reaction catalyzed by copper(I) chloride in acetonitrile [40]. When a pyridine complex of copper(I) chloride is used in methanol solution, a major product is the monomethyl ester of *cis,cis*-muconic acid **10** [41]. The same product is obtained from 1,2-dihydroxybenzene [42]. Probably both oxidations involve *o*-benzoquinone **9** as an intermediate prior to ring cleavage.

The copper(I)-catalyzed oxidations of phenols show considerable ortho-para specificity. When the ortho positions are blocked by alkyl or halo substituents, reaction occurs at the para position, even with the CuCl/pyridine complex as a catalyst. With very bulky ortho substituents such as *tert*-butyl, two phenol molecules couple to form the quinoid dimer **12** shown in Figure 7.5.

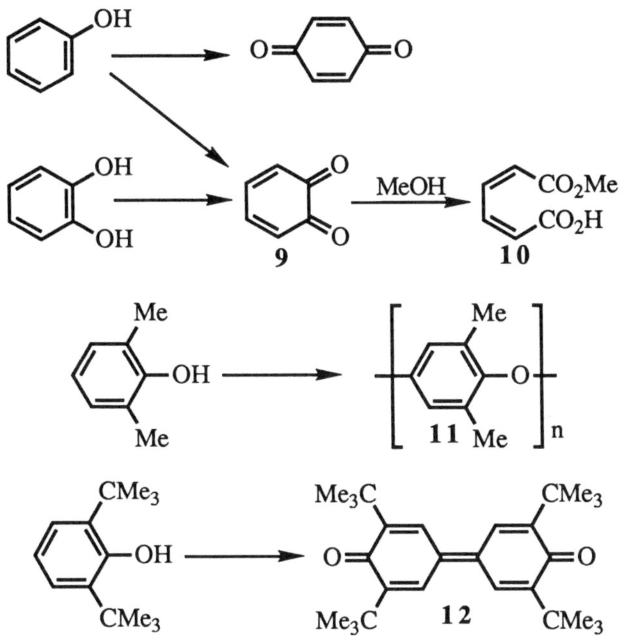

Figure 7.5 Oxidation of phenols by copper(I) chloride/amine/oxygen.

From a technological viewpoint, the most important oxidation in this class is that of 2,6-xylenol. This oxidation produces a *para*-phenylene oxide polymer 11 by coupling an oxygen of one phenol molecule to the para carbon of another. This aromatic polyether is a high melting plastic which is very resistant to heat and to water. It has found wide use as an engineering thermoplastic under the trade name PPO [43]. An even more rigid and higher melting material is obtained by the analogous oxidation of 2,6-diphenylphenol. The all-aromatic character of this material imparts outstanding thermal stability.

The oxidation of 2,6-xylenol is easy. In a semiworks experiment that may simulate commercial practice, 2,6-xylenol is reacted with oxygen in toluene at 40°C in the presence of copper(II) chloride, dibutylamine, NaBr, and a quaternary ammonium dispersing agent [44]. A high molecular weight polymer is formed in about 80 minutes. The polymerization process is reversible in the presence of catalyst. The catalyst is deactivated with a chelating agent or with HCl in order to stabilize the polymer. The latter approach is used in a laboratory synthesis of poly(2,6-dimethyl-1,4-phenylene ether) [45].

The course of the oxidation is sensitive to the amine:copper ratio in the catalyst. High ratios of amine to copper produce the polyphenylene oxide polymer as described above. However, at low ratios, the major product is a quinoid dimer 13 [46]:

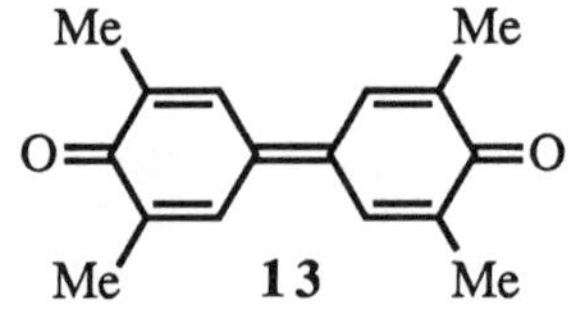

This product resembles the dimer **12** obtained from 2,6-di-*tert*-butylphenol (Figure 7.5). The basis for the divergence in the reaction pathway seems to lie in the interaction of the phenol with the catalytic copper complex.

The reaction of copper(I) chloride with oxygen in pyridine gives a polymeric copper(II) oxide, $[Cu(py)_xO]_n$ [47]. In methanol that contains pyridine, the oxidation gives a copper(II) methoxide, $[CuCl(OMe)(py)_x]$ [48]. This methoxide reacts with phenols to give phenolates which undergo the observed coupling processes.

$$CuCl(OMe) + ArOH \rightleftharpoons CuCl(OAr) + MeOH$$

Without excess pyridine, the preformed complex gives mainly quinoid dimer. It seems likely that the copper(II) phenolate gives rise to a phenolate radical which is still associated with a copper(I) ion:

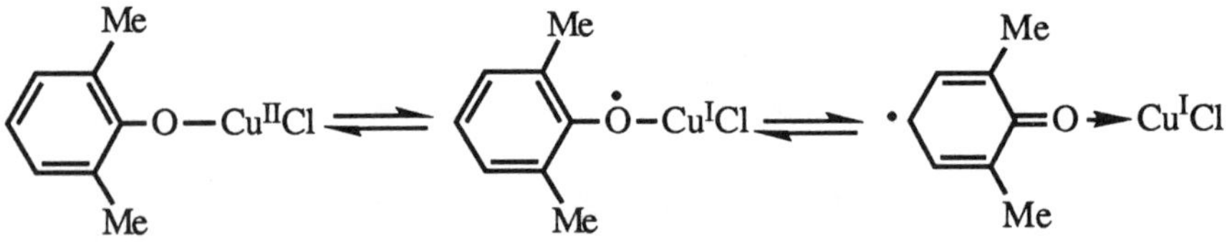

The O-Cu association inhibits O-C interactions between two radicals but leaves a clear path for C-C coupling of the *para*-quinoid form to give the observed dimer **13**.

Heating the CuCl(phenolate) complex with excess pyridine gives mainly phenylene oxide polymer [46]. This reaction is most easily explained by assuming that the pyridine coordinates to copper and displaces the phenolate ligand. The displaced ligand has the properties of a free radical. Indeed, the polymerization process is most conveniently viewed as a coupling of phenolate radicals [49]. Both monomeric and polymeric radicals have been identified in the esr spectra of polymerization mixtures [50]. The simplest chain-growth mechanism is coupling of a monomeric phenolate radical with a similar radical **14** generated from a polymer chain (Figure 7.6). Addition of the polymeric radical to the para position of a monomeric radical forms a coupling product **15** with a cyclohexadienone end group. Tautomerization of the end group to a phenol structure yields the enlarged polymer. The phenol end group can react with copper(II) again to form another polymeric radical that can undergo chain growth by a similar mechanism. Kinetic and product studies [49] indicate that many other processes including coupling of polymeric radicals are important parts of the polymerization process. The undesirable degradative processes require the catalyst deactivation step after polymerization.

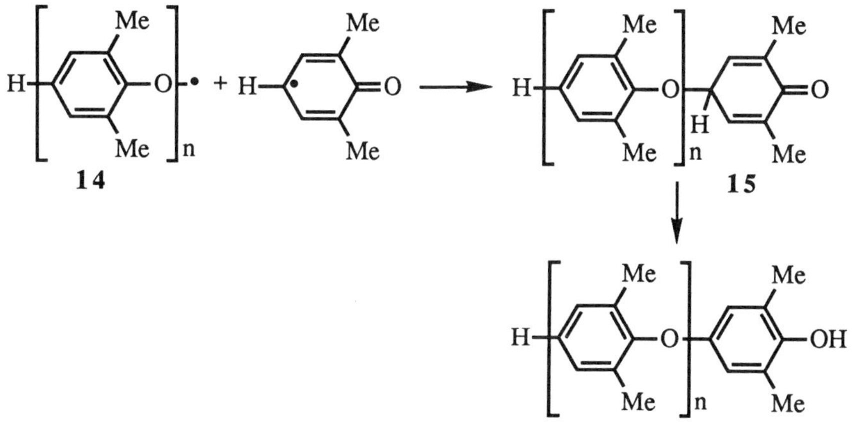

Figure 7.6 A chain-growth step in the copper-catalyzed oxidative polymerization of 2,6-xylenol.

In all the phenol oxidation processes shown in Figure 7.5 it seems likely that a copper(II) species is the actual oxidant even though a copper(I) salt is often the preferred catalyst precursor. The role of copper(II) as a one-electron oxidant can also be filled by a cobalt or manganese complex that bears a tightly bound chelating ligand. In particular, the cobalt complex **16** known as Salcomine or cobalt(salen) is very efficient for the oxidation of phenols to *p*-benzoquinones [33,51].

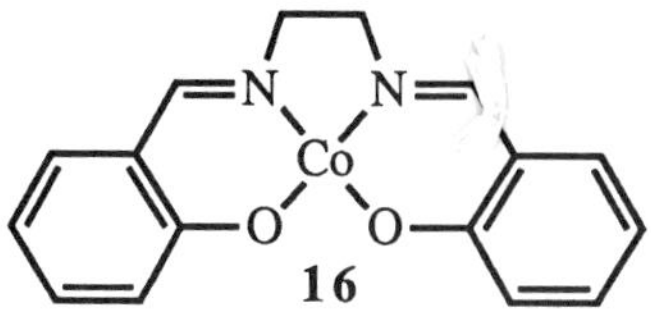

7.4 COUPLING REACTIONS OF ARYL HALIDES

Many reactions of halobenzenes are catalyzed by soluble transition metal complexes. The Ar-X bond is notoriously sluggish toward direct substitution even by powerful nucleophiles such as RS⁻. Traditionally such nucleophilic substitutions have been catalyzed by copper salts, as in the ammonolysis of *p*-chlorobenzotrifluoride to form *p*-aminobenzotrifluoride, an important industrial intermediate [52]. A more broadly useful synthesis tool is the coupling of aryl halides with organometallic compounds such as Grignard reagents. The coupling catalysts or reagents usually are complexes of copper or nickel although many other metals are active. These coupling reactions have been reviewed extensively [53-56].

Salts and complexes of Fe, Co, Ni, Pd, and Cu catalyze the reactions of halobenzenes with Grignard reagents to form alkylbenzenes and biphenyls [56-59]:

$$C_6H_5X + RMgX \longrightarrow C_6H_5\text{---}R + MgX_2$$

This reaction does not proceed well in the absence of the transition metal compound, but it becomes rapid when the proper catalyst is present. Alkyllithium, zinc, and aluminum compounds can often be substituted for the Grignard reagent [56].

Nickel complexes have received the broadest study in this reaction. The most effective catalysts are $NiCl_2(PR_3)_2$ complexes although the yields vary with the nature of the phosphine and of the Grignard reagent [56]. Generally, highest yields are obtained with complexes of chelating phosphines such as $Ph_2P(CH_2)_3PPh_2$ (DPPP). For example, o-dichlorobenzene reacts with n-butylmagnesium bromide in the presence of $NiCl_2(DPPP)$ to form o-dibutylbenzene in about 80% yield [57]:

This product would be difficult to prepare by conventional organic syntheses. Similar yields are obtained with a wide range of n-alkyl and aryl Grignard reagents. However, with sterically hindered aryl Grignard reagents such as 2,4,6-trimethyl-phenylmagnesium bromide, the nonchelate complex $NiCl_2(PPh_3)_2$ is most effective.

Although the coupling reactions are usually straightforward with the catalysts described above, seemingly minor changes in the ligands can change the course of the reaction drastically. This effect is illustrated in the coupling of $(CH_3)_2CHMgCl$ and chlorobenzene with various $NiCl_2L_2$ complexes [57].

Ligand	Product Distribution		
	$i\text{-}PrC_6H_5$	$n\text{-}PrC_6H_5$	C_6H_6
$Ph_2P(CH_2)_3PPh_2$	96	4	0
$Me_2P(CH_2)_2PMe_2$	9	84	7
$(PPh_3)_2$	16	30	54

The chelating phosphine ligand DPPP gives almost entirely the simple coupling product. However, the methyl chelate ligand leads to extensive isomerization. The triphenylphosphine complex produces extensive reduction of the chlorobenzene to benzene.

These seemingly anomalous results can be accommodated by a mechanism proposed for the coupling reaction [56]. The mechanism is illustrated in Figure 7.7. As in the stoichiometric coupling described above, a nickel(0) complex is a key intermediate and provides entry to the catalyst cycle. Oxidative addition of chlorobenzene to NiL_2 gives a σ-phenyl complex **17**. A metathetical reaction of the Grignard reagent gives a complex **18** containing both σ-aryl and σ-alkyl ligands. In the normal coupling process, reductive elimination of the two organic ligands from

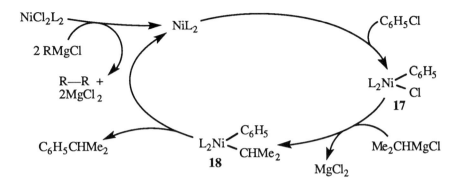

Figure 7.7 A simplified mechanism for the coupling of chlorobenzene and isopropyl-magnesium ($L = R_3P$).

nickel gives isopropylbenzene and NiL_2 to complete the catalytic cycle. However, when the phosphine ligands are small or weakly bound, other reactions can occur. β-Hydrogen elimination from the isopropyl ligand forms a hydrido olefin complex:

$$\text{Ni—CH}\begin{smallmatrix}CH_3\\|\\\\|\\CH_3\end{smallmatrix} \rightleftharpoons \text{Ni—}\begin{smallmatrix}H\\|\\||\\CH_2\end{smallmatrix}CHCH_3 \rightleftharpoons \text{Ni—CH}_2CH_2CH_3$$

As indicated in the equation, the hydrido complex is in equilibrium with both *n*- and *i*-propyl compounds. If isomerization of the alkyl group is faster than reductive elimination of the alkyl and aryl ligands, *n*-propylbenzene is the major product. Reductive elimination of the Ni-H and Ni-Ph bonds in the hydrido complex gives benzene.

The reaction pathways described above account for the observed products simply, but they do not represent a detailed mechanism. They fail to account for the observations that oxidizing agents such as O_2 and ArBr accelerate the coupling reaction and electron acceptors such as nitroarenes inhibit it [60]. The accelerating effect of simple oxidizing and alkylating agents is probably due to depletion of the phosphine ligand which, in turn, opens coordination sites on the metal ion. A more important effect, however, is probably access to odd electron species such as Ni(I) and Ni(III) that accelerate the reductive elimination process [60]. Mechanisms that involve free radicals or one-electron transfers have been suggested for catalysts based on Fe, Co, and Cu complexes.

As with nickel, phosphine complexes of palladium catalyze coupling of halobenzenes with organomagnesium and zinc reagents [58,59]. The high yields and mild conditions of these reactions often justify the use of the precious metal catalysts. The reaction pathways seem similar to those of nickel. Zero-valent complexes such as

$Pd(PPh_3)_4$ are quite effective, but the more stable $PdCl_2(PPh_3)_2$ or $Pd(Ar)(I)(PPh_3)_2$ complexes are more convenient for most purposes [58].

7.5   ARENE HYDROGENATION

Hydrogenation of aromatic compounds with soluble catalysts has received concerted scientific scrutiny only in recent years. Homogeneous catalysts for the hydrogenation of arenes have been known since the early 1950s [61], but they have received little attention because heterogeneous catalysts are extraordinarily effective for this reaction. For the organic laboratory, Adams' catalyst (brown PtO_2) hydrogenates aromatics at 25°C and 3 atmospheres pressure [62]. In commercial practice, palladium-on-carbon or high surface area nickel catalysts are used to hydrogenate benzene to cyclohexane on a very large scale. In recent years, however, two major developments in industrial benzene hydrogenation have occurred.

- A seemingly soluble nickel catalyst developed by the Institute Francais du Petrole (IFP) has been applied extensively in Europe for hydrogenation of benzene to cyclohexane [63].
- The selectivity of ruthenium catalysts for hydrogenation of benzene to cyclohexene has been increased to the point that industrial use is likely. The potentially practical catalysts are heterogeneous, but soluble ruthenium complexes provide useful models for the chemistry.

These developments as well as some work on cobalt-based catalysts are outlined below.

IFP Process

Benzene is hydrogenated to cyclohexane on a scale of millions of tons per year to provide the feedstock for making adipic acid (Chapter 10), a major intermediate in production of nylon. The process is customarily carried out with a Raney nickel heterogeneous catalyst, but the conventional technology is being displaced by IFP's soluble nickel catalyst [63]. Apparently the mechanical and thermal advantages of working with a soluble catalyst rather than a slurry outweigh the problems of catalyst separation and recycle that usually handicap homogeneous catalysts. In this instance, the volatility of cyclohexane facilitates separation of the product from the catalyst.

It has been known for many years that the Ziegler olefin polymerization catalysts (Chapter 4) catalyze the hydrogenation of arenes. Complexes prepared by reaction of triethylaluminum with a cobalt or nickel salt catalyze the hydrogenation of benzene and its derivatives. For example, benzene is reduced to cyclohexane rapidly and quantitatively at 150-190°C and about 75 atmospheres pressure with $Al(C_2H_5)_3$ and $Ni(2\text{-ethylhexanoate})_2$ as the catalyst [64]. Similarly, a combination of $Co(2\text{-ethylhexanoate})_2$ and excess alkylaluminum compound reduces the xylenes to dimethylcyclohexanes [65]. The *cis*-dimethylcyclohexanes are favored over trans by about 2:1, consistent with a predominant cis addition of hydrogen.

preference to cyclohexene. When used in the presence of aqueous NaOH, ruthenium-on-magnesia gives about 50% cyclohexene at moderate conversion of benzene [71]. Recently issued patents [72] and papers [73] point the way to even greater selectivity with aqueous slurries of metallic ruthenium catalysts.

While industrial cyclohexene production will probably use heterogeneous catalysts, research on soluble ruthenium and osmium catalysts may be instructive in understanding the operation of the metallic catalysts. Bis(hexamethylbenzene)-ruthenium(0) hydrogenates benzene rapidly at 90°C and 2-3 atmospheres pressure [74]. The reaction resembles that catalyzed by ruthenium metal in that substantial amounts of cyclohexene are formed (40-55% dimethylcyclohexenes from the xylenes). The bis(hexamethylbenzene)ruthenium catalyst differs in several respects from the allyl cobalt catalysts discussed above, but is said to operate by a similar mechanism. It differs in giving cyclohexene as a substantial product and in producing extensive H/D exchange when D_2 is the reducing agent. When xylene is treated with D_2, deuterium appears in the methyl groups of the unreduced xylene. In addition, the hexamethylbenzene ligands of recovered catalyst undergo methyl H/D exchange [74]. It was proposed that this exchange occurred via a π-benzyl intermediate. The bis(hexamethylbenzene)ruthenium(0) is interesting in that one ligand is symmetrically complexed (η^6) but the other is coordinated through only two C=C bonds (η^4) [75].

Recent research on the hydrogenation of arene complexes of osmium [76] sheds new light on the selective hydrogenation process. Osmium, like ruthenium, binds arenes tightly even when only one or two localized C=C bonds are coordinated to the metal. η^2-Arene complexes such as the anisole complex **19** have been characterized crystallographically [77]. When this complex is hydrogenated with a heterogeneous rhodium catalyst [76], the two noncoordinated double bonds are reduced and a methoxycyclohexene complex **20** is isolated:

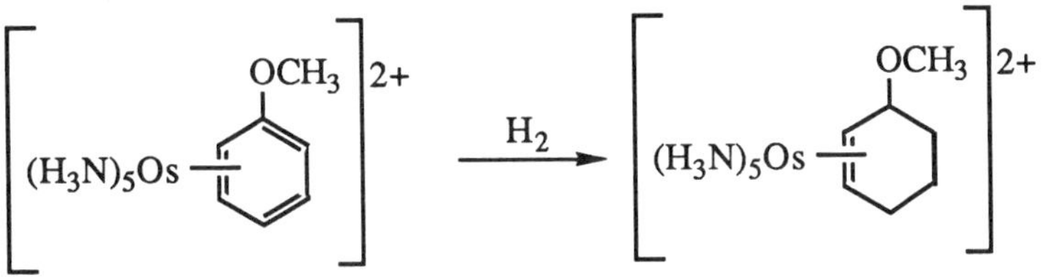

A reasonable extrapolation to the selective hydrogenation process observed with metallic ruthenium catalysts is that a benzene ring may bind η_2 to a surface ruthenium atom. In doing so, the coordinated double bond is protected from hydrogenation by other ruthenium sites that activate H_2 (or by "spillover hydrogen"). The uncoordinated double bonds are unprotected and are reduced to give a coordinated cyclohexene. A key factor in selectivity may be the competition for coordination sites between arene and cyclohexene. This data is not available for the actual hydrogenation catalysts, but further studies of ruthenium and osmium complexes may be enlightening.

7.6   AROMATIC INTERMEDIATES FOR SPECIALTY CHEMICALS

Standing in contrast to the large-scale processes for the oligomerization of phenols and the hydrogenation of benzene are some homogeneous catalytic reactions that are used to produce specialty chemicals on a scale of less than 5000 tons per year (in North America). These reactions are particularly important in the production of dyes, colored pigments, and agrichemicals [78]. Most have been developed empirically and have received little mechanistic study.

Metal Ion-Directed Ortho Substitution

One especially valuable property of metal ions is to steer an electrophile or a nucleophile to a position ortho to a substituent on a benzene ring. This effect is seen with widely varying metal ions from d_0 (Al^{+3}, Zr^{+4}) to d_{10} (Hg^{+2}). This family of reactions embraces a wide range of mechanisms, but it is generally assumed that the metal ion coordinates to an electron-donating substituent such as -OH or -NH$_2$. A second coordination site on the metal ion engages a reagent such as an olefin or a disulfide and steers it to the sterically accessible ortho site on the arene ring. This effect is illustrated in Figure 7.9, which outlines a speculative mechanism for the ortho-ethylation of aniline [79]. Aluminum alkyls react with aniline to form the trianilide, which contains Al-N bonds [80]. In the postulated mechanism, interaction of filled π-orbitals of an ethylene molecule with a vacant p-orbital on aluminum guides the olefin into a location in which a series of electron pair migrations **21** create a new

$$3\ PhNH_2\ +\ AlR_3\ \longrightarrow\ (PhNH)_3Al\ +\ 3\ RH$$

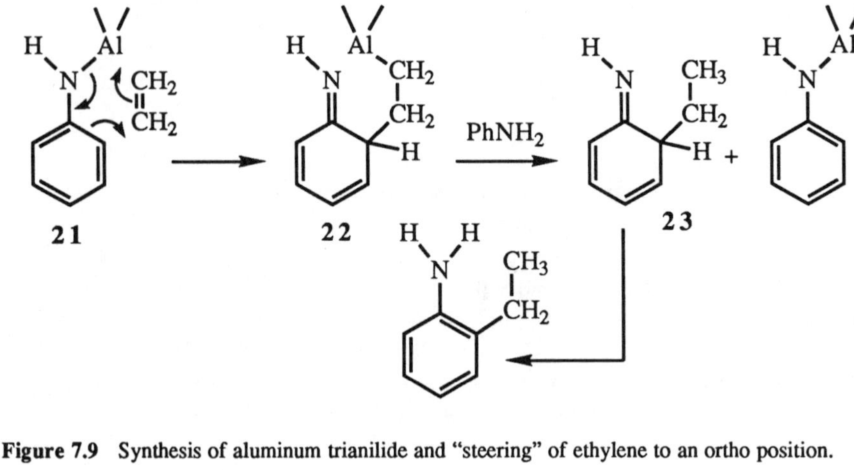

Figure 7.9   Synthesis of aluminum trianilide and "steering" of ethylene to an ortho position.

C-C bond ortho to the nitrogen substituent on the benzene ring. Subsequent aminolysis of the Al-C bond in **22** leads to regeneration of an aluminum anilide and formation of intermediate **23**, which subsequently tautomerizes to *ortho*-ethylaniline. While the postulated mechanism accounts for the observed products, it is tempting to propose an alternative in which the aluminum ion migrates from the nitrogen of the anilide to the ortho position on the benzene ring. Insertion of ethylene into the ortho C-Al bond, a well-documented reaction, would yield **22**, which then proceeds as shown in Figure 7.9 to give *ortho*-ethylaniline.

Whatever the mechanism, the ortho-ethylation of anilines has become an important industrial process for the manufacture of the widely used herbicides, Lasso® and Dual®. A critical intermediate for the production of Lasso® is 2,6-diethylaniline,

which is produced by reaction of aniline with ethylene in the presence of aluminum trianilide [79-81]. The catalyst can be generated in situ by reaction of triethylaluminum or activated aluminum turnings with excess aniline. The solution of aluminum trianilide in aniline is heated with ethylene at 300-400°C and 30-65 atmospheres pressure. Distillation of the reaction mixture gives over 90% 2,6-diethylaniline and leaves a residue of aluminum trianilide, which can be used as a catalyst for a subsequent reaction batch. A similar process is used to convert *o*-toluidine to 2-ethyl-6-methylaniline, a key for production of Dual®.

Chemistry closely analogous to that of aniline ethylation is used to produce 2,6-di-*t*-butylphenol, which is widely used in formulating antioxidants and ultraviolet stabilizers for polymers [82-85]. The reaction of isobutene with phenol in the presence of a protonic acid catalyst ordinarily gives a mixture of *t*-butyl phenyl ether and *para*-*t*-butylphenol along with minor quantities of *ortho*-*t*-butylphenol. The course of the reaction changes dramatically when aluminum phenoxide is used as the catalyst. A solution of $Al(OC_6H_5)_3$, prepared by dissolving aluminum turnings in molten phenol, reacts with isobutylene (ca. 2:1 olefin:phenol) at 100°C under autogenous pressure to give approximately 75% conversion to 2,6-di-*t*-butylphenol along with 9% each of 2-*t*-butylphenol and 2,4,6-tri-*t*-butylphenol. No significant quantity of the *para*-*t*-butylphenol is formed. The reaction can be driven through the use of excess isobutene to produce largely 2,4,6-tri-*t*-butylphenol, which is also useful as a component of stabilizing formulations for polymers.

In recent years, another ortho substitution of phenol has received attention as a potential process for making agrichemical intermediates. Many of Du Pont's sulfonylurea herbicides, such as chlorosulfuron, are characterized by strong species selectivity in controlling weeds (in addition to their innocuous relationship to life forms other than plants).

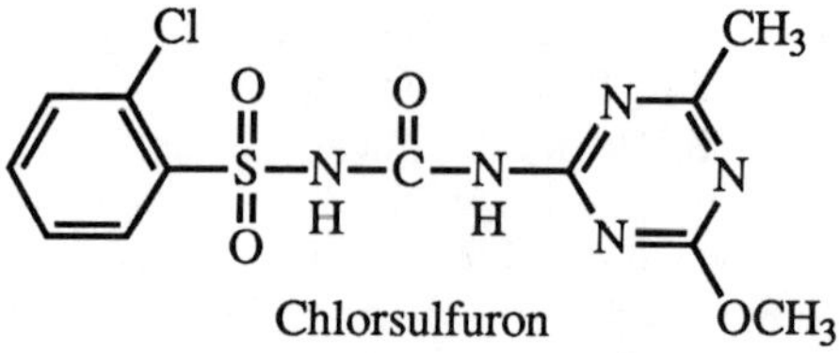

Chlorsulfuron

The selectivity for particular plant species is determined both by the nature of the nitrogen heterocycle at one end of the molecule and the substituents on the arene at the other end. It is often desirable to introduce a substituent ortho to the sulfonyl group on the benzene ring.

One approach to placing a hydroxyl group and a sulfur side by side on a benzene ring is the *ortho*-alkylthiolation of phenol. The acid-catalyzed reaction of phenol with dialkyldisulfides generally gives a mixture of ortho and para substituted products:

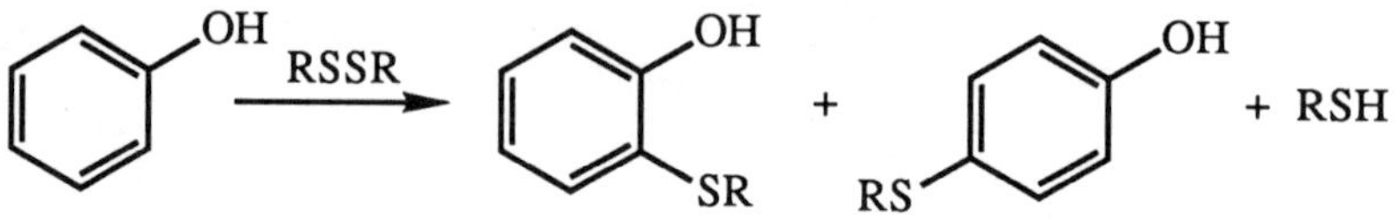

It has been observed [86] that aluminum phenoxide catalyzes the alkylthiolation of phenol with greater ortho selectivity than that attained with simple Bronsted or Lewis acid catalysts. For example, an aluminum phenoxide solution, prepared by in situ reaction of aluminum powder with excess phenol, reacts with dimethyl disulfide at 123-170°C to produce 2-(methylthio)phenol in 40% yield after distillation. The crude reaction mixture contains the *ortho*- and *para*-methylthiophenols and two bis(thiophenols) in a ratio of 17:7:3:1. More recently it has been reported [87] that zirconium phenoxide is effective in catalyzing the ortho methylthiolation of phenol. With both the aluminum and zirconium catalysts, it is plausible to suggest that the metal ion "steers" the dialkyl disulfide reagent to the ortho position on the phenol ring through an assemblage such as

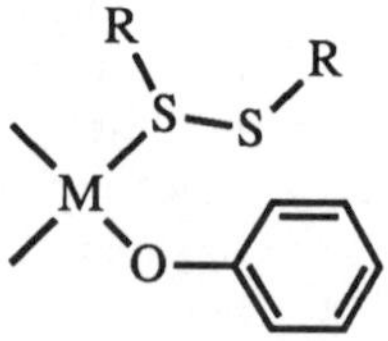

Another ortho-substitution process of industrial importance is the mercury(II)-catalyzed sulfonation of anthraquinone to produce anthraquinone-1-sulfonic acid, a major intermediate in making anthraquinone dyes, the second largest class of textile dyes. The effect of Hg^{2+} ion as a catalyst in this process is dramatic. The uncatalyzed reaction of anthraquinone with oleum containing 45% SO_3 at 150°C produced the 2-sulfonic acid almost exclusively. When the reaction is carried out in the presence of a small amount of a mercuric salt, the product is primarily anthraquinone-1-sulfonic acid [88].

Continued reaction leads to introduction of a second sulfonic acid group. The second group is directed to an ortho position (5 or 8) of the other benzenoid ring. In mercury-catalyzed reactions such as these, it is likely that the first step is mercuration of the arene ring to form a C-Hg bond. Subsequent reaction of the aryl-Hg function with another reagent (e.g., SO_3) places the incoming substituent at the site of the initial attack by mercuric ion. In the anthraquinone example, it is possible that precoordination of the metal ion to the carbonyl group directs the Hg^{2+} to attack the nearest C-H function.

The Hg^{2+}-catalyzed sulfonation of anthraquinone is valuable industrially because the SO_3H group in the 1-sulfonic acid is displaceable by OH^- and NO_2^- to form the 1-hydroxy- and 1-nitroanthraquinones. The latter can be hydrogenated to 1-aminoanthraquinone, a versatile dye precursor. The 1,5-anthraquinonedisulfonic acid, prepared as described above, can be converted similarly to 1,5-diaminoanthraquinone. This intermediate was used in making the now-obsolete dye, Vat Yellow 3 (1,5-dibenzamidoanthraquinone) [88,89].

Despite the chemical selectivity conveyed by mercury(II) catalysis, the use of this technology has declined due to health and safety problems related to the extreme toxicity of mercury(II) compounds. Other less toxic metals have been explored in seeking alternative catalysts. Palladium(II) salts, which are often involved in ortho-metallation chemistry [90,91], are reported to catalyze specific α-sulfonation of anthraquinone [92]. The reaction conditions are somewhat different from those used with mercury(II) catalysts. Both palladium(II) sulfate in tetramethylene sulfone and palladium(II) acetate in liquid SO_2 catalyze the sulfonation of anthraquinone by SO_3 to form anthraquinone-1-sulfonic acid.

SPECIFIC REFERENCES

1. G. A. Olah, *Friedel Crafts Chemistry*, Wiley Interscience, New York, 1973.
2. A. C. MacFarlane, "Monsanto's Ethylbenzene Process," in L. F. Albright and A. R. Goldsby, Eds., *Industrial and Laboratory Alkylations*, American Chemical Society Symposium Series, Vol 55, American Chemical Society, Washington, DC, 1977, pp. 341-59.
3. G. W. Parshall, *Catalysis*, 1, 334-68 (1977).
4. A. Sen, E. Gretz, T. F. Oliver, and Z. Jiang, *New J. Chem.*, 13, 755-60 (1989).
5. R. R. Schrock and G. W. Parshall, *Chem. Rev.*, 76, 243-68(1976); G. W. Parshall and J. J. Mrowca, *Adv. Organomet. Chem.*, 7, 157-209 (1968).
6. D. R. Coulson, *J. Am. Chem. Soc.*, 98, 3111-9 (1976).
7. R. F. Heck, *Palladium Reagents in Organic Synthesis*, Academic Press, New York, 1985.

8. S. F. Davison and P. M. Maitlis, "Oxidations Using Palladium Compounds," in W. J. Mijs and C. R. H. I. de Jonge, Eds., *Organic Syntheses by Oxidation with Metal Compounds*, Plenum Press, New York, 1986, pp. 469-502, esp. pp. 488-493.

9. P. M. Henry, *Palladium-Catalyzed Oxidations of Hydrocarbons*, Reidel, Dordrecht, 1980, pp. 306-338.

10. Y. Fujiwara, I. Moritani, S. Danno, R. Asano, and S. Teranishi, *J. Am. Chem. Soc.*, **91**, 7166-9 (1969).

11. R. S. Shue, *J. Catal.*, **26**, 112-7 (1972).

12. I. Moritani and Y. Fujiwara, *Synthesis*, 524-33 (1973).

13. L. M. Stock, K-T. Tse, L. J. Vorvick, and S. A. Walstrum, *J. Org. Chem.*, **46**, 1757-59 (1981).

14. R. F. Heck, *J. Am. Chem. Soc.*, **91**, 6707-14 (1969).

15. I. Moritani, Y. Fujiwara, and S. Danno, *J. Organomet. Chem.*, **27**, 279-82 (1971).

16. J. E. Plevyak and R. F. Heck, *J. Org. Chem.*, **43**, 2454-6 (1978).

17. R. van Helden and G. Verberg, *Rec. Trav. Chim. Pays-Bas*, **84**, 1263-73 (1965).

18. H. Itatani and H. Yoshimoto, *J. Org. Chem.*, **38**, 76-9 (1973); *Bull. Chem. Soc., Jpn.*, **46**, 2490-2 (1973); M. Kashima, H. Yoshimoto, and H. Itatani, *J. Catal.*, **29**, 92-8 (1973).

19. A. I. Rudenkov, G. U. Mennenga, L. N. Rachkovskaya, K. I. Matveev, and I. V. Kozhevnikov, *Kinet. Katal.*, **18**, 915-20 (1977).

20. I. V. Kozhevnikov, V. I. Kim, E. P. Talzi, and V. N. Sidelnikov, *J. Chem. Soc., Chem. Commun.* 1392-94 (1985); *Izv. Akad. Nauk SSSR, Ser. Khim.*, 2167-74 (1985).

21. A. Shiotani, H. Itatani, and T. Inagaki, *J. Mol. Catal.*, **34**, 57-66 (1986).

22. H. Itatani, A. Shiotani, and A. Yokota, U.S. Patents 4,292,435 (1981) and 4,338,456 (1982).

23. J. E. Lyons and C-Y. Hsu, Biol. Inorg. Copper Chem., Proc. Conf. Copper Coord. Chem., 1984 (Pub. 1986) **2**, 57-76; *Chem. Eng. News*, 32 (30 April 1984).

24. P. M. Henry, *J. Org. Chem.*, **36**, 1886-90 (1971).

25. L. Hornig and T. Quadflieg, U.S. Patent 3,642,873 (1972).

26. T. Jintoku, K. Takaki, Y. Fujiwara, Y. Fuchita, and K. Hiraki, *Bull. Chem. Soc. Jpn.*, **63**, 438-41 (1990).

27. L. Eberson and L. Jonsson, *Acta Chem. Scand.*, B30, 361-4 (1976); T. Itahara, *Chem. Ind.*, 599-600 (1982).

28. L. Eberson and L. Gomez-Gonzales, *Acta Chem. Scand.*, **27**, 1255-67 (1973).

29. P. M. Henry, *Tetrahedron Lett.*, 2285-7 (1968).

30. W. C. Baird, R. L. Hartgerink, and J. H. Surridge, Ger. Offenleg. 2,310,629 (1973).

31. J. J. van Venrooy, U.S. Patent 4,093,647 (1978).

32. A. Schoenberg, I. Bartoletti, and R. F. Heck, *J. Org. Chem.*, **39**, 3318-27 (1974).

33. C. R. H. I. de Jonghe, "Oxidations of Organic Compounds Catalyzed by Copper- and Cobalt-Amine Complexes," in W. J. Mijs and C. R. H. I. de Jonghe, Eds. *Organic Syntheses by Oxidation with Metal Compounds*, Plenum Press, New York, 1986, pp. 423-443.

34. A. Cairncross, J. R. Roland, R. M. Henderson, and W. A. Sheppard, *J. Am. Chem. Soc.*, **92**, 3187-9 (1970).

35. W. W. Kaeding, *Hydrocarbon Proc.*, **43**, 173-6 (Nov. 1964).

36. A. Gelbein and A. S. Nislick, *Hydrocarbon Proc.*, 125-8 (Nov. 1978).

37. W. W. Kaeding, R. O. Lindblom, and R. G. Temple, U.S. Reissue Patent 24,848 (1960).

38. W. Schoo, J. U. Veenland, J. A. Bigot, and F. L. J. Sixma, *Rec. Trav. Chim. Pays-Bas*, **80**, 134-8 (1961).

39. W. W. Kaeding and G. R. Collins, *J. Org. Chem.*, **30**, 3750-3, 3754-9 (1965).

40. E. L. Reilly, U.S. Patent 3,987,068 (1976); U.S. Patent 4,257,968 (1981).

41. M. M. Rogic and T. R. Demmin, *J. Am. Chem. Soc.*, **100**, 5472-87 (1978).

42. T. R. Demmin, M. D. Swerdloff, and M. M. Rogic, *J. Am. Chem. Soc.*, **103**, 5795-5804 (1981); D. Bankston, *Org. Syn.*, **66**, 180-3 (1988).

43. A. S. Hay, *Poly. Eng. Sci.*, **16**, 1-10 (1976); A. S. Hay and D. M. White, "Polymerizations by Oxidative Coupling," in C. E. Schildknecht and I. Skeist, Eds., *Polymerization Processes*, Wiley-Interscience, New York, 1977, pp. 537-581; *Chem. Eng. News*, 6 (15 Feb. 1982).

44. G. D. Cooper and D. E. Floryan, Ger. Offenleg. 2,754,887 (1978); *Chem. Abstr.* **89**, 75647 (1978).

45. A. S. Hay, H. S. Blanchard, G. F. Endres, and J. W. Eustance, *Macromol. Synth.*, *Coll. I*, 81-4 (1977).

46. S. Tsuruya, T. Kuse, M. Masai, and S. I. Imamura, *J. Mol. Catal.*, **10**, 285-303 (1981).

47. I. Bodek and G. Davies, *Inorg. Chem.*, **17**, 1814-8 (1978).

48. H. Finkbeiner, A. S. Hay, H. S. Blanchard, and G. F. Endres, *J. Org. Chem.*, **31**, 549-55 (1966).

49. G. D. Cooper and A. Katchman, "Polyphenylene Oxides by Oxidative Coupling," in N. A. J. Platzer, Ed., *Addition and Condensation Polymerization Processes, Advances in Chemistry Series*, **91**, American Chemical Society, Washington, DC, 1969, pp. 660-78.

50. W. G. B. Huysmans and W. A. Waters, *J. Chem. Soc. B*, 1163-9 (1967).

51. C. R. H. I. de Jonge, H. J. Hageman, G. Hoentjen, and W. J. Mijs, *Org. Synth. Coll. VI*, 412-3 (1988).

52. L. P. Seiwell, *J. Org. Chem.*, **44**, 4731-3 (1979).

53. G. Bringmann, R. Walter, and R. Weirich, *Angew. Chem., Int. Ed.* **29**, 977-991 (1990).

54. R. Noyori, "Coupling Reactions via Transition Metal Complexes," in H. Alper, Ed., *Transition Metal Organometallics in Organic Synthesis*, Academic Press, New York, 1976, pp. 83-189.

55. J. K. Kochi, "Coupling of Alkyl Groups Using Transition Metal Catalysts," in L. F. Albright and A. R. Goldsby, Eds., *Industrial and Laboratory Alkylations*, American Chemical Society Symposium Series, Vol. 55, American Chemical Society, 1977, pp. 167-85.

56. M. Kumada, *Pure Appl. Chem.*, **52**, 669-79 (1980).

57. M. Kumada, K. Tamao, and K. Sumitani, *Org. Synth., Coll. VI*, 407-11 (1988).

58. A. Sekiya and N. Ishikawa, *J. Organomet. Chem.*, **118**, 349-54 (1976).

59. E-I. Negishi, T. Takahashi, and A. O. King, *Org. Syn.*, **66**, 67-74 (1988).

60. T. T. Tsou and J. K. Kochi, *J. Am. Chem. Soc.*, **101**, 7547-60 (1979).

61. I. Wender, R. Levine, and M. Orchin, *J. Am. Chem. Soc.*, **72**, 4375-8 (1950).

62. R. Adams and J. R. Marshall, *J. Am. Chem. Soc.*, **50**, 1970-3 (1928).

63. Y. Chauvin, J. Gaillard, J. Leonard, P. Bonnifay, and J. W. Andrews, *Hydrocarbon Proc.*, 110-2 (May 1982).

64. S. J. Lapporte and W. R. Schuett, *J. Org. Chem.*, **28**, 1947-8 (1963).

65. G. Bressan and R. Broggi, *Chim. Ind. (Milan)*, **50**, 1194-9 (1968).

66. D. Durand, G. Hillion, C. Lassau, and L. Sajus, U.S. Patent 4,271,323 (1981).

67. J. M. Corker and J. Evans, *J. Chem. Soc., Chem. Commun.*, 1104-5 (1991).

68. F. K. Schmidt, Y. S. Levkovskii, V. V. Saraev, N. M. Ryutina, O. L. Kosinski, and T. I. Bakunina, *React. Kinet. Catal. Lett.*, **7**, 445-50 (1977).

69. L. S. Stuhl, M. Rakowski Du Bois, F. J. Hirsekorn, J. R. Bleeke, A. E. Stevens, and E. L. Muetterties, *J. Am. Chem. Soc.*, **100**, 2405-10 (1978).

70. J. Blum, I. Amer, K. P. C. Vollhardt, H. Schwarz, and G. Hohne, *J. Org. Chem.*, **52**, 2804-13 (1987).

71. W. C. Drinkard, U.S. Patent 3,767,720 (1973).

72. H. Nagahara and M. Konishi, U.S. Patent 4,734,536 (1988).

73. S-I. Niwa, F. Mizukami, M. Toba, T. Tsuchiya, K. Shimizu, and J. Imamura, *Shokubai*, **31**, 421-4 (1989); Y. Fukuoka and H. Nagahara, *202nd National Meeting, American Chemical Society*, New York, 25 Aug. 1991.

74. J. W. Johnson and E. L. Muetterties, *J. Am. Chem. Soc.*, **99**, 7395-6 (1977).

75. G. Huttner and S. Lange, *Acta Cryst.*, **B28**, 2049-60 (1972).

76. W. D. Harman, W. P. Schaefer, and H. Taube, *J. Am. Chem. Soc.*, **112**, 2682-5 (1990).

77. W. D. Harman, M. Sekine, and H. Taube, *J. Am. Chem. Soc.*, **110**, 5725-31 (1988).

78. G. W. Parshall and W. A. Nugent, *Chemtech*, **18**, 314-20, 376-83 (1988).

79. G. G. Ecke, J. P. Napolitano, A. H. Filbey, and A. J. Kolka, *J. Org. Chem.*, **22**, 639-42 (1957).

80. J. P. Napolitano, U.S. Patent 3,649,693 (1972).

81. R. Stroh, J. Ebersberger, H. J. Haberland, and W. Hahn, *Angew. Chem.*, **69**, 124-31 (1957).

82. A. J. Kolka, J. P. Napolitano, A. H. Filbey, and G. G. Ecke, *J. Org. Chem.*, **22**, 642-6 (1957).

83. C. W. Matthews, E. S. Batman, and J. F. King, U.S. Patent 4,929,770 (1990).

84. R. Stroh, R. Seydel, and W. Hahn in W. Foerst, Ed., *New Methods of Preparative Organic Chemistry*, Academic Press, New York, 1963, pp. 337-59.

85. G. G. Knapp, W. H. Coffield, J. P. Napolitano, H. D. Orloff, and C. J. Worrel, *Proceedings of the 7th World Petroleum Congress*, Mexico City, Vol. 5, 1967, pp. 403-13.

86. P. F. Ranken and B. G. McKinnie, *Synthesis*, 117-9 (1984).

87. P. W. Wojtkowski, U.S. Patent 4,599,451 (1986); P. W. Wojtkowski and W. A. Nugent, Lecture, International Conference on the Organic Chemistry of Sulfur, Odense, Denmark, August 1988.

88. C. W. Maynard in J. A. Kent, Ed., *Riegel's Handbook of Industrial Chemistry*, 8th Ed., Van Nostrand Reinhold, New York, 1983, pp. 809-61.

89. K. Venkataraman, Ed., *The Chemistry of Synthetic Dyes*, Academic Press, New York, Vol. 2, 1952, pp. 834-91; Vol. 6, 1972, p. 297.

90. G. R. Newkome, W. E. Puckett, V. K. Gupta, and G. E. Kiefer, *Chem. Rev.*, **86**, 451-89 (1986).

91. I. Omae, *Organometallic Intramolecular-coordination Compounds*, Elsevier, New York, 1986.

92. I. Ogata, T. Hayashi, and H. Iida, Japan Kokai 76,100,063 and 76,100,064 (1976); *Chem. Abstr.*, **86**, 43449-50 (1977).

8 | REACTIONS OF ACETYLENES

Acetylene was once a major starting material for the organic chemical industry. Commercial processes for vinyl acetate, vinyl chloride, acetaldehyde, acrylonitrile, acrylates, and chloroprene were largely based on acetylene in the years immediately following World War II. Subsequently, development of technology for synthesis of these materials from ethylene, propylene, or butadiene made acetylene-based processes obsolete. The major chemical advantage of acetylene, the large amount of energy stored in the C≡C function became a disadvantage economically. The "energy crisis" of the 1970s made high-energy materials such as acetylene ($\Delta F° = 50.84$ kcal/mole) extremely expensive as feedstocks. Acetylene-based processes have survived mainly in Eastern Europe and in developing countries where capital to replace obsolescent technology is not readily available. According to Chemical Week [1], the USSR was the world's largest producer of calcium carbide-derived acetylene. It produced about 250,000 m.t./year which was used mainly to make vinyl chloride, acetaldehyde, acetic acid, and vinyl acetate by processes described in Section 8.3.

In western Europe, Japan, and the United States, the largest remaining uses of acetylene as a feedstock are to produce acrylic acid, acrylate esters, and butyne-1,4-diol, a precursor of tetrahydrofuran and 1,4-butanediol. Even in these uses, however, there is a strong economic incentive to replace acetylene-based processes with those based on olefins and alkanes. Oxidation of propylene appears promising as a route to acrylates. Chlorination of ethylene is replacing acetylene chlorination to make highly chlorinated ethylenes and ethanes (Section 12.1). The oxidation of butane is being widely implemented to make maleic anhydride, a precursor to tetrahydrofuran and 1,4-butanediol, but new acetylene-based facilities continue to be built in Germany.

Nearly all the major acetylene processes are catalytic. The first major industrial application of homogeneous catalysis appears to be the $FeCl_3$-catalyzed chlorination of acetylene to tetrachloroethane, which was commercialized about 1910 [2]. This innovation was followed by the development of practical catalysts for several other reactions of acetylene. Both soluble and heterogeneous catalysts were used. In practice, it is sometimes difficult to determine if a relatively insoluble catalyst such as copper acetylide acts in solution. When it is used as a slurry in a liquid reaction mixture, it is quite possible that catalysis occurs on the surface of the solid.

Generally, reactions of the acetylenic C-H bond are catalyzed by copper salts, which presumably form copper acetylides, $Cu(C{\equiv}CR)_n$. Additions to the $C{\equiv}C$ function are catalyzed by copper or mercury salts. Group VIII metal compounds are used to catalyze carbonylation of C_2H_2 to acrylate esters. The Group VIII metals also yield some remarkable dimers, trimers, and tetramers of acetylenes. Recently, there has been considerable work on the polymerization of acetylene because polyacetylene exhibits potentially useful electronic and optical properties. These five classes of reactions are discussed separately after a brief introduction to the coordination chemistry of acetylene.

8.1   COORDINATION CHEMISTRY OF ACETYLENES

Acetylenes interact with transition metal ions in many different ways [3-6]. The π-orbitals of the $C{\equiv}C$ bond can donate electrons to vacant metal orbitals and the π^* orbitals can accept electron density from filled metal d-orbitals. The simplest situation, π-bonding to a single metal atom, is shown in structure **1** in Figure 8-1 [7]. The presence of two mutually orthogonal π-systems in the acetylene bond also permits simultaneous π-bonding to two metal atoms. This situation, which is quite common, is illustrated by a cobalt carbonyl complex of acetylene **2**. The two cobalt atoms are sufficiently close to permit formation of a Co-Co bond [8]. In fact, the compound may be regarded as a derivative of a pseudotetrahedral C_2Co_2 cluster.

The bonding situation can be more complex for acetylenes which bear hydrogen substituents. The acetylenic hydrogens are modestly acidic and form acetylide complexes with many metals [3,9,10]. The acetylide ion, C_2^{2-}, is isoelectronic with N_2 and CN^-. Like these two ligands, acetylides often prefer to bond "end-on" to a transition metal ion. The anionic acetylide ligands are good σ-donors and, in this respect, resemble cyanide ion. Many anionic polyacetylide complexes, analogous to polycyano complexes, have been prepared. For example, copper(I) forms a tris(phenylacetylide) complex, $[Cu(C{\equiv}CPh)_3]^{2-}$ [11].

Many acetylide complexes, especially the catalytically important copper compounds, are polymeric [9]. The acetylide ligands in these complexes form σ-bonds to one metal atom and form π-bonds between the $C{\equiv}C$ function and a second metal atom. The crystal structure **3** of the seemingly simple complex, $[Cu(C{\equiv}CPh)(PMe_3)]_4$, shows that two phenylacetylides are σ-bonded to two coppers and two other acetylide ligands are π-bonded to the same two metal atoms [12]. Such complexity is typical of copper and silver acetylide complexes and is involved in catalysis of acetylene reactions by these compounds. For example, a structure containing four copper ions bound to the acetylide has been reported in a study of the oxidative coupling of phenylacetylene [13].

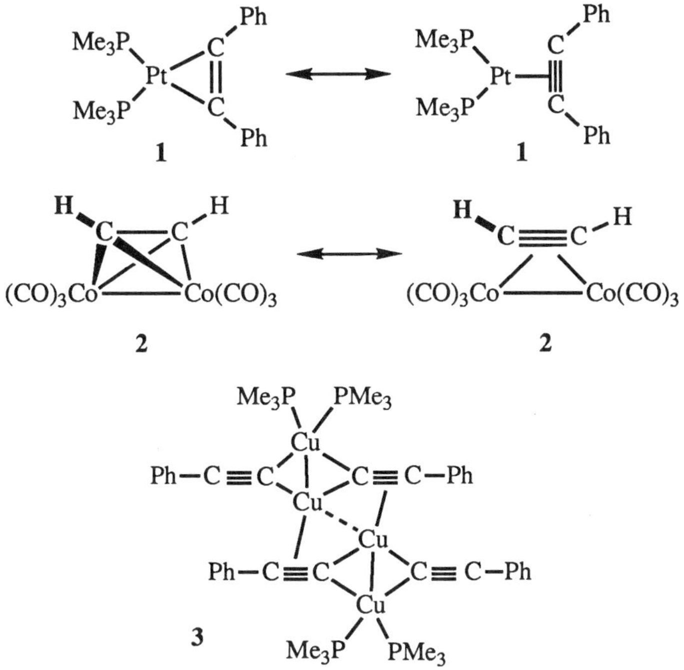

Figure 8.1 Typical bonding modes found in acetylene and acetylide complexes.

The simple acetylene complexes are usually stable in the absence of oxygen, but the acetylide complexes are often violently explosive. This danger is prevalent for the copper(I) acetylide complexes used as catalysts for many commercial reactions of acetylene. Gaseous acetylene reacts with ammoniacal aqueous solutions of copper(I) chloride to form yellow, red, or brown precipitates of Cu_2C_2. These precipitates detonate on heating (120-123°C in air) or mechanical disturbance. Explosive deposits form when oxidized copper surfaces are exposed to acetylene. Consequently, copper fittings should never be used to handle acetylene. Acetylene itself is treacherously explosive, especially in the liquid state. Consequently, acetylene is usually shipped in cylinders as a solution in acetone. A scrubbing process is necessary to obtain acetone-free acetylene. Detailed directions for handling acetylene have been compiled [14].

The reactions of mercury salts with acetylenes are also important catalytically and are just as complex as those of copper salts. Discrete complexes of mercury(II) with alkylacetylides are known. These presumably have complex structures analogous to those of the copper acetylides. The presence of positive charge on an ion such as [XHgC≡CR] should render the C≡C function susceptible to nucleophilic attack, as was noted for olefin complexes earlier. The situation is complicated, however, by the tendency of mercuric salts to add to the acetylenic triple bond. For example, $HgCl_2$ adds to acetylene to form *cis*- or *trans*-chlorovinyl mercury derivatives, depending on reaction conditions:

$$HgCl_2 + C_2H_2 \longrightarrow ClCH=CH\text{-}HgCl$$

Similar reactions probably occur with copper(II) chloride and are important in catalysis of chlorination and other additions to acetylenes.

8.2 ACETYLIDE-CATALYZED REACTIONS

A major family of catalytic reactions of acetylene is based on reactions of copper acetylides. These reactions include oxidative coupling to diynes and the addition of the Cu-C bond to aldehydes and ketones. The addition to C≡C was an especially important process for the manufacture of vinylacetylene, an intermediate in neoprene production.

Oxidative Coupling

The oxidation of acetylenes to give diacetylenes,

$$2R\text{-}C\equiv C\text{-}H \longrightarrow R\text{-}C\equiv C\text{-}C\equiv C\text{-}R$$

is a useful synthetic procedure [15]. It is not used commercially, but it illustrates some principles of copper acetylide catalysis. The oxidative coupling discovered by Glaser in 1869 involves reaction of terminal acetylenes with copper(I) salts to give copper(I) acetylides which are oxidized by air to give diacetylenes. For example, trimethylsilyl acetylene is coupled in good yield by bubbling oxygen through a solution of the acetylene in the presence of a copper(I) chloride-tetramethylethylenediamine complex [16]:

$$2\,Me_3Si\text{-}C\equiv CH \xrightarrow{\;O_2\;} Me_3Si\text{-}C\equiv C\text{-}C\equiv C\text{-}SiMe_3$$

In a variant of this procedure which is sometimes more convenient, an amine solution of a copper(II) salt is used as the oxidant. This alternative is used in the oxidative coupling of three molecules of 1,5-hexadiyne to form an 18-membered ring as a precursor to [18]-annulene [17].

Both the catalytic Glaser coupling and the stoichiometric oxidation with copper(II) salts seem to involve the same mechanism [15]. As illustrated in Figure 8.2, the catalytic cycle begins with the formation of a copper(I) acetylide **4** from a copper(I) salt and a terminal acetylene. This acetylide is presumed to have a complex structure like those found in the crystal structures of two copper phenylacetylides [12,13]. Oxidation of **4** by a copper(II) salt is proposed to give an unstable dimeric copper(II) acetylide (**5**). Decomposition of **5** forms the observed diacetylene product and completes the catalytic cycle by regeneration of copper(I) ions.

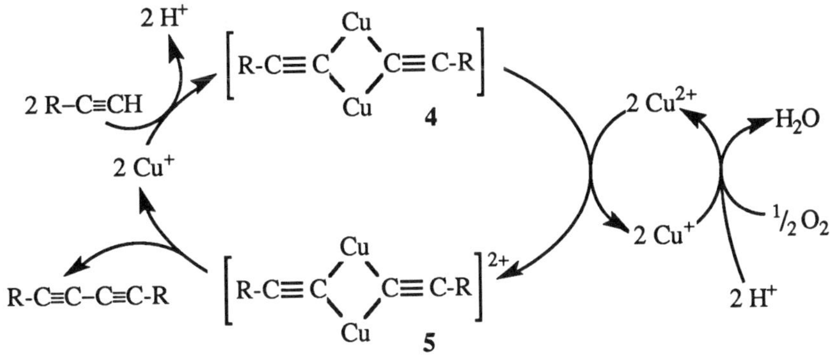

Figure 8.2 A proposed mechanism for catalytic oxidative coupling of acetylenes (amine ligands omitted for simplicity.)

In the catalytic Glaser coupling, the copper(II) oxidant is supplied by oxidation of copper(I) by air or oxygen (right-hand cycle in Figure 8.2). The reaction of copper(I) chloride with oxygen in pyridine has been shown to form $CuCl_2(Py)_2$ and a soluble copper(II) oxide polymer [18]. The latter species is proposed to be the reagent for oxidative coupling of phenols (Chapter 7), anilines, and acetylenes.

Addition to Aldehydes and Ketones

The largest remaining use of acetylene as a chemical intermediate is in the synthesis of butanediol and tetrahydrofuran via the sequence

$$C_2H_2 + 2\ CH_2O \longrightarrow HOCH_2C{\equiv}CCH_2OH \xrightarrow{\ H_2\ } HO(CH_2)_4OH \xrightarrow{\ H^+\ }$$

Similar additions of the C-H bonds of acetylene to higher aldehydes and ketones are also carried out commercially. For example, acetone is condensed with acetylene to give

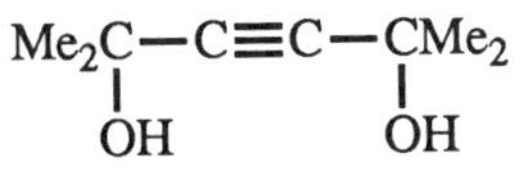

$$\underset{OH}{Me_2C}-C{\equiv}C-\underset{OH}{CMe_2}$$

Such condensations of acetylene with carbonyl compounds can be carried out by reaction of sodium acetylide ($NaC{\equiv}CH$) or lithium acetylide with the aldehyde or ketone [19,20]. These stoichiometric reactions are the most convenient way to effect condensation on a laboratory scale. However, industrial practice employs copper acetylide to catalyze the reaction of acetylene itself with the carbonyl compound.

The condensation of acetylene and formaldehyde is usually carried out as a heterogeneous catalytic reaction [20]. A particulate form of copper acetylide may be prepared by adding gaseous acetylene to a suspension of basic copper oxide in 20-35% aqueous formaldehyde [21]. Minerals such as silica, magnesium silicate, and kieselguhr may be used as supports [22]. The acetylene reacts with the copper oxide and formaldehyde to give a complex acetylide:

$$CuO \xrightarrow{\text{HCHO}} Cu_2O \xrightarrow{C_2H_2} Cu_2C_2 \xrightarrow{C_2H_2} Cu_2C_2 \cdot nC_2H_2$$

Typical conditions for the acetylene-formaldehyde condensation involve reaction of aqueous formaldehyde with acetylene at 1.5 atmospheres pressure and 95°C [21]. Acetylene and formaldehyde solution are fed continuously as required to a slurry of the catalyst. The reaction proceeds in high conversion to give over 90% 1,4-butynediol with a little HC≡CCH₂OH as a byproduct.

The reaction of formaldehyde with acetylene may also be directed to form propargyl alcohol as the major product:

$$HC \equiv CH + HCHO \longrightarrow HC \equiv CCH_2OH$$

if the amount of acetylene is carefully controlled. This procedure, which uses a mixture of copper acetylide and Fuller's earth as the catalyst, has been recommended as a laboratory synthesis of the alcohol [20,23]. Some butynediol is formed as a byproduct because formaldehyde reacts rapidly with the acetylenic C-H in propargyl alcohol. The reactions of higher aldehydes and ketones with acetylenic C-H bonds proceed much more slowly. Usually polar organic solvents such as dimethylformamide are used to attain good solubility of both reactants in laboratory-scale preparations [20].

As described in Chapter 2, ethynylation of a ketone is widely used in the commercial synthesis of fragrances and of Vitamin A [24]. A key step in several of these processes is the reaction of acetylene with a methylheptenone to form dehydrolinalool:

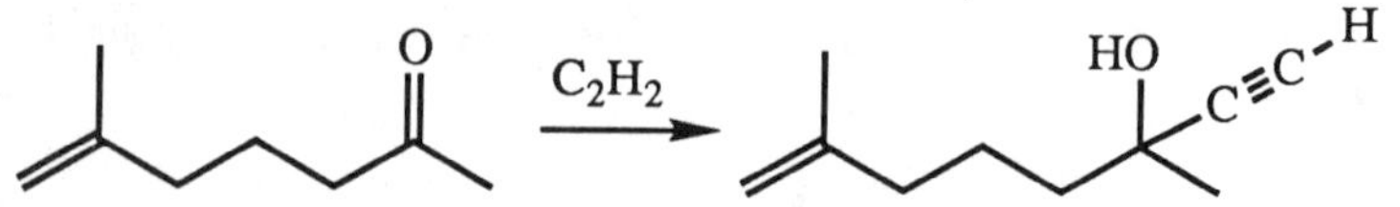

The catalytic addition of acetylene to a carbonyl function appears to be a straightforward organometallic reaction although the kinetics are complex, for example, 0.4th order in formaldehyde in the butynediol synthesis [22]. One can visualize the following steps:

$$RC \equiv C-Cu + H_2C=O \longrightarrow RC \equiv C-CH_2-O-Cu$$

$$RC \equiv CCH_2OCu + RC \equiv CH \longrightarrow RC \equiv CCH_2OH + RC \equiv C-Cu$$

The addition of the C-Cu bond to the carbonyl group resembles the addition of a Grignard reagent to an aldehyde or ketone. In this instance, the alkoxide is protonated by the modestly acidic acetylenic C-H (K_a 10^{-22}). Although such a weak acid will not ordinarily protonate an alkoxide, precipitation of the insoluble copper(I) acetylide provides a driving force for this step. Overall, the addition of C-H to C=O is thermodynamically favorable.

Chloroprene Synthesis

The dimerization of acetylene was a key step in an obsolete synthesis of chloroprene, the monomer for neoprene rubber [25,26]:

$$2\ C_2H_2 \longrightarrow CH_2=CHC{\equiv}CH \xrightarrow{\ HCl\ } CH_2=CHCCl=CH_2$$

Both acetylene dimerization and the addition of HCl to vinylacetylene are catalyzed by copper(I) chloride. In contrast to butynediol synthesis, the catalysts in these processes are soluble in the reaction mixtures. Even though the normally insoluble copper(I) acetylides probably form in these solutions, the copper remains in solution. Coordination to chloride ions and to excess acetylene, both of which are good ligands for copper(I) ions, inhibits formation of acetylide polymers.

The dimerization of acetylene is a formal addition of an acetylenic C-H bond to the C≡C function of a second acetylene molecule. The process probably involves nucleophilic attack of an acetylide ion on a C≡C bond that is activated by coordination to a copper ion. In commercial operation, acetylene is fed continuously to a dilute aqueous HCl solution of copper(I) and potassium chlorides. At 55-75°C, dimerization is rapid and vinylacetylene is swept out of the mixture into a drying system. Distillation removes the major byproduct, divinylacetylene, an acetylene trimer which is an isomer of benzene. The divinylacetylene is a treacherously explosive material, a major hazard in the operation of this process.

The addition of hydrogen chloride to vinylacetylene is conceptually similar to the Cu_2Cl_2-catalyzed addition of HCl to acetylene itself (Section 8.3). However, in contrast to vinyl chloride synthesis, it occurs in two steps, both of which are catalyzed by copper(I) chloride:

$$CH_2=CHC{\equiv}CH \longrightarrow [ClCH_2CH=C=CH_2] \longrightarrow CH_2=CHCCl=CH_2$$

The initial addition of HCl occurs in a 1,4-manner to form 4-chloro-1,2-butadiene. This compound is isomerized in the reaction mixture to give chloroprene. This 1,3-shift of chlorine resembles that in the Cu_2Cl_2-catalyzed isomerization of dichlorobutenes (Section 12.1). The overall HCl addition to vinylacetylene to give chloroprene occurs under mild conditions which can be carried out easily in the laboratory [27]. The reaction of vinylacetylene with concentrated aqueous HCl, Cu_2Cl_2, and NH_4Cl at room temperature gives chloroprene in 97% yield at 94% conversion of vinylacetylene.

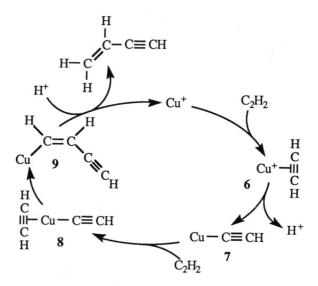

Figure 8.3   The role of copper(I) ion in the dimerization of acetylene.

Both the copper-catalyzed dimerization of acetylene and the HCl addition appear to involve addition to a C≡C bond activated by coordination to a copper(I) ion [28]. In the dimerization, the metal ion also serves to activate the C-H bond. The latter aspect resembles its role in butynediol synthesis. These two functions are illustrated in Figure 8.3. Acetylene coordinates to the metal ion to form complex **6**, which dissociates a proton to form the acetylide complex **7**. Although acetylene is a very weak acid, dissociation is assisted by coordination to the metal ion. In the presence of excess acetylene, the copper acetylide coordinates another molecule of C_2H_2. Insertion of the coordinated C_2H_2 into the Cu-acetylide bond in **8** forms the dimer as a copper complex **9**. Protonation of **9** yields vinylacetylene. It has been proposed [28] that all these reactions take place in a cluster complex, $[Cu_4Cl_4C≡CH]^-$. Yellow chlorocopper(I) acetylide complexes of this composition were delineated by spectroscopic studies of simulated reaction mixtures.

A simple model for the acetylide addition is the reaction of lithium dialkylcuprates with acetylenes [29], for example:

$$Li[CuBu_2] + HC≡CH \longrightarrow Li[Cu(CH=CHBu)_2]$$

In this reaction, the acetylene must coordinate to the cuprate ion before insertion into the Cu-C bond. In a competitive experiment with a $[C_7H_{15}-Cu-C≡CBu]^-$ salt, acetylene inserts into the Cu-alkyl bond in preference to the Cu-acetylide function.

A similar coupling of a nucleophile and an acetylene is proposed to account for the addition of HCl to vinylacetylene [30]. Coordination of the C≡C bond to a dichlorocuprate (-1) ion activates the molecule for attack by chloride ion.

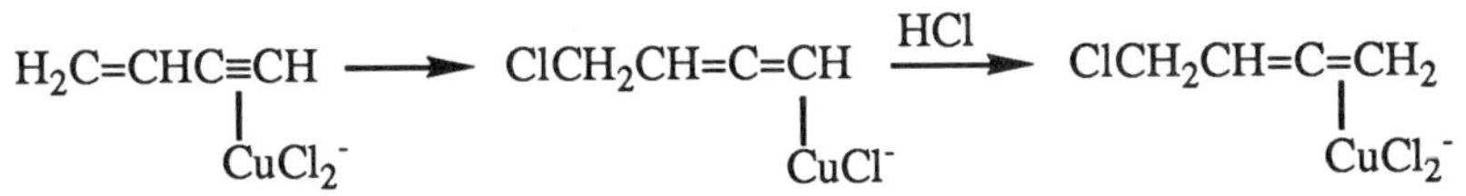

$$\text{H}_2\text{C}=\text{CHC}\equiv\text{CH} \underset{\overset{|}{\text{CuCl}_2^-}}{} \longrightarrow \text{ClCH}_2\text{CH}=\text{C}=\text{CH} \underset{\overset{|}{\text{CuCl}^-}}{} \xrightarrow{\text{HCl}} \text{ClCH}_2\text{CH}=\text{C}=\text{CH}_2 \underset{\overset{|}{\text{CuCl}_2^-}}{}$$

A subsequent addition of chloride ion to the copper-activated allene is then followed by chloride elimination in a mechanism proposed to account for the isomerization of the allene to chloroprene [31]. The pattern of attack of a coordinated nucleophile on a complexed acetylene is seen repeatedly in the following section.

8.3 ADDITIONS TO ACETYLENES

In the 1950s, the synthesis of vinyl monomers was based largely on reactions in which HX molecules (X = Cl, OAc, CN) were added to the triple bond of acetylene [2]. Similarly acetaldehyde production was accomplished by addition of water to acetylene, presumably via vinyl alcohol as a transient intermediate. These processes are now obsolete, but the chemistry of these additions continues to be of interest. Chlorination of acetylene to make tetrachloroethane is still practiced on a significant scale.

Most such additions were carried out with soluble salts of copper(I) or mercury(II) as the catalysts. A general mechanistic picture of these reactions has emerged. Acetylenes form π-complexes with these salts which activate the C≡C bond toward nucleophilic attack to give σ-vinyl complexes:

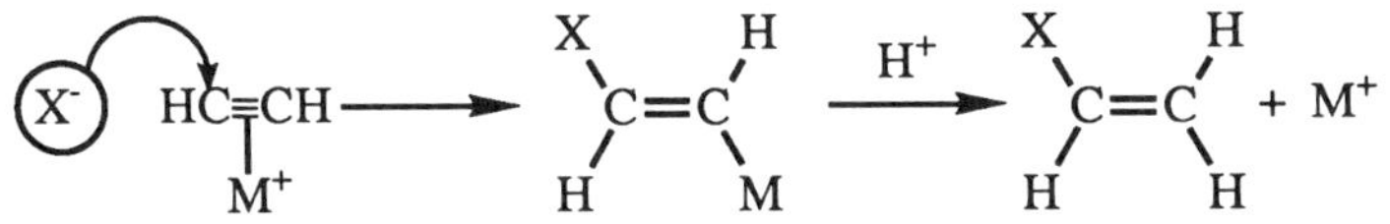

This catalysis of nucleophilic attack by a metal cation closely resembles that for cationic olefin complexes. Electron density is transferred from the acetylene to the metal. This depletion of electron density on the acetylenic carbons makes them susceptible to nucleophilic attack. In contrast to the situation with the alkylmetal complexes arising from olefins, the vinylmercury and -copper compounds protonate cleanly to give $\text{H}_2\text{C}=\text{CHX}$ compounds and regenerate the catalytic metal ion.

Acetaldehyde Synthesis

Before the development of the Wacker process (Section 6.1), most acetaldehyde was made by hydration of acetylene. Typically gaseous acetylene was passed through a sulfuric acid solution of HgSO_4 and FeSO_4 at 95°C and approximately 2 atmospheres pressure [2,32]. About 55% of the acetylene was converted in the reactor. The unchanged acetylene swept out the acetaldehyde before it underwent deleterious side reactions. Fractionation of the volatile products gave acetaldehyde in about 95% yield. Copper(I) chloride also catalyzes the hydration of acetylene to acetaldehyde, but this process was not used commercially.

Similar hydrations of higher acetylenes have been studied extensively. The addition of water typically occurs in Markovnikov fashion. For example, terminal acetylenes give largely methyl ketones rather than aldehydes. The hydration of 1-heptyne gives a 94% yield of 2-heptanone with a catalyst obtained by reacting mercury(II) oxide with a perfluoroalkylsulfonic acid resin [33].

Substantial evidence has accumulated for the formation of acetylene complexes with mercury(II) ions in the early stages of the addition process [34]. When a substituted acetylene is reacted with a mercury salt that contains no coordinating anions, a discrete 2:1 complex is formed. With phenylacetylene and $Hg(ClO_4)_2$, this complex can be observed to form and subsequently decay as the acetylene is hydrated to acetophenone. Presumably, the hydration occurs by nucleophilic attack of water on the coordinated acetylene:

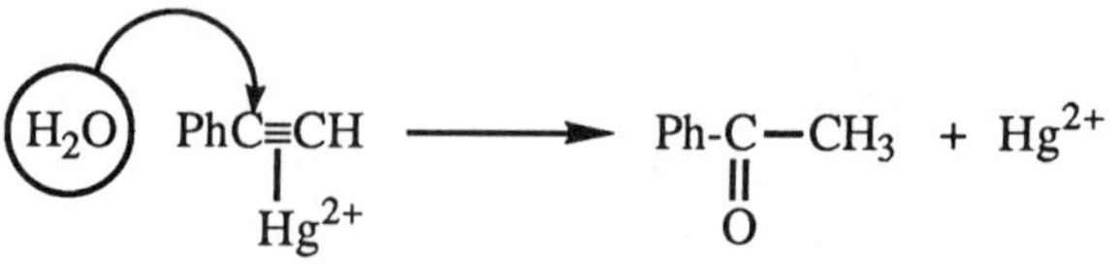

Strongly coordinating anions such as chloride compete with the acetylene for sites on the metal ion and prevent spectroscopic detection of the acetylene complex. An acetylene complex evidently forms with $HgCl_2$, but it reacts rapidly with chloride ion to give a chlorovinyl complex.

Vinyl Acetate Synthesis

The early syntheses of vinyl acetate were based on addition of acetic acid to acetylene with a homogeneous catalyst [2]. This technology was displaced by a vapor-phase process that used a carbon-supported zinc acetate catalyst [35]. This process, in turn, has been largely displaced by a heterogeneous catalytic acetoxylation of ethylene (Section 6.2).

The liquid phase HOAc addition closely resembles other additions to the acetylenic triple bond. Typically, acetylene is swept through an acetic acid solution of $Hg(OSO_2OAc)_2$ at 35°C. Rapid reaction takes place to give a mixture of vinyl acetate and ethylidene diacetate [2]:

$$C_2H_2 + HOAc \longrightarrow CH_2=CHOAc + CH_3CH(OAc)_2$$

Rapid gas passage and low conversions of acetylene are used to minimize formation of the latter product. Yields of vinyl acetate up to about 70% are obtained. Similar technology could also be used to prepare other volatile vinyl esters. Nonvolatile esters are best prepared by a mechanistically related transvinylation [36]:

$$CH_2=CHOAc + RCOOH \xrightleftharpoons{HgSO_4} CH_2=CHOOCR + HOAc$$

The mechanism of acetic acid addition to acetylenes is similar to that for hydration. It appears that a π-complex forms from the C≡C bond and a mercury(II) ion. The π-complex then isomerizes to β-acetoxyvinylmercury acetate which reacts with a proton to give vinyl acetate. Diphenylacetylene, which is known to form a complex with mercury(II) ion [34], readily undergoes the π-σ transformation [37]:

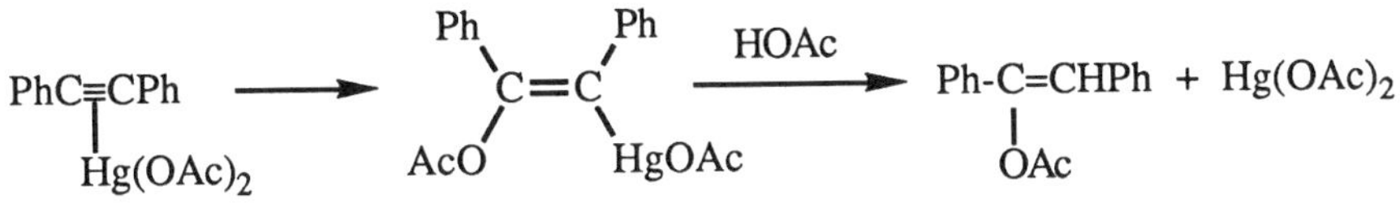

$$PhC{\equiv}CPh \atop Hg(OAc)_2 \quad \longrightarrow \quad {Ph \atop AcO}C{=}C{Ph \atop HgOAc} \quad \xrightarrow{HOAc} \quad Ph\text{-}C{=}CHPh \atop OAc \quad + \quad Hg(OAc)_2$$

In contrast to the hydration process, the direction of acetic acid addition is often anti-Markovnikov. Methylacetylene gives

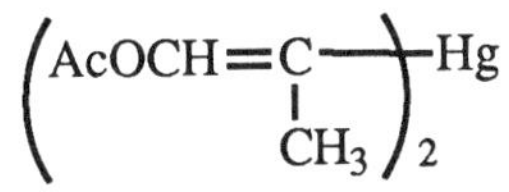

$$\left(AcOCH{=}C{\overset{\displaystyle |}{\underset{\displaystyle CH_3}{}}}{-}Hg \right)_2$$

Hydrolysis of this adduct produces propionaldehyde rather than acetone, which is formed by direct hydration of methylacetylene.

Vinyl Chloride Synthesis

The addition of hydrogen chloride to acetylene is less facile than addition of water or acetic acid. In the industrial process used before oxychlorination of ethylene was developed, HCl was added to acetylene over a carbon-supported $HgCl_2$ catalyst at 100-200°C [2]. Homogeneous catalysts were relatively ineffective, but the mechanism of hydrochlorination has been studied extensively in solution with both $HgCl_2$ and Cu_2Cl_2 as catalysts [28]. As in the additions of water and acetic acid, it seems likely that a π-complex of acetylene rearranges to 2-chlorovinyl derivative which is protonated to give vinyl chloride. The $HgCl_2$-catalyzed addition of HCl to acetylene resembles the CuCl-catalyzed addition to vinylacetylene discussed earlier.

Hydrocyanation

Until about 1960, the major industrial synthesis of acrylonitrile was the addition of hydrogen cyanide to acetylene:

$$HC{\equiv}CH + HCN \quad \longrightarrow \quad CH_2{=}CHCN$$

Like most acetylene-based processes, it was replaced by an olefin reaction, the Sohio process for cooxidation of propylene and ammonia. The acetylene-based process appears to remain in use in eastern Europe.

The acrylonitrile synthesis from acetylene was carried out in aqueous solution [2]. Both the process and the catalyst closely resembled those used for synthesis of vinylacetylene [28]. Dilute acetylene and hydrogen cyanide streams are fed to a solution of copper(I) and ammonium chlorides at 70-90°C and 1.3 atmospheres pressure. A recent Russian patent [38] suggests that yields are improved by pulsing the reaction temperature. The products are swept out of the reactor by unconverted acetylene (HCN conversion is quantitative). A complex purification scheme is required to give polymer grade acrylonitrile, but the overall yield is 80-90%. As might be expected from the similarity of the catalyst to those used in other processes, major byproducts are vinylacetylene and acetaldehyde. The catalyst for hydrocyanation of acetylene is also similar to one described [39] for the hydrocyanation of butadiene (Section 3.6).

The general similarity between dimerization and hydrocyanation of acetylene also extends to mechanism [40]. The similarity between $[C\equiv CH]^-$ and CN^- as ligands suggests that these two nucleophiles add to acetylene similarly. The steps in hydrocyanation shown in the following equations parallel those of the dimerization mechanism of Figure 8.3.

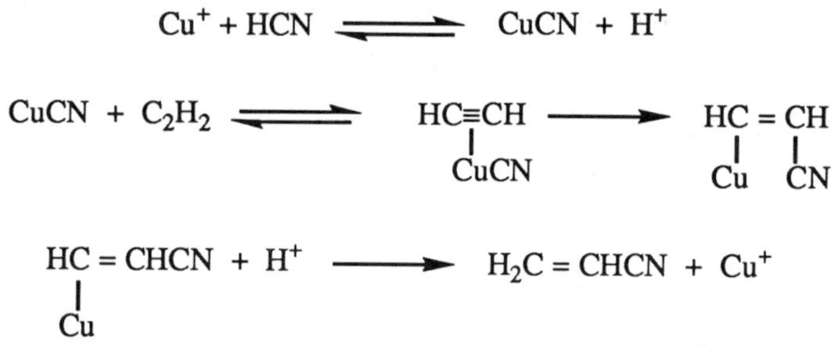

$$Cu^+ + HCN \rightleftharpoons CuCN + H^+$$

$$CuCN + C_2H_2 \rightleftharpoons \underset{\underset{CuCN}{|}}{HC\equiv CH} \longrightarrow \underset{\underset{Cu\quad CN}{|\quad\;|}}{HC = CH}$$

$$\underset{\underset{Cu}{|}}{HC = CHCN} + H^+ \longrightarrow H_2C = CHCN + Cu^+$$

Indeed copper(I) ions form an extensive series of cyanocuprate complexes. In the hydrocyanation mixture, the presence of chloride ions further complicates the structure of the catalytic species. The presence of a strong acid like HCl is required for catalytic activity, probably to cleave the Cu-C bond in the cyanovinylcopper σ-complex.

Chlorination

As noted earlier, the liquid phase chlorination of acetylene may have been the first large-scale homogeneous catalytic process. The chlorination gives 1,1,2,2-tetrachloroethane, an intermediate in the production of trichloroethylene, a widely used solvent.

$$C_2H_2 + 2\,Cl_2 \longrightarrow CHCl_2CHCl_2 \xrightarrow{\;-HCl\;} Cl_2C=CHCl$$

Although mixtures of acetylene and chlorine explode in the presence of air or light, the pure gases do not react significantly in the dark. A catalyst is necessary to produce a controllable reaction to form tetrachloroethane. In the commercial process used since

1910, the catalyst is anhydrous iron(III) chloride [2]. Acetylene is fed to a solution of chlorine and $FeCl_3$ in tetrachloroethane at 80-90°C under reduced pressure. Rapid chlorination occurs and tetrachloroethane is continuously distilled out of the system. Yields exceed 95%.

The uncontrollable air- or light-catalyzed reaction of chlorine with acetylene is almost certainly a radical chain process. It seems likely that the liquid-phase chlorination involves electrophilic catalysis because the best catalysts are Lewis acids like $SbCl_5$, $AlCl_3$, and $FeCl_3$. Although $SbCl_5$ forms a 1:1 complex with acetylene, a major factor in the catalysis is the activation of chlorine. Lewis acids assist heterolytic dissociation of molecular chlorine, for example,

$$Cl_2 + SbCl_5 \rightleftharpoons Cl^+ + SbCl_6^-$$

Incipient Cl^+ ion is a potent electrophile capable of attack on arenes, olefins, or, in this case, acetylene.

8.4 ACETYLENE REACTIONS WITH CO

Carbon monoxide and acetylenes react with transition metal complexes in an extraordinary number of ways [41]. Some of these reactions are useful in synthesis. In this section, we cover three reactions that had considerable industrial potential when acetylene was a viable feedstock. These are the nickel-catalyzed synthesis of acrylate esters, the cobalt-catalyzed synthesis of "bifurandione," and an iron- or ruthenium-catalyzed synthesis of hydroquinone.

Acrylate Synthesis

Until recently, most of the acrylic acid and acrylate esters used in polymers were made by carbonylation of acetylene [42]:

$$HC \equiv CH + CO + ROH \longrightarrow CH_2 = CHCOOR$$

This process has been displaced by syntheses based on propylene oxidation in the United States, but is used extensively elsewhere. Two major variants of this process are practiced. One is a true catalytic process which uses a nickel salt as a catalyst. The other is semicatalytic in that about 20% of the acrylate is formed in a stoichiometric reaction of nickel tetracarbonyl with acetylene. This reaction provides a catalyst for the other 80% of the carbonylation that occurs in the same vessel.

In the catalytic process for acrylic acid, an acetylene solution in tetrahydrofuran reacts with CO, water, and $NiBr_2$ at about 200°C and 60 atmospheres pressure [43]. Acrylic acid is formed in about 90% yield based on acetylene. The major byproducts are acrylic acid oligomers. The semicatalytic process [44] is carried out very much as originally developed by Reppe in the 1940s [45]. In the synthesis of ethyl acrylate, acetylene, CO, ethanol, $Ni(CO)_4$, and HCl are fed to a reactor at 30-50°C. Acrylate yields of about 85% based on acetylene are attained. Careful control of temperature and reactant ratios is necessary to keep the rather unstable catalyst system effective. In both processes, it seems likely that carbonyl nickel halides are the active catalytic species. A $PdCl_2/CuCl_2$ catalyst system has been shown to give good yields of methyl acrylate from acetylene, CO, and methanol under mild conditions [46].

The mechanisms of these processes are not firmly established [47]. One plausible reaction sequence for the semicatalytic process closely parallels the chemistry of olefin carboxylation (Section 4.4). Catalyst formation was proposed to occur via protonation of $Ni(CO)_4$ to give an unstable hydride [48]:

$$Ni(CO)_4 + HCl \longrightarrow HNi(CO)_2Cl + 2\ CO$$

Similar species may be generated from NiX_2 in the reducing environment of the catalytic process. (The water-gas shift reaction provides a source of hydrogen.)

$$CO + H_2O \rightleftharpoons CO_2 + H_2$$

Once the hydride is formed, the analogy to the olefin reaction seems straightforward:

$$HC{\equiv}CH + H{-}Ni{-}CO \longrightarrow H_2C{=}CH{-}Ni{-}CO \longrightarrow \underset{\underset{O}{\overset{\|}{}}}{CH_2{=}CHC}{-}Ni$$

Cleavage of the acyl-Ni bond by water gives acrylic acid. Presumably HCl gives acrylyl chloride and alcohols give the acrylate esters. It is reported [49] that ester synthesis does not occur under anhydrous condition, but this limitation may reflect failure to form the necessary catalytic species.

Similar conditions also bring about the carbonylation of higher alkynes [47,49], but yields tend to be much lower than for acetylene. Markovnikov addition seems to be the rule for alkylacetylenes. For example, methylacetylene gives a 50% yield of methyl methacrylate in a stoichiometric reaction with $Ni(CO)_4$, HCl, and methanol. Disubstituted acetylenes give exclusively cis addition of the H and COOR moieties to the triple bond. This result is consistent with the Ni-H addition to the $C{\equiv}C$ bond that was suggested in the mechanistic scheme above.

Both iron and cobalt carbonyls can be used as catalysts for acrylate synthesis, but yields are low [47]. These metals tend to direct the reaction of acetylenes with CO to form more complex products as described below.

Bifurandione Synthesis

Cobalt carbonyls catalyze the reaction of acetylene and CO to form two isomeric bifurandiones [50,51].

$$2\ C_2H_2 + 4\ CO \longrightarrow$$

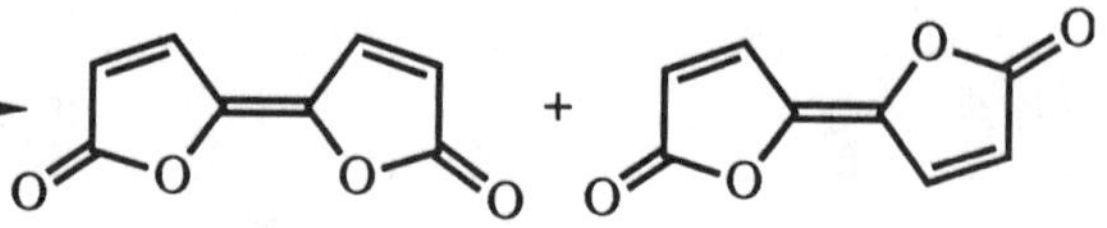

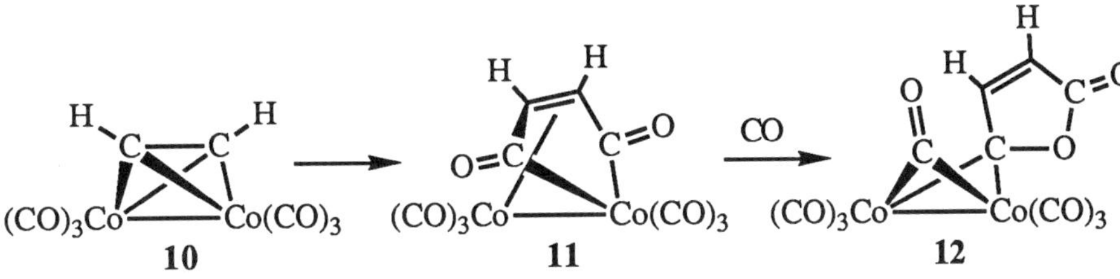

Figure 8.4 Steps in lactone synthesis from acetylene and CO.

This remarkable reaction was investigated extensively because the products could be hydrogenated to give useful polymer intermediates. Hydrogenation with a copper chromite catalyst at 160-190°C gives octanedioic acid and at 250-300°C gives 1,8-octanediol [52].

The synthesis of bifurandiones is typically carried out by adding acetylene and carbon monoxide to a solution of $Co_2(CO)_8$ and heating the mixture to 90-100°C. High pressures (over 200 atmospheres) are generally used, but yields of bifurandione can be fairly good (50-70%) if CO is in excess [50]. Equally good yields can be obtained at lower pressures by adding tributylphosphine or trimethyl phosphite as a catalyst modifier [53]. The trans isomer usually predominates, but the cis isomer is the only product when tetramethylurea is the solvent. Monoalkylacetylenes undergo the same sort of reaction to give complex mixtures of isomeric dialkylbifurandiones.

The mechanism of formation of bifurandione is unclear, but some possible intermediates in the process have been characterized. Both acetylene and disubstituted acetylenes react with cobalt carbonyls to form acetylene-bridged dicobalt complexes like **10** in Figure 8.4, some of which have been characterized crystallographically [8]. It has been suggested [47] that insertion of CO into two of the acetylene-cobalt bonds in **10** gives a product such as **11**. Isomerization of **11** and further CO absorption gives **12**, which has also been characterized crystallographically [54]. The lactone unit in **12** may be regarded as a carbenoid ligand analogous to CO or C≡NR. Such pseudocarbenes are known to bridge metal atoms in cluster compounds. Coupling of two carbene-like lactone ligands from **12** would form the observed bifurandiones, although the mechanism of coupling is not known.

In support of the double CO insertion that converts **10** to **11**, the reaction of acetylene, CO, and water in the presence of cobalt salts gives good yields of succinic acid [55]. Hydrolysis of **11** directly would give maleic acid, but cobalt carbonyls are good catalysts for hydrogenation of olefinic double bonds. Hydrogen is readily available in this system from the water gas shift reaction, noted above. Carbon dioxide is observed in the gaseous byproducts.

Hydroquinone Synthesis

Hydroquinone is formed in iron- and ruthenium-catalyzed reactions of acetylene, CO, and water [47,56]:

$$2\ C_2H_2 + 3\ CO + H_2O \longrightarrow HO-\!\!\!\left\langle\!\!\!\bigcirc\!\!\!\right\rangle\!\!\!-OH + CO_2$$

Other alkynes react under similar conditions to form substituted quinone derivatives. These reactions have created much interest in industry because hydroquinone is widely used as a reducing agent, especially in photography.

Perhaps the best results are obtained when $Ru_3(CO)_{12}$ is the catalyst and moist 2-butanone is the solvent [47]. Acetylene and excess carbon monoxide react at 250°C and 200 atmospheres pressure to give hydroquinone in 73% yield. Roughly one mole of carbon dioxide forms per mole of hydroquinone. Tetrahydrofuran and other ether and ketone solvents give the best yields and rates. Careful control of temperature and reactant levels is necessary for good yields and to avoid acetylene decomposition.

For a potential commercial process, a non-noble metal catalyst is much more attractive than $Ru_3(CO)_{12}$ or a ruthenium salt. Iron-based systems have received a great deal of attention. As in acrylate synthesis, one of the best procedures is semicatalytic [56]. Acetylene, $Fe(CO)_5$, and water react stoichiometrically at 50-80°C to form hydroquinone in fair yields. However, under high CO pressure (600-700 atmospheres), the reaction becomes partially catalytic and yields up to 70% are obtained. Substituted acetylenes react under similar conditions to form the corresponding alkylhydroquinones. A mixture of methyl- and dimethylacetylenes produces trimethylhydroquinone, a vitamin E precursor:

$$CH_3C{\equiv}CCH_3 + CH_3C{\equiv}CH \xrightarrow{\ CO\ } HO-\!\!\!\left\langle\!\!\!\bigcirc\!\!\!\right\rangle\!\!\!-OH$$

The desired trimethyl derivative forms 55% of the hydroquinone products along with 22% tetramethylhydroquinone and 23% mixed dimethyl derivatives.

The mechanism of hydroquinone synthesis is not clear. Two possible mechanisms are illustrated in Figure 8.5. Several potential intermediates such as **13-15** have been characterized, but the reactions of acetylene with iron carbonyls are extremely complex [41], and it has not been possible to define the mechanistic pathways conclusively. Intermediates analogous to those in Figure 8.5 (particularly **16**) are found in the reactions of higher alkynes with iron carbonyls.

8.5 ACETYLENE OLIGOMERIZATION

In the absence of carbon monoxide, Group VIII metal complexes catalyze some remarkable dimerizations, trimerizations, and tetramerizations of acetylenes. Other unsaturated substrates such as olefins and dienes can also be incorporated into the products. Even nitriles react to give aromatic heterocycles. None of these processes are used commercially, but several are useful in the synthesis of cyclic compounds that would be difficult to prepare in any other way.

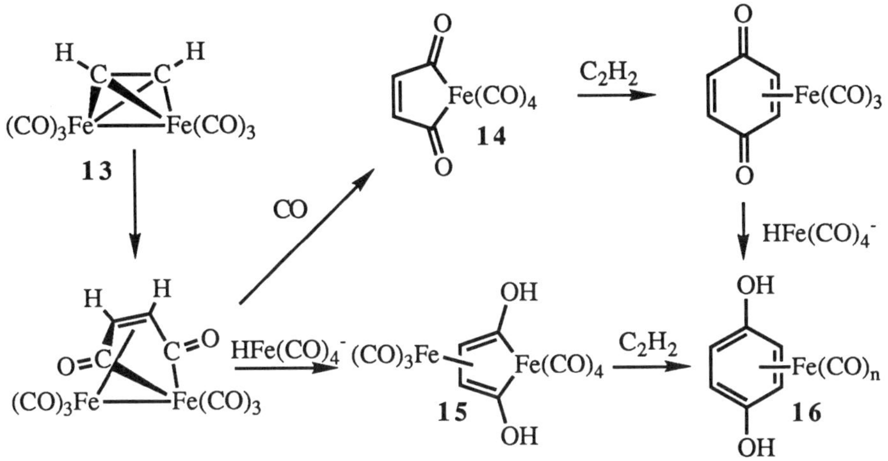

Figure 8.5 Two possible schemes for $Fe(CO)_5$ catalyzed synthesis of hydroquinone (as an iron carbonyl complex).

The mechanism of alkyne cyclization has been controversial for 30 years [57,58]. One early proposal, a "π-complex multicenter mechanism" [59], was that all the alkyne molecules that are to be assembled into a ring gather on a single metal atom. The alkynes then link to form a cyclobutadiene, a benzene, or a cyclooctatetraene ligand. The number of alkyne molecules that are present on the metal determines the size of the ring. This theory was appealing because it related several processes very simply. Empirically, nickel complexes that bear no strongly bonded ligands produce cyclooctatetraene from acetylene. This result fits the hypothesis that a "naked nickel" [58] should have four empty coordination sites and should assemble four acetylene molecules. Similarly, a nickel complex that bears one strongly bound ligand such as triphenylphosphine yields benzene, the expected product from the assembly of three acetylene molecules. The formation of cyclobutadiene ligands by the assembly of two acetylene molecules also fits this pattern.

A more commonly accepted proposal is that two π-bonded acetylenes join together with the metal to form a metallacyclopentadiene ring. This ring can rearrange to a cyclobutadiene ligand or can expand by addition of alkyne to form benzene or cyclooctatetraene. This mechanism is discussed in more detail in connection with trimerization. Recent work [60] with ^{13}C-labeled acetylene has failed to distinguish conclusively between the two mechanisms, but has eliminated other possibilities.

Cyclooctatetraene Synthesis

The preparation of this acetylene tetramer was studied intensively as a potential commercial process [56,57]. Reppe originally observed that nickel salts catalyze cyclotetramer formation [61].

$$4 \ HC\equiv CH \longrightarrow$$

For example, a nickel(II) cyanide slurry in anhydrous tetrahydrofuran slowly absorbs acetylene at 60-70°C and 15-20 atmospheres. The product is about 70% 1,3,5,7-cyclooctatetraene along with some benzene and an acetylene polymer known as cuprene.

Functionally substituted acetylenes also take part in this reaction. Conditions like those discovered by Reppe are used to form cotetramers from acetylene and acetylenic alcohols [62]. Propargyl alcohol ($HC\equiv CCH_2OH$) reacts with excess acetylene to form hydroxymethylcyclooctatetraene in about 20% yield. Zero-valent nickel catalysts such as $Ni(CO)_4$ or $Ni(1,5\text{-cyclooctadiene})_2$ are also effective catalysts [63]. These systems convert $HC\equiv CCMe_2OH$ into mixtures of isomeric tetrakis(2-hydroxy-2-propyl)cyclooctatetraenes in yields of 85-90%. Bis(acrylonitrile)nickel(0) is less effective as a catalyst and gives only stoichiometric yields (based on Ni) of cyclooctatetraene and benzene from acetylene [59].

In contrast to the trimerization reaction discussed below, the cyclooctatetraene synthesis seems to work well only with nickel catalysts. This observation fits the proposal that four coordination sites disposed in a tetrahedral array are needed to make the cyclic tetramer. It does not distinguish between concerted and stepwise processes for joining four acetylene molecules into a ring.

Trimerization

The joining of three molecules of acetylene to give one of benzene is favored thermodynamically but requires a catalyst to proceed well. The uncatalyzed trimerization discovered over a hundred years ago requires high temperatures (300-400°C) and gives poor yields. In the 1940s Reppe discovered that some nickel complexes produce benzene from acetylene in good yield under mild conditions. More recently the trimerization has become a valuable tool for the synthesis of complex organic molecules [57,64], especially when cobalt complexes are used as catalysts.

In the original Reppe synthesis [65], acetylene was heated at 60-70°C and 15 atmospheres pressure with an acetonitrile solution of $Ni(CO)_2(PPh_3)_2$. Benzene formed in 88% yield along with 12% styrene. The latter product probably arose from cotrimerization of acetylene with vinylacetylene, which is formed under reaction conditions. Under similar conditions, propargyl alcohol ($HC\equiv CCH_2OH$) trimerizes to a mixture of 1,2,4- and 1,3,5-tris(hydroxymethyl)benzenes in high yield. The $Ni(CO)_2(PPh_3)_2$ catalyst trimerizes a great variety of monosubstituted acetylenes to trisubstituted benzenes [66]. The catalyst tolerates many functional groups such as ester, ketone, hydroxyl, ether, olefin, and nitro. Disubstituted acetylenes generally do not react, but 1,4-butynediol gives hexakis(hydroxymethyl)benzene in good yield. Generally, phosphine- or phosphite-substituted nickel(0) complexes give benzenoid products preferentially.

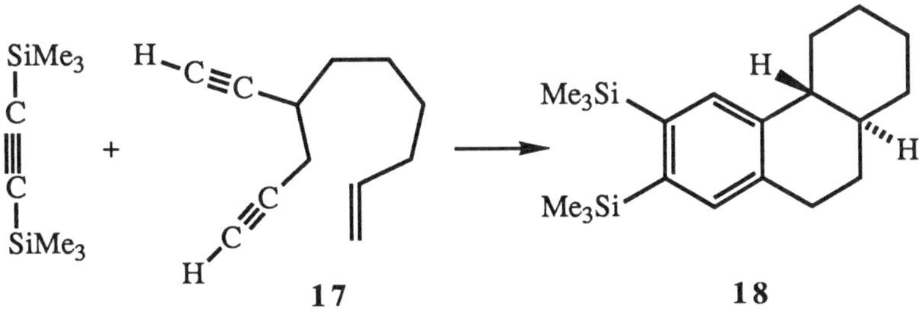

Figure 8.6 A CpCo(CO)$_2$-catalyzed synthesis of an estrone precursor by acetylene trimerization.

For synthesis of benzene, in contrast to cyclooctatetraene, complexes of cobalt rhodium and palladium are also very useful. With these metals [64,67,68] trimerization seems to occur by a stepwise mechanism which can be used to good advantage in the synthesis of unsymmetrical benzenes. Cobalt catalysts are extraordinarily tolerant of steric strain. For example, 1,5-hexadiyne reacts with monoalkylacetylenes in the presence of Co(CO)$_2$(C$_5$H$_5$) to form monoalkylbenzocyclobutenes [64].

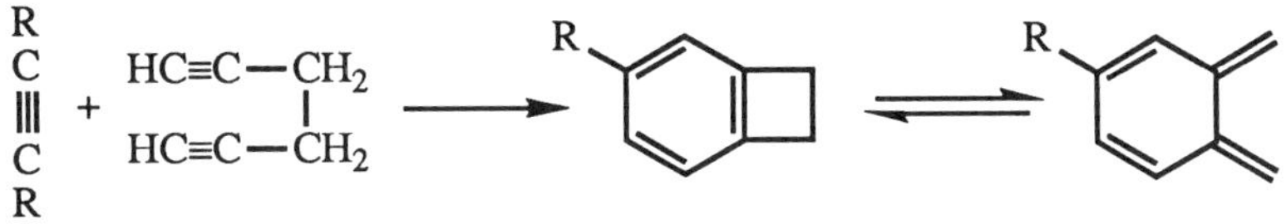

The benzocyclobutene formed in this way can ring open thermally and undergo subsequent diene reactions. An elegant application [64,69] of the acetylene trimerization is the synthesis of an estrone precursor **18** from the bis(alkyne) **17** and an acetylene that bears two bulky substituents (Figure 8.6). The steric bulk of the monoalkyne favors unsymmetrical cotrimerization. After formation of the steroid A ring by trimerization, cyclobutene opening and a Diels-Alder reaction create rings B and C.

Palladium chloride complexes catalyze oligomerization of diphenylacetylene [68,70]:

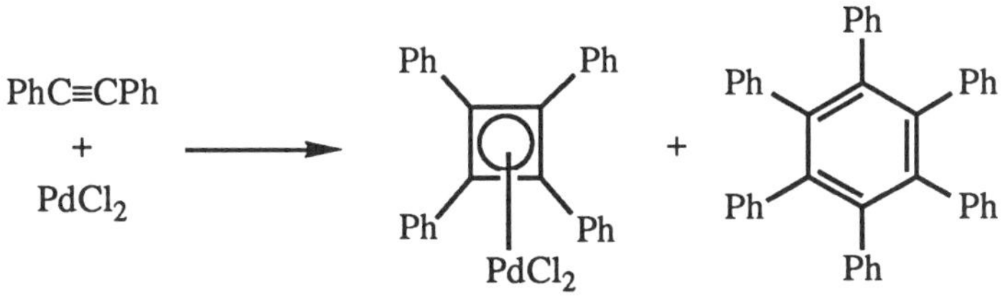

In nonhydroxylic solvents, hexaphenylbenzene is formed in 80-85% yield . Similarly mono-*tert*-butylacetylene gives largely 1,3,5-tri-*tert*-butylbenzene. A heterogeneous catalyst, palladium-on-carbon, is also effective.

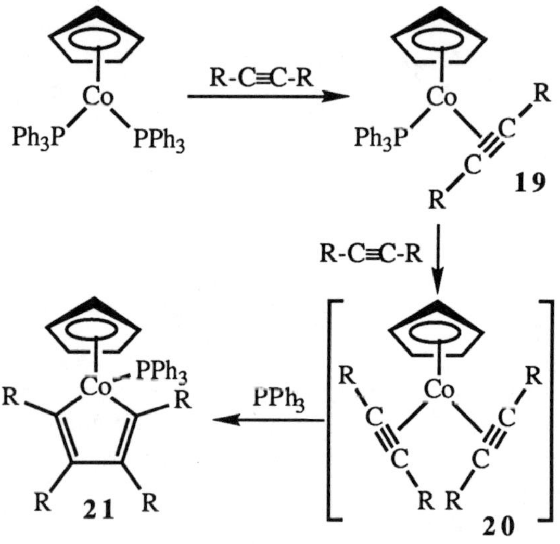

Figure 8.7 Formation of a cobaltacyclopentadiene complex by cobalt-centered coupling of two acetylene ligands.

A metallacyclopentadiene mechanism seems well established for cobalt, rhodium, and iridium catalysts and is also likely to operate with zero-valent nickel and palladium catalysts. The distinctive feature is replacement of two labile ligands to form a bis(alkyne) complex which rearranges to a metallacyclopentadiene. Perhaps the best studied case involves the reaction of $C_5H_5Co(PPh_3)_2$ with disubstituted acetylenes [64,71,72]. As shown in Figure 8.7, a monoalkyne complex **19** and a metallacycle **21** are isolable even though the bis(alkyne) complex **20** is not.

The subsequent addition of a third alkyne molecule to form the cyclic trimer could occur by several mechanisms. Three likely paths are shown in Figure 8.8. The simplest is reaction of the metallacyclopentadiene **22** with an acetylenic dienophile in Diels-Alder fashion to form **23**, which unfolds to a π-bonded benzene complex **24**. This pathway seems to operate when a good dienophile such as dimethyl acetylenedicarboxylate is present [72].

With less electrophilic acetylenes, precoordination of the third acetylene to the metallacycle is essential for trimerization to occur. In the cobalt system, it is necessary to replace the phosphine ligand in **21** to give the alkyne complex equivalent to **25** in the figure. At this point, two paths are available. A pseudo-Diels-Alder reaction within the coordination sphere of the metal can give **23** and **24** as discussed above. However, insertion of coordinated alkyne into a M-C bond of the metallacyclo-pentadiene seems more likely. The resulting metallacycloheptatriene **26** can undergo reductive elimination by coupling the M-C bonds to form the benzene complex **24**.

The relative stability of the metallacyclopentadienes such as **22** facilitates selective syntheses of unsymmetrical benzenes as noted above. The versatility also extends to incorporation of substrates other than alkynes. The cobaltacyclopentadiene **21** reacts with nitriles to form pyridines and with olefins to yield 1,3-cyclohexadienes [64].

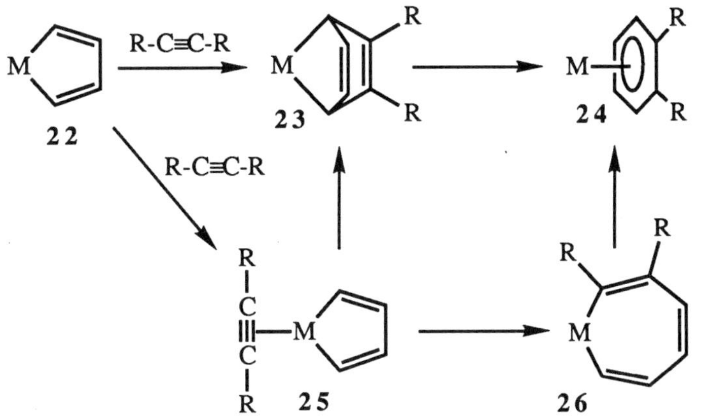

R-C≡C-R

22   23   24   25   26

R-C≡C-R

M

Figure 8.8 Three pathways for conversion of a metallacyclopentadiene to a coordinated benzene. Unreactive ligands are omitted for simplicity.

A major development in organic synthesis is based on a mechanism analogous to that in acetylene trimerization. While the process is not catalytic, the dimerization of acetylenes on a metal to form a metallacyclopentadiene provides a versatile intermediate for synthesis [73]:

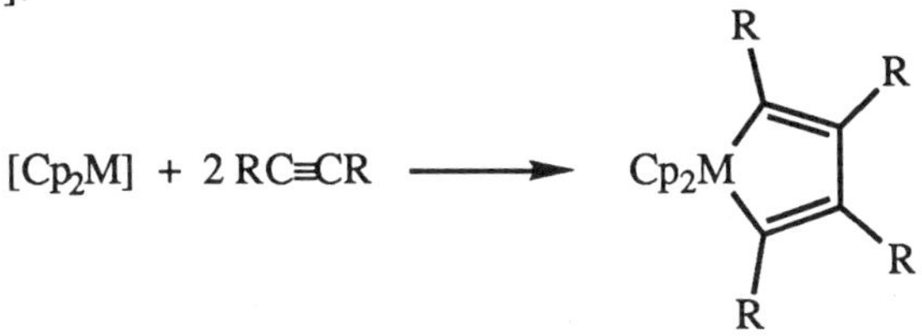

$$[Cp_2M] + 2\,RC\equiv CR \longrightarrow Cp_2M \overset{R}{\underset{R}{\diagup}} \overset{R}{\underset{R}{\diagdown}}$$

The metallocene Cp_2M in this chemistry can be generated conveniently by reduction of Cp_2TiCl_2 or Cp_2ZrCl_2 in situ. The metallacyclic intermediate can be used to produce a broad range of carbocycles and heterocycles [74]. Somewhat analogous chemistry with a palladium-centered catalyst has provided a key step in synthesis of a sesquiterpene [75].

The palladium chloride-catalyzed trimerization of acetylenes follows a different mechanism [68,76] which is characterized by a series of acetylene insertion steps. As in the copper- and mercury-catalyzed reactions discussed earlier, an acetylene complex of $PdCl_2$ readily rearranges to a chlorovinyl derivative by insertion.

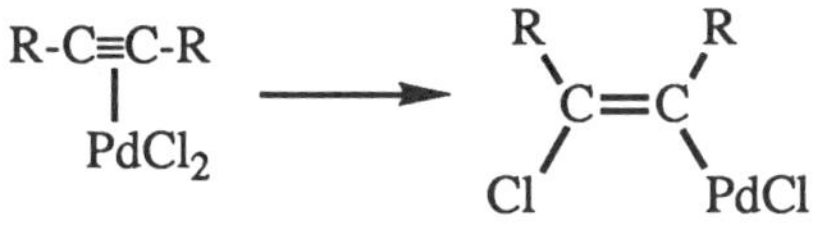

$$\underset{PdCl_2}{R-C\equiv C-R} \longrightarrow \underset{Cl}{\overset{R}{\diagup}}C=C\underset{PdCl}{\overset{R}{\diagdown}}$$

In contrast to the Cu and Hg systems, further insertion into a Pd-C bond occurs readily [76]. The resulting butadienyl complex can rearrange to a cyclobutadiene ligand by transfer of the carbon-bound chlorine back to palladium:

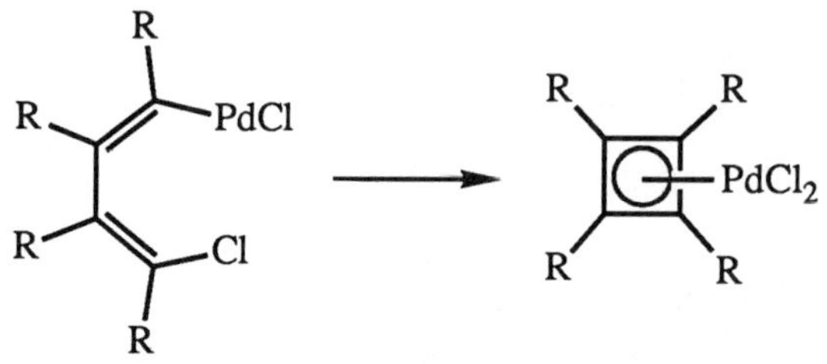

This reaction is the basis of a versatile synthesis of cyclobutadiene complexes. If steric factors do not dictate ring formation from the butadienyl ligand, insertion of a third alkyne molecule may form a 6-chlorohexatrienyl ligand. Rearrangement of such ligands to benzenes by chlorine transfer to palladium should be easy, but there is strong evidence that a less direct path involving a cyclopentadienylmethyl ligand is more favorable [76].

Cotrimerization Reactions

The catalytic trimerization of alkynes can also be modified by the addition of other substrates. In the cobalt-catalyzed trimerization of acetylene, the presence of acetonitrile gives 2-methylpyridine as a major product [77-79].

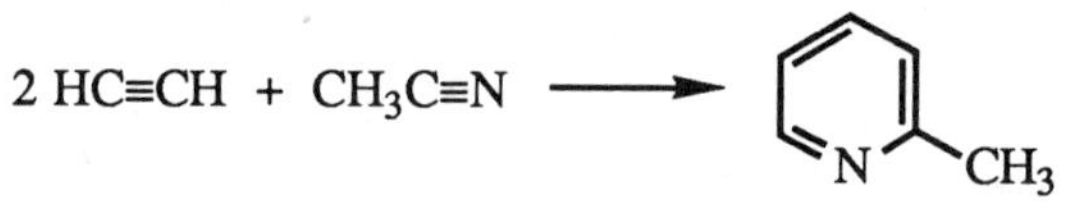

$$2 \ HC{\equiv}CH \ + \ CH_3C{\equiv}N \longrightarrow$$

Higher nitriles react even better. The chelating agent, 2,2'-bipyridine can be prepared in 95% yield by reaction of acetylene with 2-cyanopyridine using the 1,5-cyclooctadiene complex $Co(C_5H_5)(COD)$ as the catalyst. α,ω-Dinitriles give similarly high yields of alkylene-bridged bipyridines. 1-Alkynes can also be used. Acetonitrile and 1-butyne give an 84% yield of a mixture of 3,6- and 4,6-diethyl-2-methylpyridines. The main byproducts are triethylbenzenes.

8.6 POLYMERIZATION

Polyacetylene has attracted wide attention as a material that should have interesting and useful electronic and optical properties [14]. The extended conjugated chain of the polymer is the prototype of many theoretical models for molecular electronic devices. Attempts to test the predicted properties of polyacetylene have been hampered by difficulties in synthesizing and fabricating the polymer. Some of the best polyacetylene samples have been produced by indirect means such as metathesis polymerization (Chapter 9) of cyclooctatetraene [80] or benzvalene [81] or by pyrolysis of a polybutenamer produced by olefin metathesis [82]. Nevertheless, good samples of polyacetylene have also been prepared by polymerization of acetylene [14,83].

Ziegler-Natta Mechanism

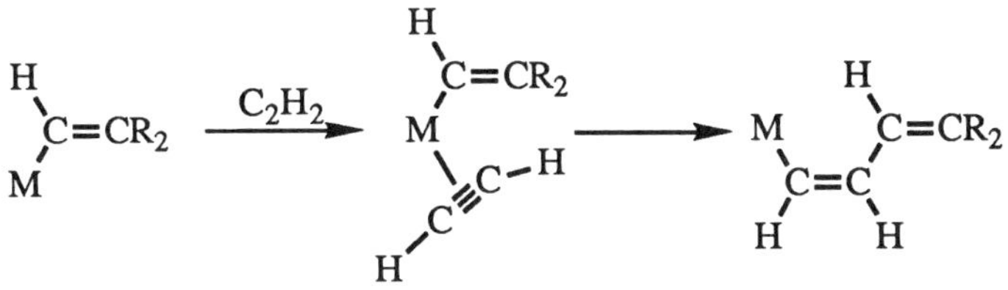

Metathesis Mechanism

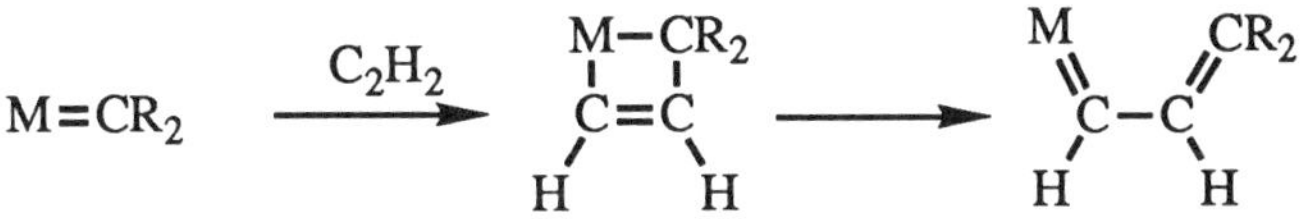

Figure 8.9 Two mechanisms for catalysis of acetylene polymerization.

Two major classes of catalyst for polymerization of acetylene have been found. They seem to function by distinctly different mechanisms. Catalysts [84] analogous to the Ziegler-Natta catalysts for olefin polymerization (Chapter 4) appear to operate by a mechanism closely analogous to that found in polymerization of olefins [85]. The second major mechanistic pathway involves a metathesis mechanism [86] analogous to the metathesis polymerization of olefins (Chapter 9). The two mechanisms are sketched in Figure 8.9.

The metathesis mechanism can be initiated either by M=C or M≡C complexes. A classic Fischer-type carbyne complex, $PhC\equiv W(CO)_5Br$, works well with both acetylene and substituted acetylenes [87]. The Schrock complex, $Me_3CC\equiv W(OR)_3(dme)$, also polymerizes acetylenes [88], but some of the best controlled polymerizations of acetylene have been accomplished with alkylidene complexes such as $M(NAr)(CH\text{-}t\text{-}Bu)(O\text{-}t\text{-}Bu)_2$ (M=Mo, W) [89,90]. Reaction of a solution of $W(NAr)(CH\text{-}t\text{-}Bu)(O\text{-}t\text{-}Bu)_2$ with acetylene gives black, insoluble polyacetylene. When quinuclidine is added as a catalyst modifier, a controlled oligomerization takes place to form yellow-to-red oligomeric $t\text{-}Bu(CH=CH)C\equiv W$ species. This material appears to be a "living" polymer because addition of another monomer such as norbornene leads to formation of a block copolymer. This chemistry provides an elegant tool for production of polyenes with control of structure and properties.

GENERAL REFERENCES

P. J. T. Morris, *Chem. Ind.*, 710-715 (1983) provides a short, readable account of the use of acetylene as a chemical feedstock.

H. G. Viehe, Ed., *Chemistry of Acetylenes*, Marcel Dekker, New York, 1969.

S. Patai, Ed., *The Chemistry of the Carbon-Carbon Triple Bond*, Wiley, New York, 1978.

SPECIFIC REFERENCES

1. A. Wood, *Chem. Week*, **145**(23), 17-18 (1989).
2. S. A. Miller, *Acetylene*, Vol. 2, Academic Press, New York, 1966.
3. A. C. Hopkinson, "Acidity, Hydrogen Bonding and Complex Formation," in S. Patai, Ed., *The Chemistry of the Carbon-Carbon Triple Bond*, Part 1, Wiley, New York, 1978, pp. 75-136.
4. R. Nast, *Coord. Chem. Rev.*, **47**, 89-124 (1982).
5. A. J. Carty, *Pure Appl. Chem.*, **54**, 113-30 (1982).
6. A. A. Cherkas, L. H. Randall, S. A. MacLaughlin, G. N. Mott, N. J. Taylor, and A. J. Carty, *Organometallics*, **7**, 969-77 (1988).
7. D. L. Packett, A. Syed, and W. C. Trogler, *Organometallics*, **7**, 159-66 (1988).
8. J. J. Bonnet and R. Mathieu, *Inorg. Chem.*, **17**, 1973-6 (1978).
9. A. M. Sladkov and L. Yu. Ukhin, *Russ. Chem. Rev.*, **37**, 748-63 (1968).
10. C. Bianchini, D. Masi, A. Meli, M. Peruzzini, J. A. Ramirez, A. Vacca, and F. Zanobini, *Organometallics*, **8**, 2179-89 (1989).
11. R. Nast and W. Pfab, *Chem. Bev.*, **89**, 415-21 (1956).
12. P. W. R. Corfield and H. M. M. Shearer, *Acta Cryst.*, **21**, 957-65 (1966).
13. M. G. B. Drew, F. S. Esho, and S. M. Nelson, *J. Chem. Soc., Chem. Commun.*, 1347-8 (1982).
14. H. W. Gibson and J. M. Pachan, "Acetylenic Polymers," in J. I. Kroschwitz, Ed., *Electrical and Electronic Properties of Polymers*, Wiley, New York, 1988, pp. 1-44.
15. P. Cadiot and W. Chodkiewicz, "Couplings of Acetylenes," in H. G. Viehe, Ed., *Chemistry of Acetylenes*, Marcel Dekker, New York, 1969, pp. 597-643.
16. G. E. Jones, D. A. Kendrick, and A. B. Holmes, *Org. Synth.*, **65**, 52-9 (1987).
17. K. Stockel and F. Sandheimer, *Org. Synth., Coll.*, **6**, 68-75 (1988).
18. I. Bodek and G. Davies, *Inorg. Chem.*, **17**, 1814 (1978).
19. M. M. Midland, J. I. McLoughlin, and R. T. Werley, *Org. Synth.*, **68**, 14-24 (1989).
20. W. Ziegenbein, "Synthesis of Acetylenes and Polyacetylenes by Substitution Reactions," in H. G. Viehe, Ed. *Chemistry of Acetylenes*, Marcel Dekker, New York, 1969, pp. 169-256.
21. J. R. Kirchner, U.S. Patent 3,560,576 (1971).
22. S. S. Kale, R. V. Chaudhari, and P. A. Ramachandran, *Ind. Eng. Chem.*, **20**, 309-15 (1981).
23. W. Reppe und Mitarbeiten, *Liebig's Ann. Chem.*, **596**, 1-38 (1955).
24. W. Reif and H. Grassner, *Chem. Ing. Tech*, **45**, 646-52 (1973).
25. C. A. Hargreaves and D. C. Thompson, "2-Chlorobutadiene Polymers," in *Encyc. Poly. Sci. Tech.*, **3**, 705-30 (1965).
26. G. T. Martirosyan and A. Ts. Malkhasyan, *Zh. Vses. Khim., O-Va. im. D. I. Mendeleeva*, **30**, 263-7 (1985).
27. W. H. Carothers, G. J. Berchet, and A. M. Collins, *J. Am. Chem. Soc.*, **54**, 4066 (1932).
28. O. A. Chaltykyan, *Copper-Catalytic Reactions*, Consultants Bureau, New York, 1966, pp. 39-57; O. N. Temkin and R. M. Flid, *Catalytic Conversions of Acetylenic Compounds in Solutions of Metal Complexes*, Nauka, Moscow 1968; O. N. Temkin, R. M. Flid, G. F. Tikhonov, G. K. Shestakov, A. Yermakova, L. I. Yarovaya, and V. G. Michal'chenko, *Kinet. Katal.*, **10**, 1004, 1230 (1969).
29. A. Alexakis, J. Normant, and J. Villieras, *Tetrahedron Lett.*, 3461-2 (1976).
30. G. K. Shestakov, F. I. Bel'skii, S. M. Airyan, and O. Temkin, *Kinet. Katal.*, **19**, 334 (1978).

31. G. K. Shestakov, S. M. Airyan, F. I. Bel'skii, and O. N. Temkin, *Zh. Org. Khim.*, **12**, 2053 (1976).

32. D. F. Othmer, K. Kon, and T. Igarashi, *Ind. Eng. Chem.*, **48**, 1258-62 (1956).

33. G. A. Olah and D. Meidar, *Synthesis*, 671-2 (1978).

34. W. L. Budde and R. E. Dessy, *J. Am. Chem. Soc.*, **85**, 3964-70 (1963).

35. B. A. Morrow, *J. Catal.*, **86**, 328-32 (1984).

36. D. Swern and E. F. Jordan, *Org. Synth., Coll.*, **4**, 977 (1963).

37. A. N. Nesmeyanov, A. E. Borisov, I. S. Saveleva, and M. A. Osipova, *Izv. Akad. Nauk, SSSR, Otd. Khim. Nauk*, 1249-52 (1961).

38. N. R. Yusupbekov and M. L. Tsatkin, USSR Patent 1,049,477 (1983); *Chem. Abstr.*, **100**, 52177.

39. D. Y. Waddan, U.S. Patent 3,869,501 (1975).

40. O. N. Temkin, R. M. Flid, G. F. Tikhonov, L. V. Melnikova, T. G. Sukhova, G. K. Shestakov, A. A. Khorkin, and L. N. Reshetova, *Katal. Reakts. Zhidk. Faze, Tr. 2nd Vses. Konf*, Alma-Ata, 1966, pp. 49-57; *Chem. Abstr.*, **69**, 76328 (1968).

41. W. Hubel, "Organometallic Derivatives from Metal Carbonyls and Acetylene Compounds," in I. Wender and P. Pino, Eds., *Organic Syntheses via Metal Carbonyls*, Vol. 1, Wiley-Interscience, New York, 1968, pp. 273-342.

42. K. Weissermel and H-J. Arpe, *Industrial Organic Chemistry*, Verlag Chemie, Weinheim, 1978, pp. 254-8.

43. T. Toepel, *Chim. Ind. (Paris)*, **91**, 139-45 (1964).

44. F. T. Maher and W. Bauer, "Acrylic Acid and Its Derivatives," in J. J. McKetta and W. A. Cunningham, *Encyclopedia of Chemical Processing and Design*, Marcel Dekker, New York, Vol 1, 1976, p. 401.

45. W. Reppe, *Liebig's Ann. Chem.*, **582**, 1-21 (1953).

46. H. Alper, B. Despeyroux, and J. B. Woell, *Tetrahedron Lett.*, **24**, 5691-4 (1983).

47. P. Pino and G. Braca, "Carbon Monoxide Addition to Acetylenic Substrates," in I. Wender and P. Pino, Eds., *Organic Syntheses via Metal Carbonyls*, Vol. 2, Wiley-Interscience, New York, 1977, p. 419.

48. R. F. Heck, *J. Am. Chem. Soc.*, **85**, 2013-4 (1963).

49. J. Falbe, *Carbon Monoxide in Organic Synthesis*, Springer-Verlag, Berlin, 1970, pp. 87-96.

50. J. C. Sauer, R. D. Cramer, V. A. Engelhardt, T. A. Ford, H. E. Holmquist, and B. W. Howk, *J. Am. Chem. Soc.*, **81**, 3677-81 (1959).

51. G. Albanesi and M. Tovaglieri, *Chim. Ind. (Milan)*, **41**, 189 (1959); G. Albanesi and E. Gavezzotti, *ibid.*, **47**, 1322-4 (1965).

52. H. E. Holmquist, F. D. Marsh, J. C. Sauer, and V. A. Engelhardt, *J. Am. Chem. Soc.*, **81**, 3681-6 (1959).

53. G. Varadi, I. T. Horvath, J. Palagyi, T. Bak, and G. Palyi, *J. Mol. Catal.*, **9**, 457-60 (1980).

54. O. S. Mills and G. Robinson, *Proc. Chem. Soc.*, 156-7 (1959).

55. G. Natta and G. Albanesi, *Chim. Ind. (Milan)*, **48**, 1157-61 (1966).

56. W. Reppe, N. von Kutepow, and A. Magin, *Angew. Chem. Int. Ed.*, **8**, 727-33 (1969); W. Reppe and H. Vetter, *Liebig's Ann. Chem.*, **582**, 133-42 (1953).

57. C. Hoogzand and W. Hubel, "Cyclic Polymerization of Acetylenes by Metal Carbonyl Compounds," in I. Wender and P. Pino, Eds., *Organic Syntheses via Metal Carbonyls*, Vol. I, Wiley-Interscience, New York, 1968, p. 343-71.

58. P. W. Jolly and G. Wilke, *The Organic Chemistry of Nickel II*, Academic Press, New York, 1975, p. 94.

59. G. N. Schrauzer, *Chem. Ber.*, **94**, 1403-9 (1961); *Adv. Organomet. Chem.*, **2**, 1 (1964).
60. R. E. Colborn and K. P. C. Vollhardt, *J. Am. Chem. Soc.*, **108**, 5470-77 (1986).
61. W. Reppe, O. Schlichting, K. Klager, and T. Toepel, *Liebig's Ann. Chem.*, **560**, 1 (1948).
62. A. C. Cope and D. F. Rugen, *J. Am. Chem. Soc.*, **75**, 3215 (1953); A. C. Cope and R. M. Pike, *ibid.*, 3220 (1953).
63. P. Chini, N. Palladino, and A. Santambrogio, *J. Chem. Soc.*, C, 836-40 (1967).
64. K. P. C. Vollhardt, *Angew. Chem., Int. Ed.*, **23**, 539-56 (1984).
65. W. Reppe and W. J. Schweckendiek, *Liebig's Ann. Chem.*, **560**, 104-13 (1948).
66. L. S. Meriwether, E. C. Colthup, G. W. Kennerly, and R. N. Reusch, *J. Org. Chem.*, **26**, 5155 (1961).
67. R. Grigg, R. Scott, and P. Stevenson, *Tetrahedron Lett.*, **23**, 2691-2 (1982).
68. P. M. Maitlis, *Acc. Chem. Res.*, **9**, 93-9 (1976).
69. R. L. Funk and K. P. C. Vollhardt, *J. Am. Chem. Soc.*, **101**, 215-7 (1979).
70. A. T. Blomquist and P. M. Maitlis, *J. Am. Chem. Soc.*, **84**, 2329-34 (1962).
71. Y. Wakatsuki, O. Nomura, K. Kitaura, K. Morokuma, and H. Yamazaki, *J. Am. Chem. Soc.*, **105**, 1907-12 (1983).
72. D. R. McAlister, J. E. Bercaw, and R. G. Bergman, *J. Am. Chem. Soc.*, **99**, 1666-8 (1977).
73. W. A. Nugent, D. L. Thorn, and R. L. Harlow, *J. Am. Chem. Soc.*, **109**, 2788-96 (1987).
74. P. J. Fagan and W. A. Nugent, *J. Am. Chem. Soc.*, **110**, 2310-2 (1988).
75. B. M. Trost, P. A. Hipskind, J. Y. L. Chung, and C. Chan, *Angew. Chem. Int. Ed.*, **28**, 1502-4 (1989).
76. B. E. Mann, P. M. Bailey, and P. M. Maitlis, *J. Am. Chem. Soc.*, **97**, 1275-6 (1975).
77. R. A. Clement, U.S. Patent 3,829,429 (1974).
78. H. Bonnemann, *Angew. Chem. Int. Ed.*, **17**, 505-15 (1978).
79. H. Bonnemann and W. Brijoux, "The Cobalt-Catalyzed Synthesis of Pyridine and Its Derivatives," in R. Ugo, Ed., *Aspects of Homogeneous Catalysis*, Vol. 5, D. Reidel, Dordrecht, 1984, pp. 75-196.
80. C. B. Gorman, E. J. Ginsburg, S. R. Marder, and R. H. Grubbs, *Angew Chem., Int. Ed., Adv. Mater.*, **28**, 1571-3 (1989).
81. T. M. Swager and R. H. Grubbs, *J. Am. Chem. Soc.*, **111**, 4413 (1989).
82. J. H. Edwards, W. J. Feast, and D. C. Bott, *Polymer*, **25**, 395-8 (1984); K. Knoll and R. R. Schrock, *J. Am. Chem. Soc.*, **111**, 7889-8004 (1989).
83. W. Liang and C. R. Martin, *J. Am. Chem. Soc.*, **112**, 9666-7 (1990).
84. T. Ito, H. Shirakawa, and S. Ikeda, *J. Polym. Sci., Polym. Chem. Ed.*, **12**, 11-20 (1974).
85. T. C. Clarke, C. S. Yannoni, and T. J. Katz, *J. Am. Chem. Soc.*, **105**, 7787-8 (1983).
86. D. J. Liaw, C. Lucas, A. Soum, M. Fontanille, A. Parlier, and H. Rudler, in R. P. Quirk, Ed., *Transition Metal Catalyzed Polymerizations*, Cambridge Press, Cambridge, 1988, pp. 671-687.
87. T. J. Katz, T. H. Ho, N-Y Shih, Y-C. Ying, and V. I. W. Stuart, *J. Am. Chem. Soc.*, **106**, 2659 (1984).
88. J. H. Freudenberger, R. R. Schrock, M. R. Churchill, A. L. Rheingold, and J. W. Ziller, *Organometallics*, **3**, 1563-73 (1984).
89. R. Schlund, R. R. Schrock, and W. E. Crowe, *J. Am. Chem. Soc.*, **111**, 8004-6 (1989).
90. I. A. Weinstock, R. R. Schrock, and W. M. Davis, *J. Am. Chem. Soc.*, **113**, 135-44 (1991).

9 | CARBENE COMPLEXES IN OLEFIN METATHESIS AND RING-FORMING REACTIONS

Many reactions such as olefin metathesis and cyclopropanation are catalyzed by transition metal species that either activate or stabilize a "carbene" ligand. Several of these reactions are used industrially with either homogeneous or heterogeneous catalysts. Recent studies of well-characterized alkylidene complexes have provided considerable mechanistic insight into the nature of the olefin metathesis. The combination of new science and new technology has made this field one of the most exciting in organometallic chemistry in the past decade [1]. Commercial applications range from polymers used in snowmobile bodies to very complex organic fragments which are intermediates in the synthesis of insecticides. Olefin metathesis plays a critical role in Shell's large-scale synthesis of higher olefins. While not commercial, metathesis has been applied to a very elegant indirect synthesis of polyacetylene.

This chapter discusses the role of alkylidene complexes in olefin metathesis and in ring-forming reactions such as addition of a carbenoid fragment across a C=C bond. Both classes of reactions have established themselves as important technologies for the manufacture of specialty chemicals [2]. The two classes differ strongly in the types of metals utilized and the way in which the "carbene" ligand is generated. Olefin metathesis is catalyzed by complexes of elements in the Ti, V, Cr, Mn, and Fe columns of the periodic table while cyclopropanation is typically carried out with Cu(+1) and Rh(+2) complexes.

One of the most exciting developments in recent years has been the synthesis of discrete complexes that are efficient catalysts for olefin metathesis. Typically these compounds contain Mo, W, or Re, the same metals that are used in almost all commercial metathesis processes. Perhaps the best example is the well-characterized tungsten complex **1** [3]. It and the analogous molybdenum compound catalyze a wide range of ring-opening metathesis polymerizations.

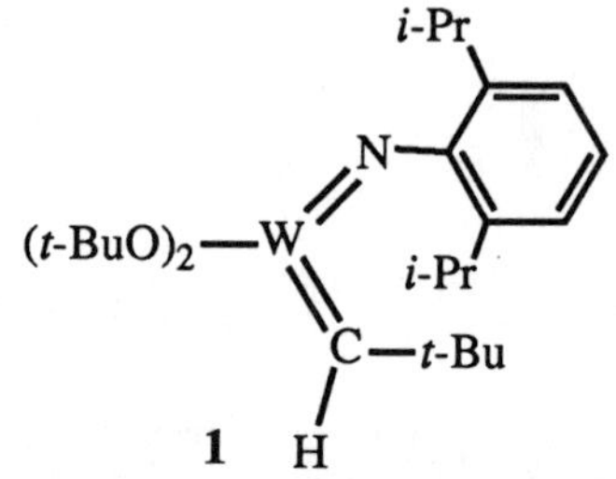

The structural factors that combine to give **1** stability as well as catalytic activity have been defined [4]:

- "Four coordination allows a relatively small substrate to attack the metal to give a five-coordinate intermediate metallacyclobutane complex."
- "Bulky alkoxide and imido ligands prevent intermolecular reactions." "The diisopropylphenylimido ligand almost certainly plays a major role in slowing down bimolecular decomposition reactions and in directing steric interactions throughout these molecules."
- The alkoxy groups have a dramatic effect on reaction rates. The *t*-butoxy complex **1** will metathesize *cis*-2-pentene at a rate of only ~2 turnovers per hour at 25°C. In contrast, the analogous complex bearing two $(CF_3)_2(CH_3)CO^-$ ligands will metathesize *cis*-2-pentene at a rate of ~10^3 turnovers per *minute*! The difference is explained in terms of an electrophilic attack on the olefin by the metal. The attack is enhanced by the electrophilicity of the metal bearing the fluorinated alkoxy ligands.

The role of these complexes in metathesis is discussed in Section 9.3. An important recent development is the isolation of rhenium complexes somewhat analogous to **1** [5]. Like the heterogeneous metathesis catalysts based on rhenium, the new Re complexes tolerate a range of organic functional groups. This property promises to extend the useful scope of the olefin metathesis reaction.

9.1   OLEFIN METATHESIS AND ITS APPLICATIONS

The nature of olefin metathesis is nicely illustrated by the reaction of ethylene with stilbene to form styrene:

$$H_2C=CH_2 + PhHC=CHPh \rightleftharpoons \begin{array}{c} H_2C \\ \parallel \\ PhHC \end{array} + \begin{array}{c} CH_2 \\ \parallel \\ CHPh \end{array}$$

In simplest terms, the methylene and benzylidene groups of the two olefins recombine to produce an equilibrium mixture of the three olefins. The other C-C bonds and the C-H bonds in the olefins are unaffected. This remarkable reaction has stimulated much research in both industrial and academic laboratories and the mechanism is now well established (Section 9.3). A broad range of applications are emerging [6], most of which employ heterogeneous catalysts.

Most simple olefins undergo metathesis although some catalysts are ineffective with terminal olefins. Like most catalytic reactions, metathesis gives an equilibrium mixture of products. For example, in the "Phillips Triolefin Process" [7], propylene is converted to ethylene and butene:

$$2 \; CH_3HC{=}CH_2 \; \rightleftharpoons \; H_2C{=}CH_2 \; + \; CH_3HC{=}CHCH_3$$

The initial product is an equilibrium mixture of the three olefins, but the volatility of ethylene permits its selective removal from the process. Like the volatility of ethylene, the stability of the six-membered ring can be a strong driving force. Cyclohexene is a poor substrate for metathesis but its formation can be a strong driving force for the reaction, for example,

Metathesis catalysts have been prepared from almost all the transition metals, but best results are obtained with Mo, W, and Re catalysts. The soluble catalysts,which have received extensive study in both academic and industrial laboratories, are of three major types. One family might be designated Ziegler catalysts because they are based on combinations of alkyl aluminum, magnesium, or lithium reagents with transition metal chlorides such as WCl_6 and $WOCl_4$. It seems well established [8] that these systems contain alkylidene metal complexes like that in **1**. The discrete alkylidene complexes of Mo, W, and Ti have been very effective in the ring-opening metathesis polymerization reactions described below. Another family is derived from metal carbonyls, especially $Mo(CO)_6$ and $W(CO)_6$ and ligand-substituted derivatives such as $[W(CO)_5Cl]^-$. The metal carbonyl catalysts usually require activation by photolysis or by addition of a Lewis acid. The metal carbonyls also can be used to prepare heterogeneous catalysts. The reaction of $Mo(CO)_6$ with silica, alumina, or magnesia gives surface species such as -O-Mo(CO)_n which are catalysts for metathesis. Finally [9], it has been found that metathesis can be carried out in water with catalysts derived from aquoruthenium (+2) ions. This system appears to be closely related to the alcoholic $RuCl_3$ catalyst reportedly [10] used to produce polynorbornene, a major industrial use of a soluble metathesis catalyst.

The largest industrial applications of olefin metathesis use heterogeneous catalysts. These generally take the form of molybdenum, tungsten, or rhenium oxides that have been deposited on silica or alumina [11]. Although the oxide is usually applied in its maximum oxidation state such as MoO_3, reaction with the olefin at elevated temperature brings about reduction. It seems likely that Mo at the catalyst site has an oxidation state in the range +2 to +4 whether the molybdenum is deposited as MoO_3 or $Mo(CO)_6$.

Many industrial applications of metathesis have been seriously considered. The styrene synthesis from stilbene mentioned above has been considered for commercialization by Monsanto [12,13], but never used on a large scale. (The stilbene would be produced by oxidative coupling of the methyl groups of toluene in a heterogeneous catalytic process.) The "Phillips Triolefin Process" mentioned above was operated when propylene prices were low and butene was desired as a feedstock for butadiene production [7].

By far the largest application of olefin metathesis is a key step in the SHOP (Shell Higher Olefins Process) process described in Section 4.3. Several hundred thousand tons per year of mixed olefins in the C_4-C_{10} and C_{20+} ranges are metathesized to produce mixtures containing substantial amounts of the desirable C_{10}-C_{20} olefins [14-16]:

$$\begin{array}{c} CH_3CH \\ \parallel \\ CH_3CH \end{array} + \begin{array}{c} HC-C_{10}H_{21} \\ \parallel \\ HC-C_{10}H_{21} \end{array} \rightleftharpoons 2\ CH_3CH = CHC_{10}H_{21}$$

Internal olefins are preferred feedstocks for this process because terminal olefins reduce the activity of the catalyst through nonproductive degenerate metatheses. The product olefins are hydroformylated with a special cobalt catalyst (Section 5.5) to produce long-chain alcohols used in making surfactants and plasticizers.

In the SHOP metathesis step [15], a liquid mixture of C_4-C_{10} and C_{20+} olefins are fed to a fixed bed of alumina-supported metathesis catalyst. In the course of reaction at 80-140°C and approximately 13 atmospheres pressure, the olefins equilibrate to a mixture containing about 10-15% of the desired C_{10}-C_{14} olefins. These desired products are separated by distillation and the high and low boiling olefins are recycled to the isomerization and metathesis reactors, respectively. The catalyst appears to be a cobalt-promoted molybdenum oxide in a high surface area alumina hydrogel [16].

The number of applications which use olefin metathesis in making specialty intermediates is growing [6]. Phillips produces neohexene, a monomer for specialty plastics, by metathesis of ethylene and diisobutylene [17,18]:

$$\begin{array}{c} (CH_3)_2C = CHC(CH_3)_3 \\ + \\ H_2C = CH_2 \end{array} \rightleftharpoons \begin{array}{c} (CH_3)_2C \\ \parallel \\ H_2C \end{array} + \begin{array}{c} CHC(CH_3)_3 \\ \parallel \\ CH_2 \end{array}$$

The catalyst, a mixture of magnesia and WO_3 on silica [11], is maintained at about 370°C while a mixture of diisobutylene and excess ethylene is passed through it under 25-30 atmospheres pressure [18]. The products are separated by distillation and the isobutylene is recycled to the diisobutylene synthesis unit.

Another commercial application of metathesis to the production of specialty monomers and intermediates is the Shell FEAST (Further Exploitation of Advanced Shell Technology) process [19,20]. α,ω-Diolefins are produced by metathesis of ethylene with cycloolefins such as with cycloctene and cyclooctadiene. Phillips has run metathesis of cyclododecene with ethylene to make 1,13-tetradecadiene [21].

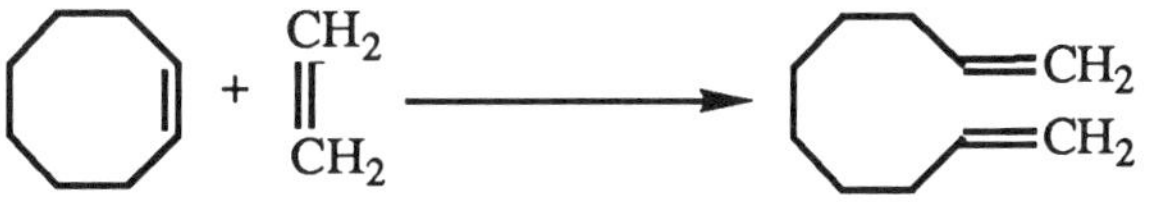

The reactions are conducted by passage of a solution of ethylene in the cycloolefin through a fixed bed of Re_2O_7 on basic alumina catalyst at 50-90°C and 15-40 atmospheres [22]. The product mixtures are complex because the olefinic products can also undergo metathesis. Metathesis of the α,ω-diolefins can give rise to higher molecular weight oligomers. Recycle of these high boiling products through the ethylene reaction, however, can cleave them to form desirable products. Overall yields to the 1,9-decadiene **2** and 1,5-hexadiene **3** are said to be high .

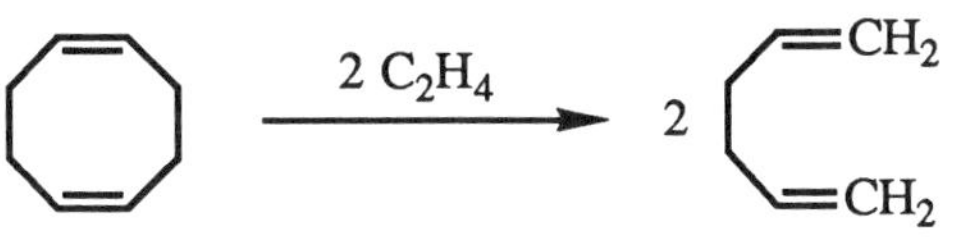

In addition to these processes aimed at producing monomers, the olefin metathesis reaction has been employed extensively as an approach to making polymers. Great scientific interest has been stirred by the use of metathesis to make polyacetylene either directly by polymerization of acetylene (Chapter 8) or cyclooctatetraene or indirectly from polycyclic polyenes (next section). Less exotic polymers obtained by metathesis polymerization of commercially available cyclic olefins have been produced industrially for 10-15 years as described below.

9.2   RING-OPENING METATHESIS POLYMERIZATION (ROMP)

When a cycloolefin is treated with a metathesis catalyst, the ring is cleaved at the double bond and many molecules are coupled together to form a polymer as shown with cyclooctene:

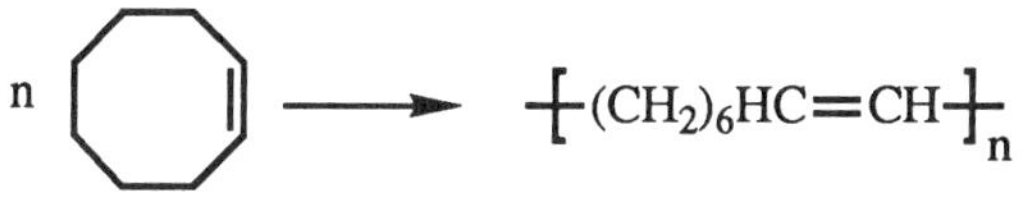

This reaction is quite general for cycloolefins other than cyclohexene, which is inert, as noted earlier. The commercial processes for olefin metathesis polymerization usually employ soluble catalysts, in contrast to the heterogeneous catalysts used in the SHOP process and the specialty olefin syntheses. The commercial ROMP catalysts tend to be mixtures of transition metal complexes rather than single molecular entities, but they provide reasonable control of polymer properties. The discrete molecular catalysts, such as the alkylidenetungsten complex **1** described earlier, give a "living polymer" in which the catalytically active metal species remains attached to the growing end of the polymer chain [1,4]. In common with other living polymers, the ROMP products can have a narrow molecular weight distribution; that is, the dispersity of molecular weights is only slightly above 1.0. The "living" character of the polymerization also facilitates the preparation of block copolymers.

The product from cyclooctene shown above is referred to generically as polyoctenamer. The commercial product sold by Huls as Vestenamer® contains 55-85% trans double bonds in the polymer chain [23]. It finds wide use as an elastomer in demanding applications such as automotive hoses and gaskets and as rollers for printing presses [6,24]. The product is made by treatment of cyclooctene with a Ziegler-type catalyst prepared from WCl_6, excess $EtAlCl_2$, and an activator such as ethanol, 2,6-di-*t*-butyl-4-methylphenol, or an allyl aryl ether [25,26]. Organic halides like 1,2-dichloroethylene are also excellent activators [27]. The reaction appears to be carried out in an inert solvent like hexane at ambient temperature and pressure.

Commercial applications have been sought for ROMP polymers of many cycloolefins. In general, the physical properties of the polymers have not been sufficiently desirable to justify the added cost vs. currently available polymers. For example, polypentenamer from potentially available cyclopentene has been studied extensively, but its properties are too much like those of polybutadiene or isoprene to justify the investment required for commercialization [6].

Two ROMP polymers that have been commercial successes are derived from strained-ring polycyclic polyenes. Polynorbornene and poly(dicyclopentadiene) are produced industrially [6,28]. Polynorbornene is of historical interest as the first ROMP polymer to be discovered [29] and the first to be produced commercially [10]. In the original example, a MoO_3 on alumina catalyst was reduced with hydrogen and was allowed to react with $LiAlH_4$ and norbornene in benzene solution at room temperature. The C=C bond of the cycloolefin opened to produce a polymer:

This type of polymer has been produced for 15 years by CdF Chimie and is sold as Norsorex®. It is a high molecular weight material which, after addition of plasticizers and fillers, can be crosslinked to produce a soft, tough rubber useful for many of the same applications as polyoctenamer. The crosslinking or vulcanization needed to produce the desired physical properties is facilitated by the C=C and allylic C-H bonds in the polymer chain. It is interesting that the polymerization is reported [10] to be carried out with a soluble catalyst ($RuCl_3$ in an alcoholic solvent) rather than a typical heterogeneous catalyst. The very high trans isomer content suggests that a Ziegler-type catalyst is used currently.

Poly(dicyclopentadiene) has been commercialized by Hercules as Metton® [30] and by B. F. Goodrich as Telene® [31,32] in formulations for reaction injection molding (RIM). In the RIM technique, an object such as an automobile bumper is formed by injecting a reactive fluid into a mold under conditions such that it polymerizes and crosslinks to a tough solid during the molding process. Dicyclopentadiene is attractive for this technology because it is cheap and because it has two double bonds that differ in chemical reactivity.

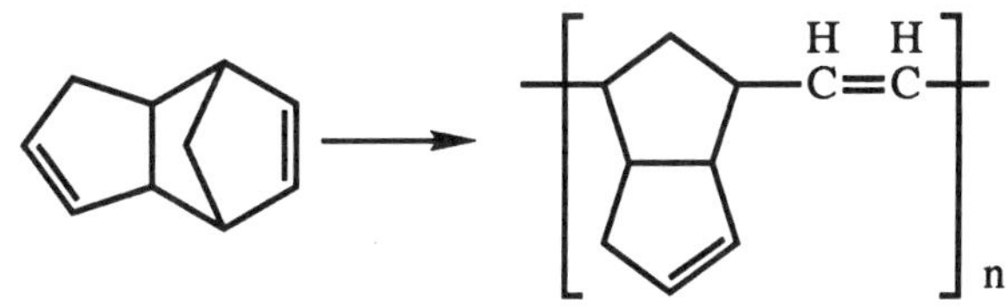

Presumably the more strained C=C bond, which resembles that in norbornene, polymerizes first to form the polymer shown. The cis double bond in the cyclopentene ring is somewhat less reactive, but still adequate for crosslinking of the polymer. The high concentration of crosslinkable bonds in the polymer leads to a very tough product.

In the Hercules process [30], two reactant streams are mixed in the RIM mold where both polymerization and crosslinking occur. One stream is typically a solution of WCl_6 in dicyclopentadiene. The other is a dicyclopentadiene solution of an activator such as $EtAlCl_2$. When they mix in the heated (ca. 60°C) mold, reaction is rapid and exothermic. A tough, strong object is formed in minutes. Fillers such as minerals, glass fibers, or butadiene-styrene elastomers may be added to either reactant stream. Hindered phenols and tin compounds are added as catalyst promoters and stabilizers.

One of the most elegant applications of the ROMP chemistry is an indirect synthesis of polyacetylene [4,28,33]. The ROMP polymerization of 4 in Figure 9.1 proceeds through opening of C=C bond in the highly strained cyclobutene ring. This reaction gives the polymer 5, which can be purified and has been well characterized. On heating, a reverse Diels-Alder reaction occurs and bis(trifluoromethyl)benzene is eliminated.

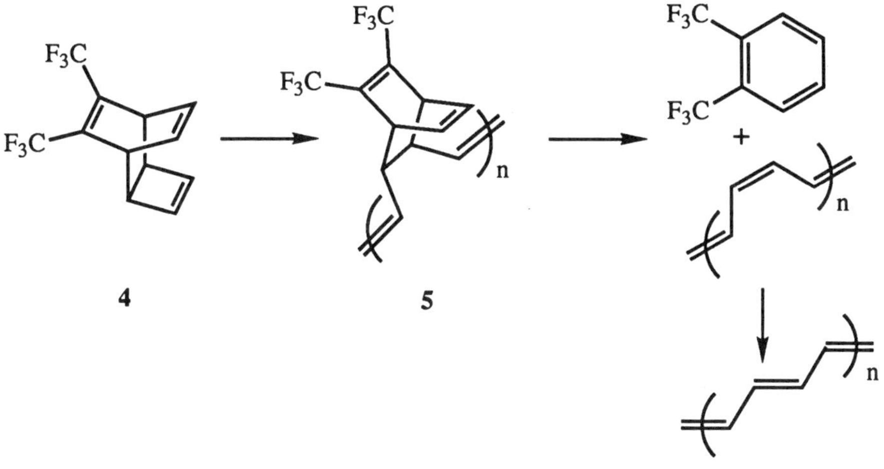

Figure 9.1 Synthesis of polyacetylene by way of a precursor polymer made by ROMP.

The residual polymer is a soluble polyacetylene containing both cis and trans C=C bonds. It can be cast into films and then heated to convert it to the familiar black, insoluble *trans*-polyacetylene (Chapter 8). The virtue of this indirect synthesis is that it produces a highly pure polymer suitable for study of its electronic and optical properties.

Amorphous hydrocarbon polymers containing alicyclic structures offer promise as substrates for optical disks and as aspherical lenses for optical disk reading systems. The characteristics which contribute to their functionality are low birefringence, low water absorption, transparency, and thermal resistance. The first and best established product is produced by Mitsui Petrochemical and is based upon Ziegler-Natta copolymerization of ethylene and cyclic monomers (see Chapter 4). Subsequent products have been based upon ROMP followed by hydrogenation. Nippon Zeon has commercialized Zeonex® [34-35], a product based upon a substituted norbornene [36].

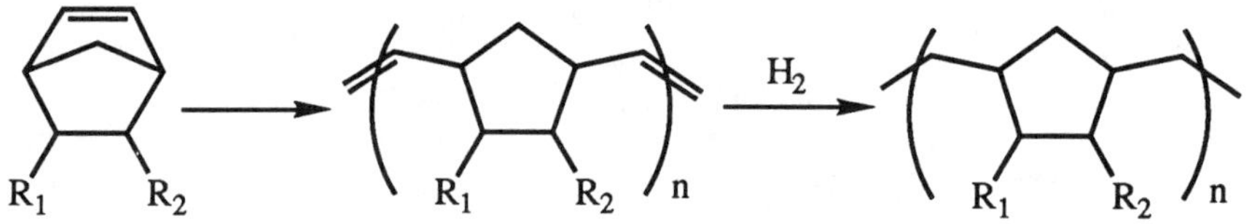

Hydrogenation is used to remove unsaturation. Japan Synthetic Rubber has commercialized Arton®, based upon ROMP of Diels-Alder adducts of dicyclopentadiene with acrylates or possibly methacrylates [37,38].

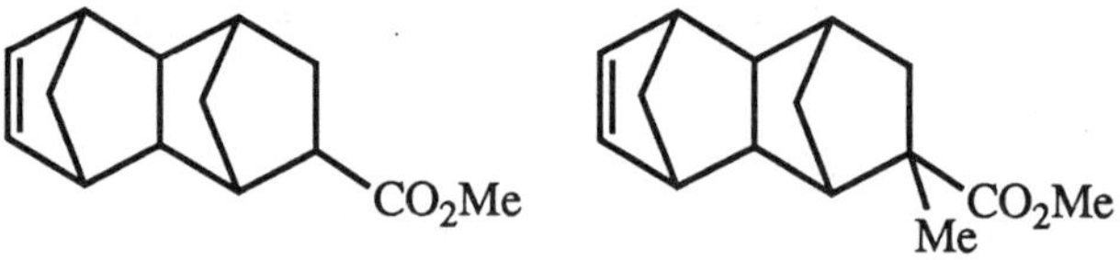

Their patents also include copolymers of dicyclopentadiene with these substrates. Their catalyst system is generally WCl_6 with $AlEt_2Cl$, and again, the polymer is hydrogenated. The functional groups are said to improve adhesion of inorganic layers to the compact disk substrates so this polymer may be used as an adhesive layer between a purely hydrocarbon substrate and the metallized layer.

9.3   OLEFIN METATHESIS MECHANISMS

Olefin metathesis is so unlike other catalytic reactions of olefins (Chapters 2-6) that its mechanism has attracted intense study. Initial thoughts centered on concerted reaction paths. These generally involved exchange of alkylidene fragments between two olefins via cyclobutane-like intermediates such as **6**. Despite the intellectual appeal of a concerted process, the evidence now firmly supports the stepwise mechanism discussed below.

$$
M + \begin{array}{c} H_2C{=}CH_2 \\ PhHC{=}CHPh \end{array} \rightleftharpoons \begin{array}{c} M \\ H_2C\cdots|\cdots CH_2 \\ \vdots \qquad \vdots \\ PhHC\cdots\text{-}\text{-}\text{-}CHPh \end{array} \rightleftharpoons \begin{array}{c} M \\ H_2C \quad + \quad CH_2 \\ \| \qquad \| \\ PhHC \qquad CHPh \end{array}
$$

$$\mathbf{6}$$

The currently accepted mechanism is illustrated by the stilbene-ethylene metathesis in Figure 9.2. This scheme, based on a proposal by Chauvin [39], invokes a metallacyclic intermediate. The catalyst cycle begins with an alkylidene complex **7** that may be formed by α-hydrogen elimination from a W-CH$_3$ group. Coordination of stilbene to **7** yields the olefin complex **8**. Intramolecular cycloaddition of C=C to W=C produces the metallacycle **9**. The cycloaddition process seems to be readily reversible. Fission of the ring can either regenerate **8** or it can produce the new olefin complex **10**. Displacement of styrene from **10** by ethylene gives **11**, which can again undergo cycloaddition and ring fission to complete the catalytic cycle. This mechanism helps explain the apparent inactivity of some catalysts in the metathesis of terminal olefins. The combination of WCl$_6$, C$_2$H$_5$AlCl$_2$, and ethanol catalyzes one of the most rapid homogeneous catalytic reactions. With internal olefins such as 2-butene, rates of 100 catalyst cycles per second may be attained, but little or no metathesis is detected with terminal olefins. This anomaly is explained by labelling studies. Terminal olefins rapidly exchange CH$_2$ groups with one another, but CHR groups do not exchange. This result may be rationalized by assuming that a methylene complex such as **7** coordinates a terminal olefin in such a way that the metallacycle

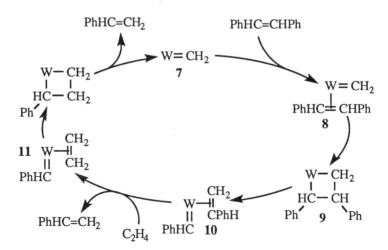

Figure 9.2 A metallacyclic mechanism for ethylene/stilbene metathesis. All the reactions are reversible.

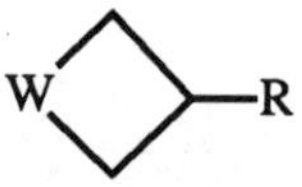

is formed selectively. Fragmentation of this ring can produce CH_2 exchange, but does not accomplish metathesis in the usual sense.

Sophisticated product analyses and labelling studies support a mechanism like that of Figure 9.2. Recently, a number of complexes like those in the catalyst cycle of the figure have been characterized. A close analog of the methylene complex **7** that initiates the catalyst cycle is the neopentylidene complex **1** described earlier. Complex **1** (Figure 9.3) is particularly significant because it is both well characterized and catalytically active [3]. It reacts with bis(trifluoromethyl)norbornadiene at low temperature to form an unstable, but isolable tungstacyclobutane **12**, which is an analog of the proposed intermediate **9** [40] . At higher temperatures, the reaction of the diene with **1** leads to formation of the trifluoromethyl-substituted poly-(norbornadiene).

The origin of the alkylidene complex **7** (Figure 9.2) that initiates metathesis seems quite varied. In organometallic systems like $WCl_6/C_2H_5AlCl_2$, α-hydrogen abstraction from an alkyl group is almost certainly the source. In this instance, the cocatalyst also serves as a Lewis acid to stabilize the alkylidene ligand. In the metal carbonyl and metal oxide catalysts, different mechanisms must be responsible. A likely mechanism for generation of the active sites in the commercial catalysts based on metal oxides is reaction of the olefin with oxometal species on the catalyst surface:

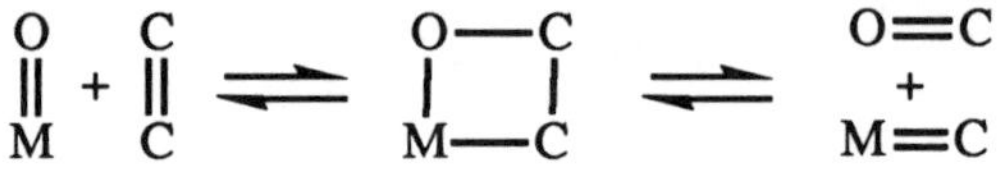

Strong inferential evidence has been developed for this mechanism in the homogeneous ROMP reaction of norbornene catalyzed by OsO_4 [41]. The olefin reacts with OsO_4 and an amine to form an osmium glycolate complex **14**, as expected from the well-known glycol synthesis described in Chapter 6. In the absence of amine, however, a significant amount of the ROMP polynorbornene is isolated. This observation was interpreted to indicate the sequence of events shown in Figure 9.4.

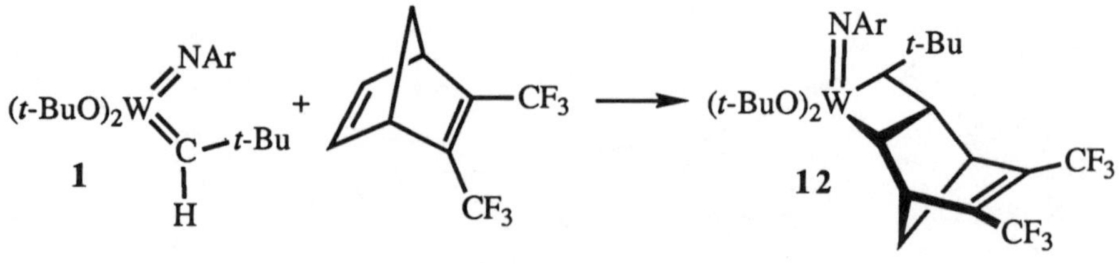

Figure 9.3 Isolation of a tungstacyclobutane complex (Ar = 2,6-diisopropylphenyl).

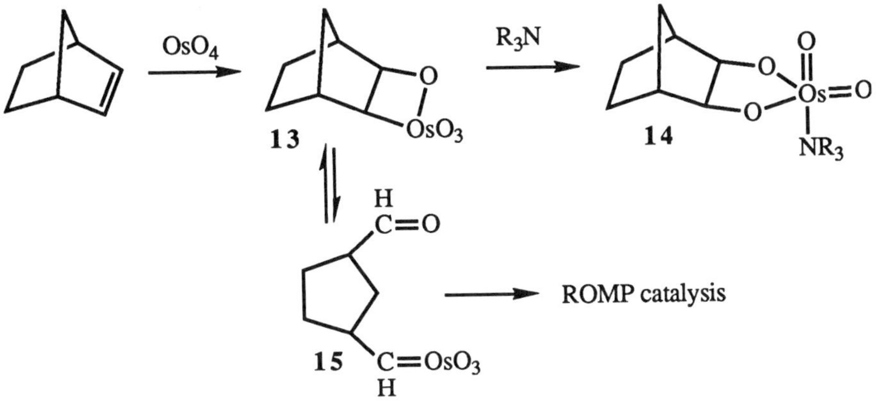

Figure 9.4 Generation of an osmium alkylidene complex from OsO₄ and an olefin.

The key intermediate is **13**, an oxaosmacyclobutane of the sort that Sharpless has postulated [42] in the osmylation of olefins. This complex can ring-expand to the glycolate, which is facilitated by the presence of an amine ligand, or it can open the C-C and Os-O bonds to form **15**. This latter process is presumed to provide the alkylidene complex necessary to initiate the ROMP reaction. The **13** ↔ **15** equilibrium is analogous to that postulated above for the heterogeneous catalysts based on metal oxides. The CH₂ or CHR complexes that take part in the metathesis cycle are relatively labile. A major decomposition pathway is dimerization of the ligands to form olefin [43].

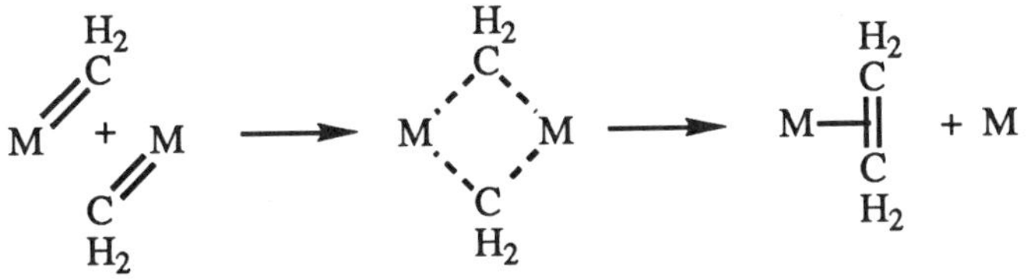

In the heterogeneous metathesis catalysts, this process is inhibited by the isolation of the alkylidene ligands at well-separated sites on the catalyst surface. As mentioned previously, many of the soluble catalysts are stabilized by coordination of the ligand to a Lewis acid. Steric hindrance also plays a major role by inhibition of the alkylidene dimerization.

9.4 RING FORMATION VIA CARBENOID SPECIES

The importance of metal-catalyzed "carbene" transfer reactions is not limited to olefin metathesis. Cyclization reactions involving carbenoid reactants have found significant commercial use in the past 15 years. These reactions may involve either addition of a carbenoid ligand to an olefinic C=C bond (cyclopropanation) or addition of an N-H bond to a divalent carbon species. In general, a complex of a late transition metal ion is used as a catalyst with three functions [44,45].

- Decomposition of a diazo compound such as N_2CHCO_2Et to produce N_2 and highly reactive carbenoid species.
- Stabilization of the "carbene" by complexation to the metal ion.
- Transfer of the carbene ligand to a C=C or N-H species, which may also be coordinated to the metal ion.

The third step may involve significant stereochemical control if the catalytic metal complex bears chiral ligands. The enantioselectivity aspect makes the technology interesting to the pharmaceutical industry and to others with a need for specific optical isomers.

In 1909, Loose [46] discovered that copper bronze catalyzes the decomposition of ethyl diazoacetate to a mixture of dimers and trimers of carboethoxycarbene:

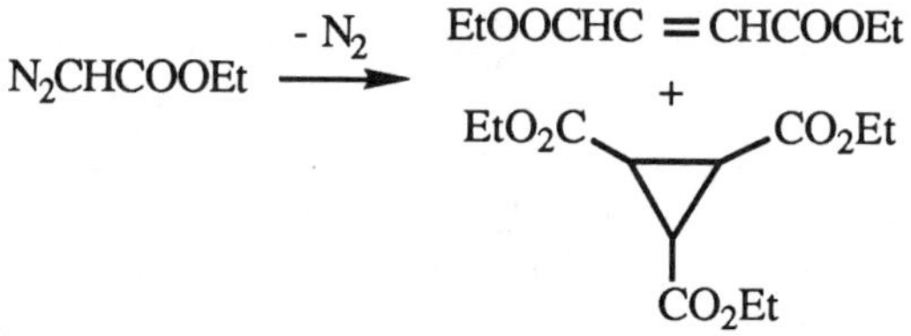

$$N_2CHCOOEt \xrightarrow{- N_2} \begin{array}{c} EtOOCHC = CHCOOEt \\ + \\ \text{(cyclopropane structure with } EtO_2C, CO_2Et, CO_2Et) \end{array}$$

It appeared that the intermediate carbenoid species, presumably a copper complex of $:CHCO_2Et$, could add across the double bond of diethyl fumarate to form a cyclopropane ring. Although many aspects of the mechanism are unknown, the copper-catalyzed reaction of N_2CHCO_2Et with olefins and dienes became a valuable procedure for making cyclopropane derivatives [47].

In recent decades [44,48] soluble complexes of copper, cobalt, and rhodium have been found to catalyze carbenoid reactions of ethyl diazoacetate analogous to those promoted by copper bronze. The soluble catalysts provide a measure of stereochemical control as described above. As a consequence, they have found several applications in the synthesis of specialty chemicals [2].

Permethric Acid

The first industrial application of these metal-catalyzed carbenoid cyclizations was the synthesis of a cyclopropanecarboxylic acid derivative, which is a central feature of an important class of insecticides. Pyrethrum, a mixture of natural compounds extracted from an East African plant, has been used extensively as a broad spectrum insecticide. It has many virtues, such as rapid kill of insects and low mammalian toxicity, but it is not very stable to light and air. To get better stability as well as to assure a supply of product not susceptible to the climatic and political variables of East Africa, extensive exploration of synthetic analogs of pyrethrum has occurred [2,49].

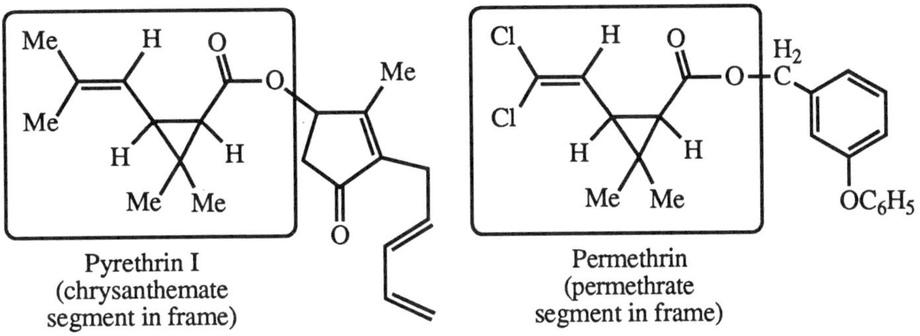

Figure 9.5 Structures of natural and synthetic pyrethroid insecticides.

Natural pyrethrum is a mixture of esters of chrysanthemic acid, typified by pyrethin I in Figure 9.5. A central feature of this molecule is a heavily substituted cyclopropane ring. This structural element is also found in many, but not all, of the synthetic analogs that exhibit the desired biological activity. A representative synthetic pyrethroid is permethrin, which is used extensively for insect control. It displays similar toxicity to insects, but is much more stable than pyrethrum. Like natural pyrethrins, it contains a vinylcyclopropane-carboxylic acid unit, but the substituents on the vinyl group and the ester are different.

A major challenge in designing an industrially useful synthesis was construction of the cyclopropane ring in the permethrate segment of permethrin (Figure 9.5). Not only is it necessary to produce the strained three-membered ring, it is also desirable to produce the isomer with the vinyl and carboxyl groups on the same side of the ring. (With permethrin, the cis isomer has maximum biological activity; in pyrethrin, the trans isomer is more active.) The activity also varies with the chirality of the carbons bearing these substituents [49].

Some commercial processes for making permethrin employ the reaction of ethyl diazoacetate with the diene **16** to form the desired cyclopropanecarboxylate **18** (ethyl permethrate):

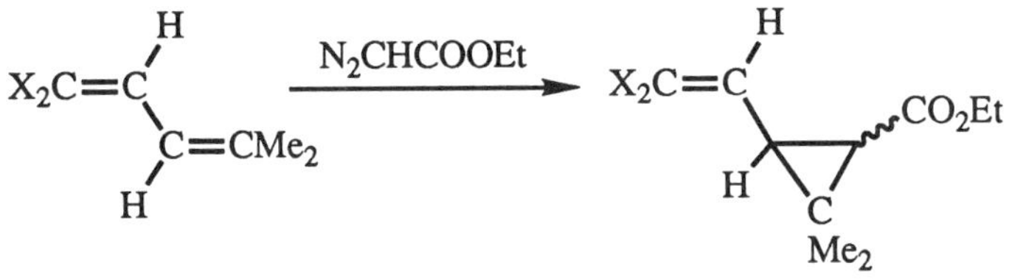

16 X = Cl
17 X = CH$_3$

18 X = Cl
19 X = CH$_3$

ICI initially employed a soluble catalyst for this process, a dimeric rhodium carboxylate, Rh$_2$(O$_2$CR)$_4$ [50]. Earlier work [51] had shown that these remarkable Rh-Rh bonded complexes catalyze the evolution of N_2 from diazoacetate esters and the addition of the :CHCO$_2$R fragment to substituted olefins. The merit of this reaction for permethrate synthesis is that the carbenoid fragment adds selectively to the electron-rich, methyl-substituted C=C bond of **16**. When ethyl diazoacetate and **16** are added to a solution of dirhodium tetrapivalate at 20°C, **18** is formed in 66% yield (based on diazo compound) and with a 60:40 ratio of cis:trans isomers.

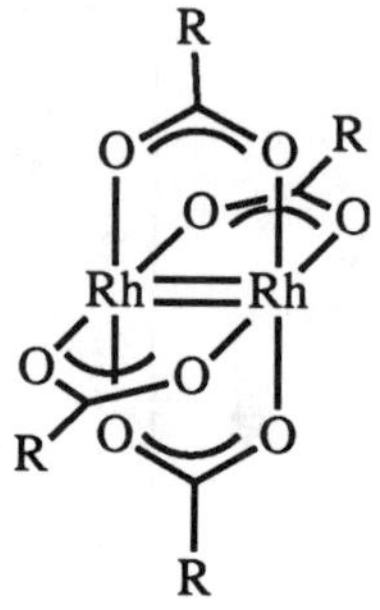

In current practice, a heterogeneous catalyst, the traditional copper bronze, is used for synthesis of **18** and ethyl chrysanthemate **19** [52]. Even though the copper catalyst produces only a 41:59 cis:trans ratio in **18** [50], the yield is 80% and it offers several practical advantages [53]. The catalyst is cheap and, because it is insoluble, it permits operation of a continuous process. In so doing, worker exposure to the toxic and explosive ethyl diazoacetate is minimized.

Extensive work continues on catalysts containing clusters of rhodium atoms [44]. Both the well-characterized $Rh_6(CO)_{16}$ and less well-defined rhodium(II) acetamide complexes show good activity. The combination of the latter catalyst with a sterically bulky diazoacetate

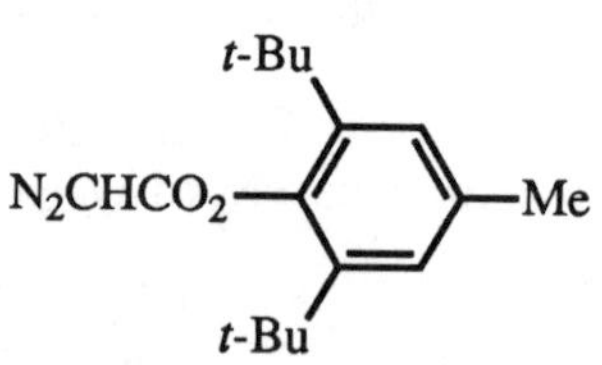

leads to extraordinarily high proportions of *trans*-cyclopropanecarboxylates [54]. With 2,5-dimethyl-2,4-hexadiene as a precursor to chrysanthemic acid, this combination gives 94-98% selectivity to the potent trans isomer.

As noted above, the chirality as well as the cis:trans ratio is important to the biological activity of the pyrethroids. Japanese work on enantioselective cyclopropanation catalysts has given high selectivity in production of the desired optical isomers. This work has led to a commercial synthesis of the drug cilastatin (next section) as well as enantioselective syntheses of permethrates and chrysanthemates. The typical catalysts are Schiff base complexes of cobalt(II) and copper(II). For example, as shown in Figure 9.6, the copper complex **20** bearing a sterically encumbered chiral ligand catalyzes the addition of ethyl diazoacetate to the olefin **21** to form **22**, a precursor of the most biologically active permethrin, the

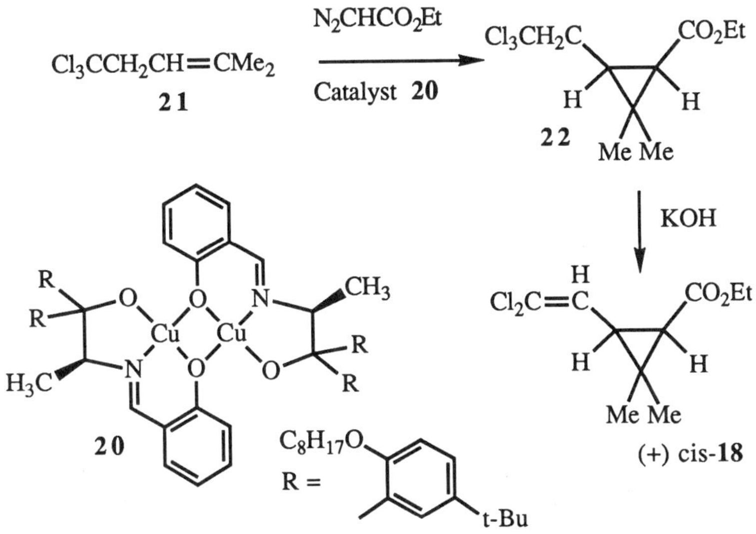

Figure 9.6 Enantioselective cyclopropanation with a chiral copper catalyst.

(+)-*cis* form [48,55]. The (S)-form of **20** gives an 85:15 cis:trans ratio with a 91% enantiomeric excess of the (+)-form of the cis isomer. Other work using cobalt(II) complexes of camphor-derived *vic*-dioximes has also given high selectivity for the desired permethrate isomer [56].

Chrysanthemic acid esters, **19**, found in natural pyrethrins, exhibit their highest biological activity in the D-*trans* form. This isomer is produced in high chemical and optical yield by reaction of L-menthyl diazoacetate with the diene **17** in the presence of the R-isomer of **20** [57].

In recent work [58], asymmetric catalysis has been achieved with dimeric rhodium complexes bearing chiral pyrrolidone ligands. These complexes, which are close analogs of the Rh$_2$(O$_2$CCH$_3$)$_4$ catalyst described earlier, impart a high degree of enantioselectivity to the intramolecular cyclopropanation of allylic diazoacetates. For example, the reaction of allyl diazoacetate itself gives a 74% yield of the chiral lactone **23** with an 88% enantiomeric excess.

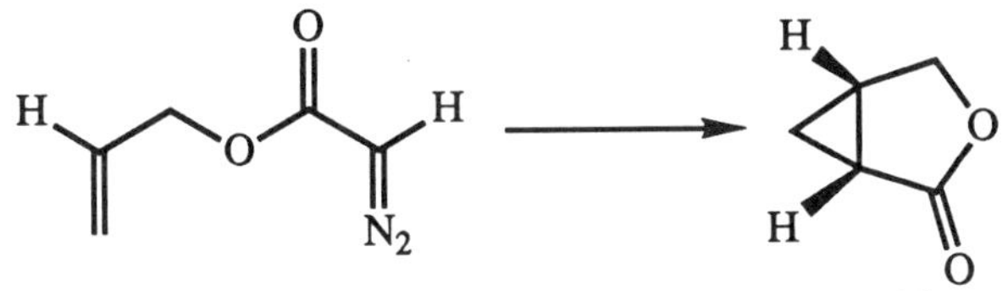

In addition to catalytic cyclopropanation, many other approaches to synthesis of permethric acid derivatives have been explored. Two of the alternatives are shown in

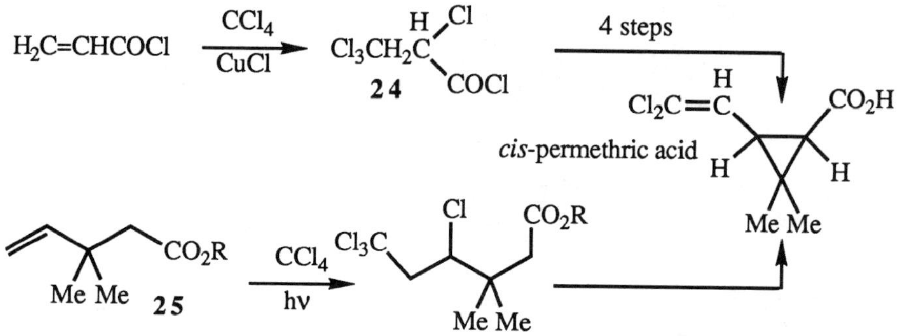

Figure 9.7 Use of Cl₃C• radical addition reactions in the Ciba-Geigy and Sagami syntheses of permethric acid.

Figure 9.7. The upper equation represents a key step in technology developed by Ciba-Geigy [59], but not commercialized. This step utilizes a homogeneous catalytic reaction that often seems overlooked. This catalysis involves the addition of a C-Cl bond to a C=C double bond via a species that may be regarded as a complexed "free radical." Copper(I) chloride and other metal ions with easily accessible one-electron transitions catalyze the addition of CCl₄ across an olefin to form a product **24** with a terminal Cl₃C-group. The formation of a 1:1 adduct with few higher telomers suggests that the intermediate Cl₃C• radical is complexed to the metal ion rather than being free. The better known, photocatalytic, free-radical addition of CCl₄ to the olefin **25** provides the basis for the industrial synthesis of permethrin commercialized by Sagami [49,60].

Cilastatin

The enantioselective cyclopropanation catalyst **20** (Figure 9.6) has been used to synthesize the drug Cilastatin, which is used extensively as an adjuvant to the antibiotic imipenem described below. A key step in the synthesis is the stereoselective assembly of the necessary cyclopropane ring (Figure 9.8).

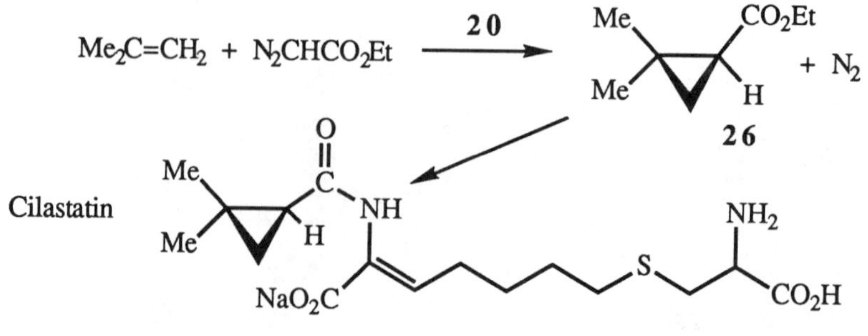

Figure 9.8 Enantioselective synthesis of the cyclopropane ring in Cilastatin.

The desired biological activity is found in the stereoisomer derived from (+) 1S-2,2-dimethylcyclopropanecarboxylic acid. The ethyl ester (**26**) of this acid is manufactured by Sumitomo through the enantioselective reaction of ethyl diazoacetate with isobutylene in the presence of a chiral catalyst. When this reaction is carried out in toluene at 40°C in the presence of a catalytic amount of **20**, the desired cyclopropanecarboxylate is formed in 92% enantiomeric excess [48,55].

Imipenem

One of the most important recent developments in antibiotic therapy was the commercialization in 1985 of the broad-spectrum antibacterial Primaxin® [61,62]. This drug is formulated from Cilastatin (described above) and imipenem, a synthetic β-lactam related to penicillin. Imipenem has extraordinary versatility and potency, but it is sensitive to hydrolysis by the kidney enzyme dehydropeptidase I. This problem is overcome by the simultaneous administration of Cilastatin, which specifically inhibits the action of the enzyme.

A key step in the manufacture of imipenem is the closure of the 5-membered ring which is fused to the β–lactam structure (Figure 9.9). This step is accomplished by rhodium-catalyzed reaction of a diazo function to generate a carbenoid intermediate that, in effect, inserts into the N-H bond of the lactam in **27**. The diazo lactam **27** [63] is heated with a CH_2Cl_2 solution of a $Rh_2(O_2CR)_4$ catalyst. (The octanoate is used commercially for solubility reasons.) Nitrogen gas is evolved and the desired penicillinate **28** is formed almost quantitatively [61,64].

It is tempting to speculate that the rhodium complex reacts with the diazo compound to produce a chelate intermediate such as **29**. Bringing the N-H and carbene functions together in the coordination sphere of the metal ion may facilitate the

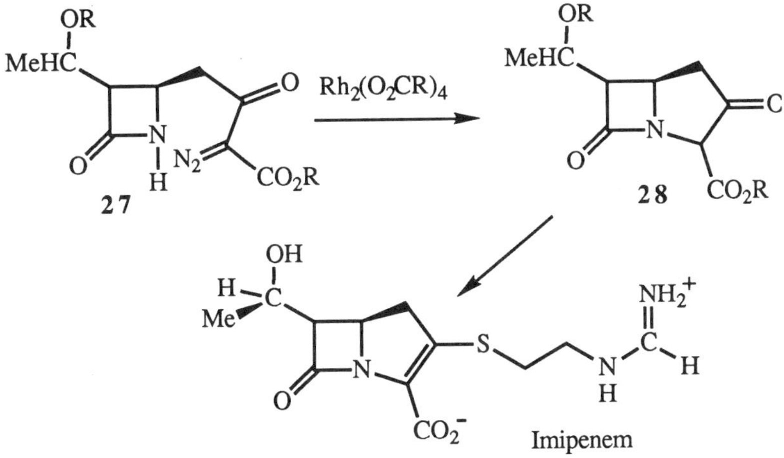

Figure 9.9 Closure of the five-membered ring in the synthesis of imipenem.

"insertion" of the carbene into the N-H bond. Extensive precedent exists for catalytic insertion of "carbenes" into N-H and C-H bonds [44,65].

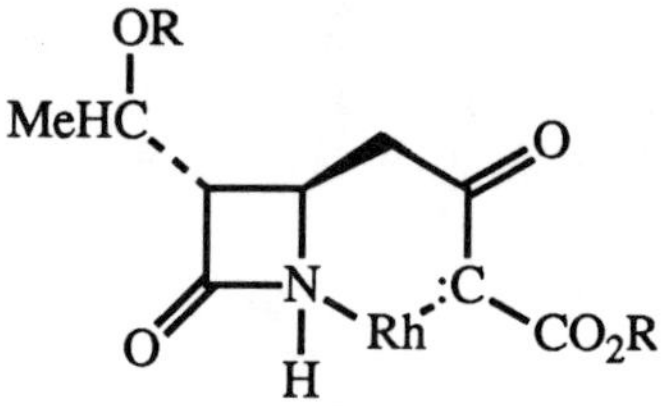

$$\underset{\textbf{29}}{\overset{\displaystyle OR}{\underset{}{}}}$$

Attempts to isolate carbenoid complexes from the metal-catalyzed decomposition of ethyl diazoacetate have generally been unsuccessful. The reaction of the diazo ester with iodorhodium(III) tetra-p-tolylporphyrin produces N_2, but the organometallic product appears to result from insertion of the carbene into the Rh-I bond [66]:

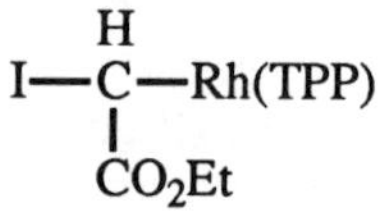

$$I\!\!-\!\!\overset{\displaystyle H}{\underset{\displaystyle CO_2Et}{C}}\!\!-\!\!Rh(TPP)$$

GENERAL REFERENCES

K. J. Ivin, *Olefin Metathesis*, Academic Press, New York, 1983.

V. Dragutan, A. T. Balaban, and M. Dimonie, *Olefin Metathesis and Ring-Opening Polymerization of Cyclo-Olefins*, Wiley, New York, 1985.

J. Mol. Catal. (special volumes on metathesis), **15** (1982); **28** (1985); **46** (1988).

R. L. Banks, "Olefin Metathesis," in B. E. Leach, Ed., *Applied Industrial Catalysis*, Academic Press, New York, 1984, pp. 215-239.

SPECIFIC REFERENCES

1. R. H. Grubbs and W. Tumas, *Science*, **243**, 907-15 (1989); J. Feldman and R. R. Schrock, *Prog. Inorg. Chem.*, **39**, 1-74 (1991).
2. G. W. Parshall and W. A. Nugent, *Chemtech*, **18**, 184-90, 376-83 (1988).
3. R. R. Schrock, R. T. DePue, J. Feldman, K. B. Yap, D. C. Yang, W. M. Davis, L. Park, M. DiMare, M. Schofield, J. Anhaus, E. Walborsky, E. Evitt, C. Kruger, and P. Betz, *Organometallics*, **9**, 2262-75 (1990).
4. R. R. Schrock, *Acc. Chem. Res.*, **23**, 158-65 (1990).
5. R. Torecki and R. R. Schrock, *J. Am. Chem. Soc.*, **112**, 2448-9 (1990).

6.   R. Streck, *J. Mol. Catal.*, **46**, 305-16 (1988); *Chemtech*, **19**, 498-503 (1989).

7.   *Hydrocarbon Proc.*, **46**, 232 (Nov. 1967).

8.   J. Kress, M. Wesolek, and J. A. Osborn, *J. Chem. Soc., Chem. Commun.*, 514-6 (1982).

9.   B. M. Novak and R. H. Grubbs, *J. Am. Chem. Soc.*, **110**, 7542-3 (1988).

10. *Inf. Chem.*, **83**, 179 (1978).

11. R. L. Banks and S. G. Kukes, *J. Mol. Catal.*, **28**, 117-31 (1985).

12. P. D. Montgomery, R. N. Moore, and W. R. Knox, U.S. Patent 3,965,206 (1976).

13. C. F. Hobbs, U.S. Patents 4,439,626; 4,439,627; 4,439,628 (1984).

14. A. J. Berger, U.S. Patent 3,726,938 (1973); P. A. Verbrugge and G. J. Heiszwolf, U.S. Patent 3,776,975 (1973).

15. E. R. Freitas and C. R. Gum, *Chem. Eng. Progr.* **75**, 73-6 (1979).

16. R. A. Kemp and D. M. Hamilton, U.S. Patent 4,754,099 (1988).

17. R. L. Banks, D. S. Banasiak, P. S. Hudson, and J. R. Norell, *J. Mol. Catal.*, **15**, 21-33 (1982).

18. D. L. Crain and R. E. Reusser, U.S. Patent 3,660,516 (1972).

19. P. Chaumont and C. S. John, *J. Mol. Catal.*, **46**, 317-28 (1988).

20. H. Short and R. Remirez, *Chem. Eng.*, 22-25 (17 Aug. 1987).

21. D. R. Fahey in P. N. Rylander and G. Greenfield, Eds., *Catalysis in Organic Syntheses*, Academic Press, New York, 1976, p. 287.

22. F. D. Mango, U.S. Patent 3,424,811 (1969); G. J. Heiszwolf and R. van Helden, British Patent 1,482,745 (1977).

23. W. A. Schneider and M. F. Muller, *J. Mol. Catal.*, **46**, 395-403 (1988).

24. K-D. Hesse, *Eur. Rubber J.*, 37-41 (Sept. 1987).

25. R. A. Streck, *J. Mol. Catal.*, **15**, 3-19 (1982).

26. R. Streck and H. Weber, U.S. Patent 3,974,092 (1976).

27. J. Finter, G. Wegner, E. J. Nagel, and R. W. Lenz, *Makromol. Chem.*, **181**, 1619-28 (1980).

28. W. J. Feast and V. C. Gibson, "Olefin Metathesis," in F. R. Hartley, Ed., *The Chemistry of the Metal-Carbon Bond*, Vol. 5, Wiley, New York, 1989, pp. 199-228.

29. H. S. Eleuterio, U.S. Patent 3,074,918 (1963); *J. Mol. Catal.*, **65**, 55-61 (1991).

30. R. P. Geer and R. D. Stoutland, *Plast. Eng.*, **41**, 41-4 (Nov. 1985); *Chem. Eng. News*, 10-11 (23 April 1990); D. W. Klosiewicz, U.S. Patent 4,400,340 (1983).

31. R. J. Minchak, T. J. Kettering, and W. J. Kroenke, U.S. Patent 4,380,617 (1983).

32. B. L. Goodall and L. F. Rhodes, U.S. Patent 4,923,936 (1990).

33. J. H. Edwards, W. J. Feast, and D. C. Bott, *Polymer*, **25**, 395-8 (1984).

34. M. Oshima, *Kino Zairyo* **10**(5), 5-10 (1990); *Chem. Abstr.*, **114**, 144944 (1990).

35. T. Natsuume, M. Oshima, and M. Yamazaki, *Mater. Res. Soc. Symp. Proc.*, **150**, 245-50, (1989); T. Natsuume, *Kogyo Zairyo* **38**(15), 69-75 (1990).

36. *The Chemical Daily* (Dec 7, 1990).

37. K. Goto, Z. Komiya, N. Yamahara, A. Iio, European Patent Application 317,262 (1989) to Japan Synthetic Rubber Co., Ltd.; *Chem. Abstr.*, **111**, 175360 (1989).

38. K. Goto, T. Komiya, N. Satoa, and A. Iio, Jpn. Kokai Tokkyo Koho 01,197,460 (1989) to Japan Synthetic Rubber Co., Ltd.; *Chem. Abstr.*, **112**, 157736 (1989).

39. J. L. Herisson and Y. Chauvin, *Makromol. Chem.*, **141**, 161-76 (1970).

40. G. C. Bazan, E. Khosravi, R. R. Schrock, W. J. Feast, V. C. Gibson, M. B. O'Regan, J. K. Thomas, and W. M. Davis, *J. Am. Chem. Soc.*, **112**, 8378-87 (1990).

41. J. G. Hamilton, O. N. D. Mackey, J. J. Rooney, and D. G. Gilheany, *J. C. S Chem. Commun.*, 1600-1 (1990).

42. S. G. Hentges and K. B. Sharpless, *J. Am. Chem. Soc.*, **102**, 4263-5 (1980).

43. R. R. Schrock and P. R. Sharp, *J. Am. Chem. Soc.*, **100**, 2389-99 (1978).

44. M. P. Doyle, *Chem. Rev.*, **86**, 919-39 (1986).

45. G. Maas, *Topics Curr. Chem.*, **137**, 75-253 (1987).

46. A. Loose, *J. Prakt. Chem.* [2], **79**, 507 (1909).

47. W. Kirmse, *Carbene Chemistry , 2nd ed.*, Academic Press, 1971.

48. T. Aratani, *Pure Appl. Chem.*, **57**, 1839-44 (1985).

49. D. Arlt, M. Jautelet, and R. Lantzsch, *Angew. Chem. Int. Ed.*, **20**, 703-22 (1981).

50. D. Holland and D. J. Milner, *J. Chem. Res. (M)*, 3734-46 (1979); British Patent 1,553,638 (1979).

51. A. J. Hubert, A. F. Noels, A. J. Anciaux, and P. Teyssie, *Synthesis*, 600-2 (1976).

52. R. Pearce and W. R. Patterson, Eds., *Catalysis and Chemical Processes*, Wiley, New York, 1981, p. 333.

53. K. S. Shim and D. J. Martin, British Patent 1,306,191; *Chem. Abstr.*, **77**, 113,878 (1972).

54. M. P. Doyle, V. Bagheri, T. J. Wandless, N. K. Harn, D. A. Brinker, C. T. Eagle, and K-L. Loh, *J. Am. Chem. Soc.*, **112**, 1906-12 (1990).

55. T. Aratani, H. Yoshihara, and G. Susukamo, U.S. Patent 4,552,972 (1985).

56. A. Nakamura, A. Konishi, V. Tatsuno, and S. Otsuka, *J. Am. Chem. Soc.*, **100**, 3443-8, (1978); A. Nakamura, A. Konishi, R. Tsujitani, M. Kudo, and S. Otsuka, *ibid.*, 3449-61 (1978).

57. T. Aratani, Y. Yoneyoshi, and T. Nagase, *Tetrahedron Lett.*, 2599-2602 (1977).

58. M. P. Doyle, R. J. Pieters, S. F. Martin, R. E. Austin, C. J. Oalmann, and P. Muller, *J. Am. Chem. Soc.*, **113**, 1423-4 (1991).

59. D. Bellus, *Pure Appl. Chem.*, **57**, 1827-38 (1985); P. Martin, E. Steiner, and D. Bellus, *Helv. Chim. Acta*, **63**, 1947-57 (1980).

60. K. Kondo, K. Matsui, and A. Negishi, "New Synthesis of the Acid Moiety of Pyrethroids" in M. Elliott, Ed., *Synthetic Pyrethroids*, American Chemical Society Symposium Series, Vol. 42, American Chemical Society, Washington, DC, 1977, pp. 128-36.

61. B. G. Christensen, *Chem. Brit.*, 371-3 (1989).

62. S. C. Stinson, *Chem. Eng. News*, 33-61 (29 Sept. 1986).

63. J. S. Amato, S. Karady, and L. M. Weinstock, U.S. Patent 4,444,685 (1984).

64. D. G. Melillo, I. Shinkai, T. Liu, K. Ryan, and M. Sletzinger, *Tetrahedron Lett.*, **21**, 2783-6 (1980).

65. D. F. Taber, "Carbon-Carbon Bond Formation by C-H Insertion," in *Comprehensive Organic Synthesis*, Pergamon, New York, 1991.

66. J. Maxwell and T. Kodadek, *Organometallics*, **10**, 4-6 (1991).

10 | OXIDATION OF HYDROCARBONS BY OXYGEN

The largest scale application of homogeneous catalysis is the oxidation of hydrocarbons by molecular oxygen. Oxidation processes and their uses are listed in Table 10.1 in the order in which they are discussed in this chapter. The remarkable scale of processes like the oxidations of *p*-xylene and cyclohexane has been indicated in the table in Chapter 1.

The oxidation of organic compounds by molecular oxygen dates to Lavoisier and stimulated much early research [1,2]. Very early on, catalysts were introduced to oxidation reactions to control their course [3]. The catalytic processes discussed in

Table 10.1   Major Industrial Oxidations of Hydrocarbons

Hydrocarbons	Oxidation Products	Applications
Cyclohexane	Cyclohexanol and cyclohexanone	Converted to adipic acid and caprolactam (polyamide precursors)
Cyclododecane	$C_{12}H_{23}OH$ and $C_{12}H_{22}O$	Oxidized to dodecanedioic acid and lauryl lactam (polyamide precursors)
Butane	Acetic acid	Solvent, vinyl acetate polymers
Toluene	Benzoic acid	Caprolactam (polyamide precursors), phenol (Section 7.3), and food preservative
	Benzaldehyde	Agrichemicals, flavorings, and fragrances
m-Xylene	Isophthalic acid	Polymers and plasticizers
p-Xylene	Terephthalic acid	Polyester fibers, films, and plastics
	Terephthalate esters	(Section 11.1), plasticizers

this chapter differ sharply from those of earlier chapters. Although soluble metal complexes enhance the yields and selectivities of these oxidations, the catalytic complex often does not interact with either O_2 or the hydrocarbon.

Dozens of metal complexes of molecular oxygen have been isolated and characterized [4,5], but such complexes do not seem to be involved in the *major* oxidation pathways. Instead, the reaction between the hydrocarbon and oxygen is a radical chain process that operates even in the absence of metal catalysts. The usual initial product of the radical chain process is an alkyl hydroperoxide, ROOH. These hydroperoxides are unstable under reaction conditions and decompose to give alcohols, ketones, or carboxylic acids. The major function of the metal complex in many oxidation processes is catalytic decomposition of hydroperoxides. In this way, the metal enhances the formation of desirable products and stimulates the production of free-radical-species that initiate the radical chain reaction between the hydrocarbon and oxygen. These two effects give substantial control over the yield and the rate of the overall oxidation process.

In the following sections, the general reactions of oxygen with metal complexes and with hydrocarbons are discussed first. The interplay between these two kinds of chemistry is illustrated in the oxidation of cyclohexane, an important hydrocarbon substrate. Finally, the oxidations of the major industrial hydrocarbons are considered individually.

10.1 REACTIONS OF O_2 WITH METAL COMPLEXES

In contrast to other simple diatomic molecules such as N_2 and F_2, dioxygen (O_2) is paramagnetic, having two unpaired electrons in the ground state. The highest occupied molecular orbitals are a pair of π^* orbitals of equal energy so that the two highest energy electrons have no driving force to spin-pair [4]. Consequently, dioxygen may be regarded as a diradical whose chemistry is dominated by one-electron steps.

$$O_2 + e^- \rightleftharpoons O_2^{\cdot -} \xrightarrow{\ e^-\ } O_2^{2-}$$

In fact, almost all of the reactions discussed in this chapter proceed by one-electron transfer processes. The most effective catalytic metal ions have two stable oxidation states related by transfer of one electron. The most important are cobalt, manganese, and copper, which undergo the following redox reactions [6]:

$$Co^{3+} + e^- \rightleftharpoons Co^{2+} \qquad E_0 = 1.8$$

$$Mn^{3+} + e^- \rightleftharpoons Mn^{2+} \qquad E_0 = 1.5$$

$$Cu^{2+} + e^- \rightleftharpoons Cu^+ \qquad E_0 = 0.17$$

These three metals vary considerably in their redox properties. As indicated by the oxidation-reduction potentials, copper(I) ion is easily oxidized by O_2 ($E_0 \approx 1.2$ at pH = 0) in aqueous solution. Cobalt(II) and manganese(II) salts are not oxidized by O$_2$ under these conditions. However, the redox potentials change substantially with changes in solvent and changes in the ligands bound to the metal ion. When six NH$_3$ or CN$^-$ ligands are attached to cobalt, the Co^{III} oxidation state becomes more stable than the Co^{II} state.

Cobalt(II) salts are the most widely used soluble catalysts for hydrocarbon oxidation reactions, but they are often used in combination with Mn, Cu, or Cr salts in the industrial processes discussed in this chapter [6]. Typically, the cobalt salt is a carboxylate. Cobalt(II) acetate is often used for oxidations in acetic acid solution. For reactions in hydrocarbon solvents, long-chain carboxylic acid salts such as naphthenic acids or 2-ethylhexanoic acid ("octoates") (see their preparation under Hydroformylation, Section 5.5) are used to gain solubility. The higher cobalt(II) carboxylates have complex, as yet unknown structures with four to six oxygen atoms arrayed about the metal atom.

Cobalt(II) salts react with O$_2$ to form labile O$_2$ complexes when certain N-containing chelate ligands are present [7]. The so-called "cobalt salen" or Salcomine gives a relatively stable O$_2$ complex in the solid state which can lose O$_2$ upon heating. Complexes of the salen family were used in oxygen storage devices for submarines during World War II and were considered as high-altitude-replenishable oxygen sources for B2 bombers [8,9].

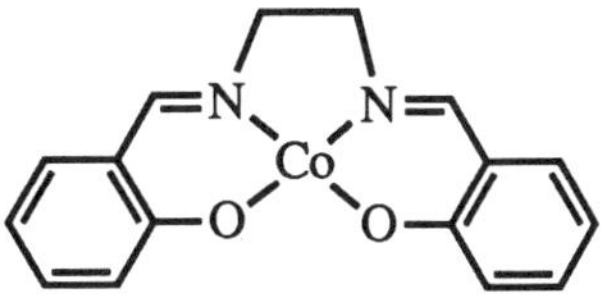

Cobalt Salen or Salcomine

The ligand array about cobalt in the starting complex is essentially square planar. When O$_2$ coordinates, it occupies an axial position in an octahedral structure. Complexation of O$_2$ is assisted by coordination of an amine to the other axial position, just as in heme porphyrin complexes. The formation of the O$_2$ complex usually involves transfer of some electron density from the metal to the O$_2$ ligand. In other words, the cobalt is partially oxidized:

$$Co^{2+} + O_2 \rightleftharpoons Co^{3+} + O_2^{\cdot -}$$

This partial oxidation is facilitated by N-donor ligands which compensate for electron withdrawal by O$_2$. The cobalt carboxylate complexes used as hydrocarbon oxidation catalysts do not react detectably with dioxygen in water, but O$_2$ complexes do form in primary amine solutions [10].

Electrons added to the coordinated O$_2$ molecule populate antibonding (π^*) orbitals and weaken the O-O bond. This effect is evident in both the O-O bond length and its dissociation energy [11]:

$$O_2 \xrightarrow{e^-} O_2^{\cdot} \xrightarrow{e^-} O_2^{2-}$$

Bond energy	118	90	50	kcal/mole
Length	1.21	1.3	1.5 Å	

The bond length and energy of cobalt-coordinated $O_2^{\cdot}$ correspond to values found in ROO$\cdot$.

In contrast to cobalt and iron, the platinum metals form "side-on" O_2 complexes [4,12]. A typical example is "Vaska's complex," which is a reversible oxygen carrier. In solution at room temperature, it absorbs O_2 but the reaction is reversed when the solution is heated. Some of these O_2-binding complexes, especially RhCl(PPh$_3$)$_3$ and Pt(PPh$_3$)$_3$, catalyze oxidation of phosphines, CO, SO$_2$, and isonitriles. However, despite much study, only modest catalytic activity in hydrocarbon oxidation has been observed because the formation of O_2 complexes is seldom involved in catalysis of hydrocarbon oxidation. The electronic factors that favor interaction with the $\cdot$O-O$\cdot$ molecule also favor reaction with the R-O-O$\cdot$ radicals present in hydrocarbon oxidation mixtures.

Two reactions of alkyl hydroperoxides with cobalt ions are especially important in catalysis [13-15]. The first involves oxidation of cobalt(II) via formation of a complex in which the hydroperoxide becomes a ligand on the cobalt ion.

$$Co^{II} + ROOH \longrightarrow [Co(HOOR)] \longrightarrow Co^{III}OH + RO\cdot$$

$$Co^{III} + ROOH \longrightarrow Co^{II} + ROO\cdot + H^+$$

Electron transfer from cobalt to oxygen occurs after complexation. Weakening the O-O bond facilitates breakdown of the complex and formation of the energetic alkoxy radical. The second reaction reduces cobalt(III) and forms the more stable alkylperoxy radical. The oxidation potentials of these reactions seem to be balanced so that the two processes occur simultaneously in solution, giving what is sometimes referred to as the Haber-Weiss cycle [16], Figure 10.1. The rapid shuttling of cobalt between the two oxidation states catalyzes the breakdown of ROOH into radicals that initiate hydrocarbon oxidation.

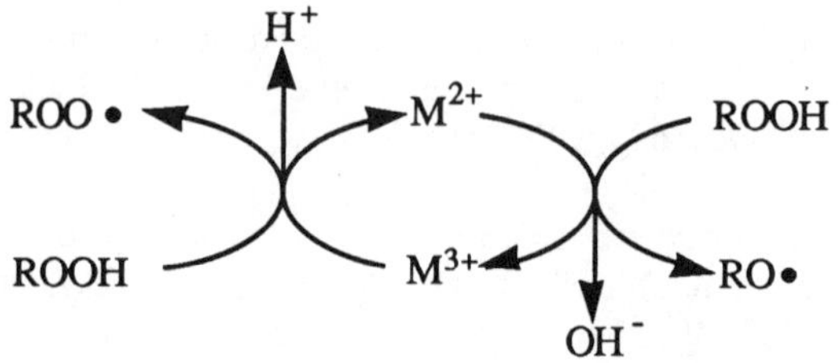

Figure 10.1 The Haber-Weiss cycle applied to the decomposition of hydroperoxides.

The radicals themselves give the oxygenated products. While cobalt is an oxidation catalyst at low concentrations, at high concentrations it actually inhibits oxidation by keeping radical concentrations so low that oxidation is not initiated [17,18].

10.2 REACTION OF O_2 WITH HYDROCARBONS

Most aliphatic hydrocarbons react with oxygen, but under noncombustion conditions, the reactions are extremely slow in the absence of a free-radical initiator. The primary products of alkane oxidations are hydroperoxides, ROOH, which can themselves be a source of radicals by O–O fission. Hence, alkane oxidation by O_2 is autocatalytic, although induction periods are very long. To achieve oxidation in a practical time period, initiators such as benzoyl peroxide may be added. Alternatively, metal ions such as cobalt(II) may be used to catalyze decomposition of ROOH and accelerate the normal autoxidation process. The latter approach is preferred in practice and is the basis for most industrial processes.

Free-radical attack on an aliphatic C-H bond precedes interaction with oxygen in the oxidation process. The attacking reagent X· abstracts hydrogen to form a new radical:

$$R—H + X\cdot \; \rightleftharpoons \; R\cdot + HX$$

This hydrogen abstraction process can be fairly selective with radical reagents such as Cl₃C· and ROO·, which are moderately stable and long-lived in solution. With these low-energy radicals, tertiary C-H bonds are attacked in preference to secondary which, in turn, are more susceptible to attack than primary. Allylic and benzylic C-H bonds are especially vulnerable. Energetic radicals such as RO· are rather indiscriminate in abstraction of hydrogen from carbon, especially at high temperatures.

An alkyl radical, once formed, combines readily with oxygen to give an alkylperoxy radical.

$$R\cdot + \cdot O—O\cdot \longrightarrow R—O—O\cdot$$

Since the process generates a species that can abstract aliphatic hydrogen, the radical initiator has started a cyclic oxidation. The decomposition of an alkyl hydroperoxide leads to many different pathways that include cleavage of O-O, O-H, C-C, and C-H bonds. Some are catalyzed by metal ions, others are simple radical chain processes, and a third group are radical chain sequences initiated by a metal ion [6]. The relative rates of these reactions are controlled by the nature of the metal ion present. Because of the great variety of ROOH reactions, those of major interest are discussed separately in the following sections. The oxidation of cyclohexane to cyclohexanol and cyclohexanone is considered first because all the C-H bonds are equivalent and no C-C bonds are to be broken; thus at first glance, it is relatively simple and reasonably well understood. In the oxidation of linear alkanes (Section 10.5), discrimination among primary and secondary C-H bonds becomes important.

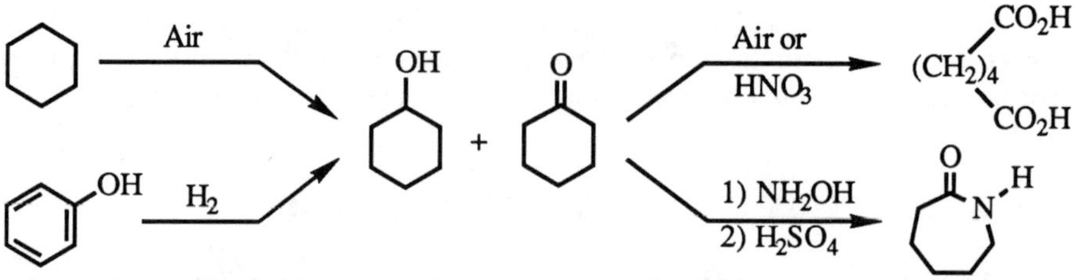

Figure 10.2 Production of the nylon intermediates, adipic acid, and caprolactam, via cyclohexanol and cyclohexanone

10.3 ADIPIC ACID SYNTHESIS

Adipic acid, an important intermediate in the production of nylon (Chapter 11), is made by oxidation of cyclohexanone and cyclohexanol [19]. These intermediates, in turn, come from air oxidation of cyclohexane or hydrogenation of phenol (Figure 10.2). They are also intermediates in the production of caprolactam, which can be polymerized to form 6-nylon directly. The two stages in oxidation of cyclohexane to adipic acid are discussed separately below.

Cyclohexane Oxidation

The commercial practice of this reaction is strongly influenced by its chemical peculiarities [19-23]. In the simplest model, cyclohexane is converted to cyclohexyl hydroperoxide and some or all of the hydroperoxide is decomposed to the observed products, cyclohexanol and cylohexanone:

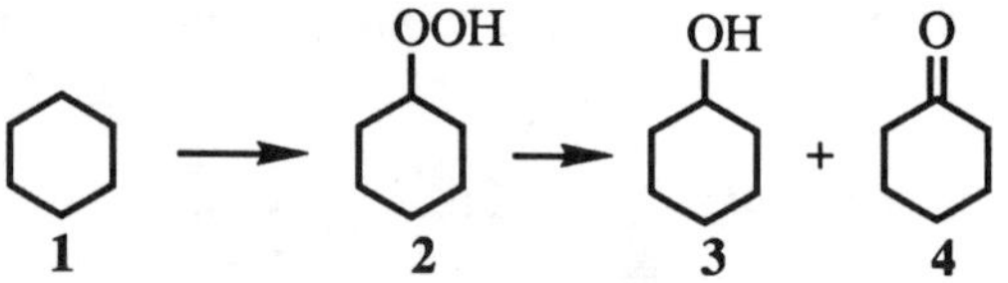

Oxidative attack on the C-H bonds of cyclohexane **1** is slow and requires vigorous reaction conditions. In contrast, the hydroperoxide **2**, alcohol **3**, and ketone **4** are easily oxidized. As a result, the reaction is generally run with low conversions of cyclohexane to avoid degradation of the desired products.

A typical industrial oxidation may be carried out by reacting air with a cyclohexane solution of a soluble cobalt(II) salt at about 140-165°C and 10 atmospheres pressure in a continuous process [21]. The residence time in the reactor is limited to achieve up to 10% conversion of the cyclohexane. Liquid reaction mixture is withdrawn continuously and is distilled; unreacted cyclohexane is recycled to the oxidation reactor. Cyclohexanol and cyclohexanone are sent to another oxidation unit for conversion to adipic acid as described below or to caprolactam (Figure 10.2). Combined yields of alcohol and ketone are 60-70% if conversion is limited to 6-9%

[19]. An alternative strategy involves maximizing the peroxide coming out of the first step by using no or minimal cobalt catalysts and passivating the walls of the reactor [24,25]. The peroxide is then decomposed under conditions which maximize the yield of cyclohexanone and cyclohexanol from the peroxide. In a process developed by Halcon, considerably higher yields can be attained at 10-12% conversion if boric acid is added to stabilize the cyclohexanol as it is formed [26-30]. The borate is believed to react with the hydroperoxide **2** to form a peroxyborate which subsequently decomposes to the borate ester of cyclohexanol [31,32]. The cyclohexyl borate protects the cyclohexanol from subsequent oxidation, thereby allowing the higher conversions. The improved yields are partially offset by the added investment and operating costs for boric acid recycle (cf. Section 10.4), but this technology is used by several major adipic acid producers.

The catalyst for most cyclohexane oxidation is a hydrocarbon soluble cobalt(II) carboxylate such as the naphthenate or 2-ethylhexanoate. Other metal ions such as manganese(II) or chromium(lII) are frequently used with the cobalt to control product distribution. The metal ions probably have no direct part in the conversion of cyclohexane to cyclohexyl hydroperoxide because this oxidation is a simple radical chain process. However, the ions have a role in controlling the conversion of the hydroperoxide to cyclohexanol and cyclohexanone. In addition, since the metal-catalyzed hydroperoxide reactions supply the free-radicals necessary to initiate and maintain the oxidation, the metal ion concentration provides some control of the overall reaction rate. The kinetic effect is not simple [33], however, since the metal ions become inhibitors of the radical chain processes when present in high concentrations [18]. Cobalt and manganese ion catalysts produce both cyclohexanol and cyclohexanone and seem to function by similar mechanisms [13]. Chromium favors the production of cyclohexanone [34].

The attack of a radical, X·, on cyclohexane initiates the oxidation process by abstracting a hydrogen atom from cyclohexane. (The nature of the initial species X·, is unknown. However, it is observed that initiation of oxidation in pure cyclohexane is difficult. On a plant scale, the cyclohexane stream is sometimes "dirtied" with cyclohexanone to help "light-off" the reaction.) The resulting cyclohexyl radical rapidly combines with O_2 to form a cyclohexylperoxy radical. When the latter encounters a cyclohexane molecule, it abstracts hydrogen from a C-H bond in an endothermic but nonetheless important, equilibrium reaction. The transfer of H from C to O produces a primary product, cyclohexyl hydroperoxide, and regenerates a cyclohexyl radical. The cyclohexyl radical is immediately trapped by oxygen, shifting the equilibrium and starting another reaction cycle.

Once oxidation begins, the process is a typical radical chain process consisting of initiation, propagation, and termination reactions. Initiation is primarily a result of the Haber-Weiss cycle decomposition of cyclohexylhydroperoxide.

$$2 \text{ CyOOH} \xrightarrow{\text{Co}} \text{CyO} \bullet + \text{CyOO} \bullet + H_2O \qquad \textit{Initiation}$$
$$\text{CyO} \bullet + \text{CyH} \longrightarrow \text{CyOH} + \text{Cy} \bullet$$

The highly energetic cyclohexyloxy radical abstracts any available hydrogen atom, primarily from the cyclohexane, which is in large excess, to give cyclohexanol.

The longer-lived cyclohexylperoxy radical is capable of abstracting a hydrogen atom from cyclohexane despite the endothermic nature of the reaction because the resulting cyclohexyl radical rapidly combines with an oxygen molecule to form a more stable cyclohexylperoxy radical, thereby driving the equilibrium to the right.

$$\text{CyOO} \bullet + \text{CyH} \xrightleftharpoons{\qquad} \text{CyOOH} + \text{Cy} \bullet$$
$$\text{Cy} \bullet + O_2 \longrightarrow \text{CyOO} \bullet \qquad \textit{Propagation}$$

This cycle provides the chain mechanism which accounts for several cyclohexylhydroperoxide molecules for each initiating event.

The major termination reaction in cyclohexane oxidations at low conversion is the bimolecular combination of two cyclohexylperoxy radicals.

The reaction occurs with little or no activation energy and is essentially independent of temperature. It is thought to proceed through a transitory tetroxide [35] which undergoes intramolecular H-atom transfer to give one cyclohexanone and one cyclohexanol. The probability of this bimolecular reaction is enhanced by the relative stability of the CyOO· radical and by coordination to manganese or cobalt [13]. The ketone is formed in a triplet state, resulting in the chemiluminescence observed during cyclohexane oxidation or cyclohexylhydroperoxide decomposition [36,37].

Because the reaction is typically run in a continuous manner at steady state, the initiation and termination reactions must occur at equal rates. The initiation gives one molecule of cyclohexanol, whereas termination gives one cyclohexanol and one cyclohexanone. This balance explains the observed cyclohexanone to cyclohexanol ratio of approximately 0.5 in cobalt catalyzed reactions.

It has become clear from microreactor studies that much of the kinetics of cyclohexane oxidation can be explained without invoking the known radical intermediates. In one model, the observed products are derived not only from consecutive oxidation of the oxidized products but also directly from the radical species derived from cyclohexane [38]. Thus, even at very low conversions, products such as hydroxycaproic acid 5 and adipic acid 6 are observed. This model clearly ignores most of the complexities of the oxidation reaction.

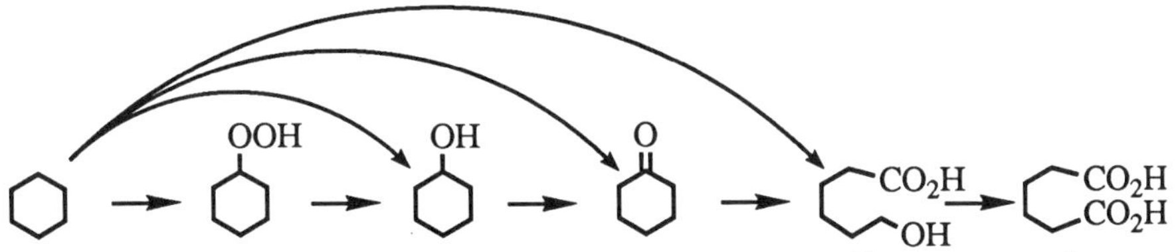

More sophisticated models containing radical intermediates and up to 150 reactions have been developed for tuning the commercial process and predicting the efficiency of new reactor configurations. Even at this level of sophistication, many families of reactions must be grouped rather than treated independently. A kinetic model which accounts for oscillations in the oxidation of one of the intermediates, cyclohexanone, requires 29 equations [39].

Cyclohexyl hydroperoxide, one of the initial products of this oxidation, has only moderate stability. Solutions can be stored for long periods at low temperatures, but decomposition is rapid when the solutions are heated above 150°C. Concentrated solutions of hydroperoxides can be treacherously explosive. As a result, the hydroperoxide is seldom isolated. Most commonly, the oxidation is carried out in the presence of a metal ion that decomposes the peroxide almost as rapidly as it is formed. This method of operation eliminates the hazard of uncontrolled peroxide decomposition and provides a source of radicals to maintain the reaction cycle. There are modes of operation, however, that provide the highest overall yields to adipic acid precursors if hydroperoxide concentrations are maximized and the decomposition is carried out in a separate reactor under milder conditions [24,25]. To obtain high peroxide concentrations, the reactions are run without catalyst and the reactor walls are passivated [40-42].

A variety of metals are effective at normal cyclohexane oxidation temperatures, but yields are higher at reduced temperatures where more effective catalysts are required to completely decompose the peroxide. Cobalt porphyrins and phthalocyanines as well as a family of cobalt bis(pyridylimino)isoindoline catalysts, **7**, are efficient [43-45].

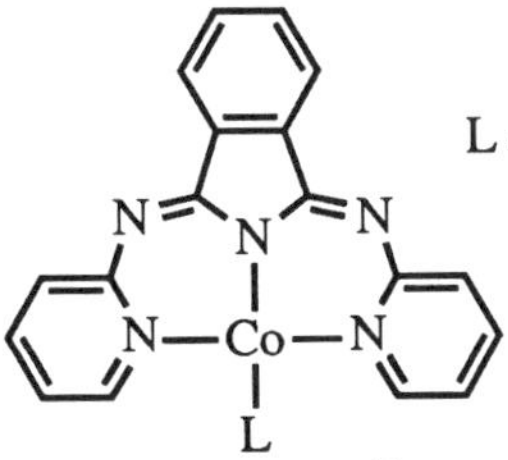

$L = RCO_2$ or a second ligand

This increased efficiency is a result more of increased catalyst lifetime than of increased catalytic rate. In contrast to Co and Mn, chromium(III) ions catalyze decomposition of CyOOH to form cyclohexanone as the major product [34,46,47]. This reaction is essentially a dehydration of the hydroperoxide:

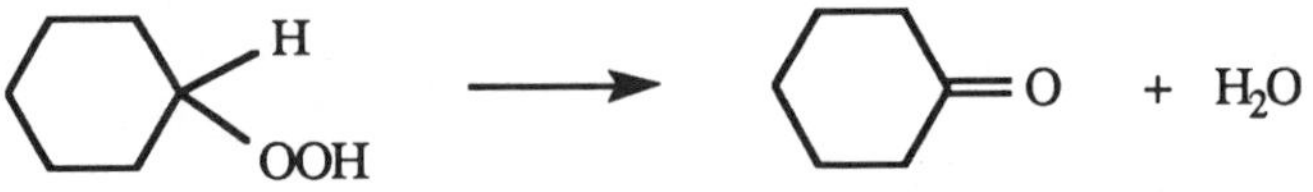

The reaction probably occurs by a nonradical mechanism in which the metal ion assists cleavage of the O-O bond through metal coordination [48]. Commonly observed byproducts, such as acetic and formic acids, can have an effect on the catalytic activity of cobalt catalysts [49].

Cyclohexylhydroperoxide can also be decomposed by strong acids or bases to give high ratios of cyclohexanone [50-52]. Though it has not been practiced commercially, hydrogenation of cyclohexylhydroperoxide gives high ratios of cyclohexanol to cyclohexanone [53-56].

In addition to these reactions which lie on the major pathway from cyclohexane to cyclohexanol and cyclohexanone, literally dozens of other reactions occur in the reaction mixture. Typical are those that involve C-C bond cleavage to form lactones and C_4-C_6 dicarboxylic acids. The α-CH_2 groups in cyclohexanone are especially susceptible to radical attack. Labelling studies implicate an α-oxy radical 8 (Figure 10.3) as a precursor to the dicarboxylic acids [38,57]. A direct oxidation of cyclohexane to adipic acid would be attractive economically [58], but yields in such processes are usually low. Present commercial operation is largely based on a separate oxidation of the alcohol-ketone mixture to adipic acid.

Chemists do not usually think of mass-transfer limitations on chemical reactions, but these are a major source of concern in cyclohexane oxidation. As a result there has been considerable work on reactor design, the coupling of physical and chemical effects, and their effects on the products of the reaction [59-61].

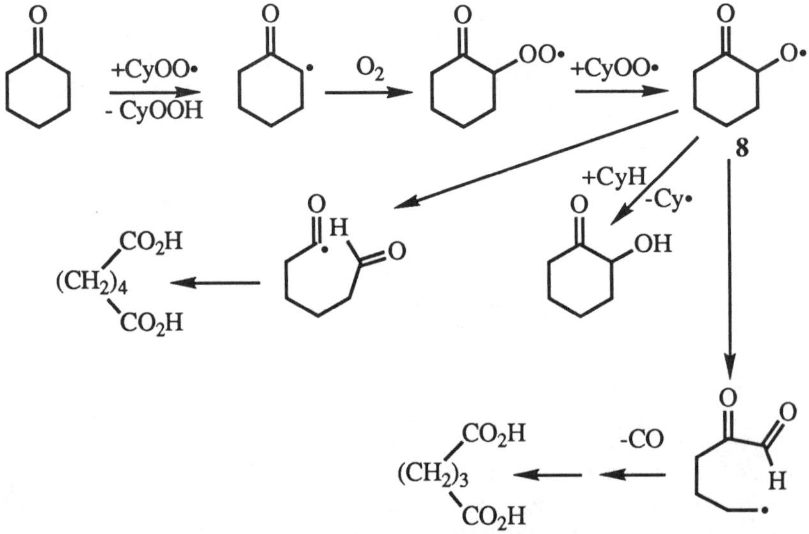

Figure 10.3 Formation of adipic and glutaric acids from cyclohexanone.

Oxidation of Cyclohexanol and Cyclohexanone

Although the cyclohexanol-cyclohexanone mixture from cyclohexane oxidation can be converted to adipic acid by a cobalt-catalyzed air oxidation, present commercial operations generally use nitric acid as the oxidant [19,62]. Even with nitric acid, however, air is the ultimate oxidant because most of the nitrogen oxide byproducts are recycled:

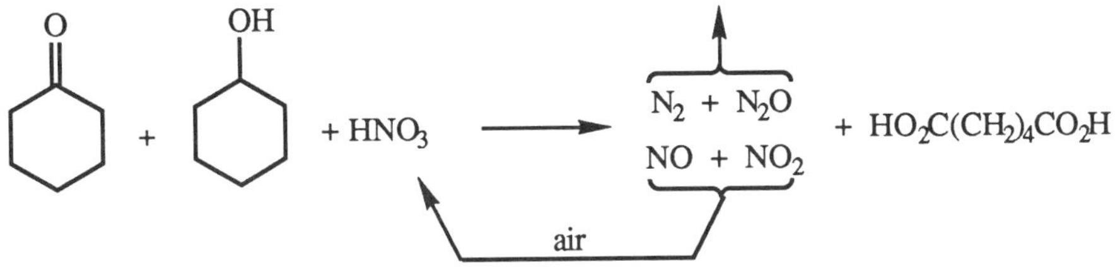

The mixture of cyclohexanol and cyclohexanone is fed continuously to a solution of $Cu(NO_3)_2$ and NH_4VO_3 in 45-50% nitric acid at 70-80°C [63]. The oxidation is complete in a few minutes. The gaseous products, mainly nitrogen oxides, are recycled to a nitric acid synthesis unit. Some nitric acid is lost to products such as N_2 and N_2O which are not reoxidized to HNO_3. The hot acid solution, which contains the organic products, is cooled to crystallize the desired adipic acid. Yields of pure adipic acid typically exceed 90%.

The chemistry in the nitric acid oxidation is very complex. Nearly all the cyclohexanol is oxidized to cyclohexanone in a noncatalytic reaction that is initiated by traces of HNO_2 [64-66]. The cyclohexanone forms adipic acid by three major pathways (Figure 10.4) [67,68]. The upper pathway involves dinitration and is observed only under high-temperature conditions. The other two, more likely to be observed under commercial conditions, begin with conversion of the ketone to its 2-nitroso derivative **9**. The mechanism of this transformation is not well established, but analogy to other oxidations of cyclohexanone suggests that the ketone enolizes in an acid-catalyzed reaction. Loss of a hydrogen atom (or an electron and a proton sequentially) gives a delocalized radical that combines with NO to form 2-nitroso-cyclohexanone **9**. The nitroso ketone forms adipic acid by a noncatalytic path and by a vanadium-catalyzed sequence [65]. The uncatalyzed reaction with HNO_3 gives a nitro-nitroso derivative **10** which hydrolyzes to "nitrolic acid" **11** and to adipic acid. Nitrous oxide, a culprit in ozone depletion, is a major nitrogen-containing byproduct [69]. In the catalytic sequence, tautomerization of the nitrosoketone **9** gives an oxime **12** that hydrolyzes to 1,2-cyclohexanedione. This diketone is stoichiometrically oxidized by two VO_2^+ ions to form adipic acid. The vanadium(IV) coproduct is easily reoxidized by nitric acid. Hence, this pathway is catalytic in vanadium.

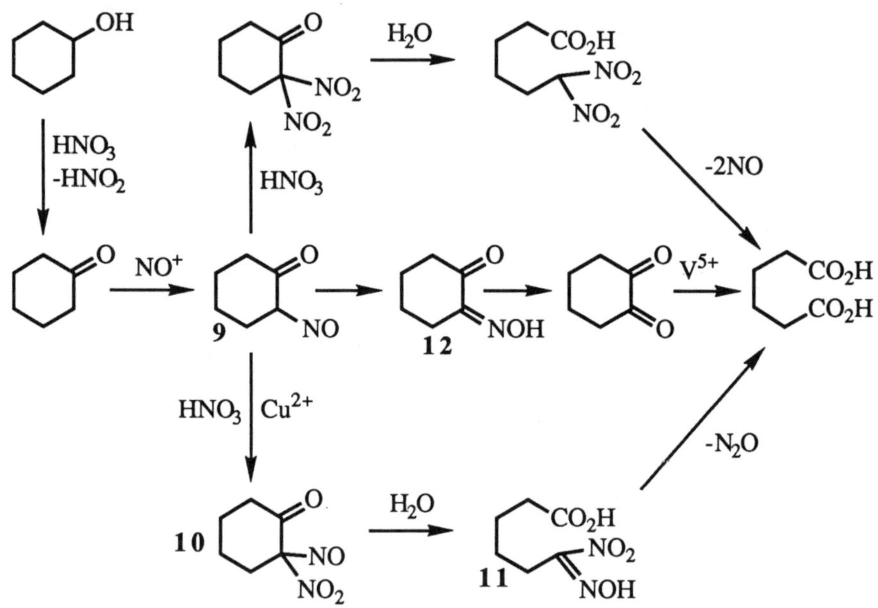

Figure 10.4 Oxidation of cyclohexanol and cyclohexanone to adipic acid in nitric acid solution.

The role of $Cu(NO_3)_2$ in the oxidation is less clear. Copper seems to increase the overall yield of adipic acid by suppressing a side reaction [65].

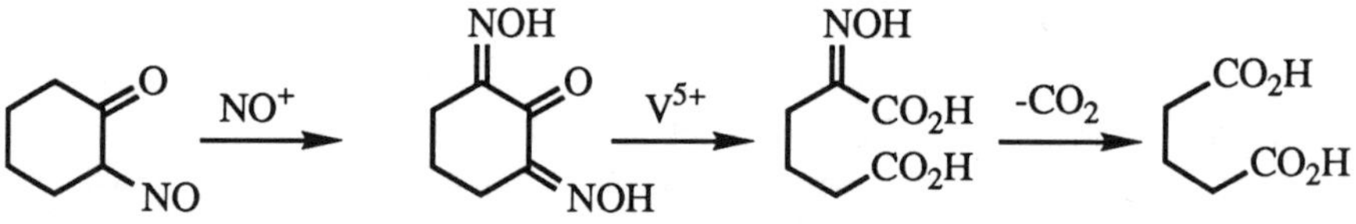

Double nitrosation of the cyclohexanone leads to an unstable intermediate which, upon further oxidation by vanadium, decarboxylates to give glutaric or succinic acids.

10.4   OXIDATION OF CYCLODODECANE

The major industrial application of the butadiene oligomerization discussed in Section 4.5 is synthesis of 1,5,9-cyclododecatriene. This compound is converted to dodecanedioic acid [70] and lauryl lactam [71] by the chemistry of Figure 10.5. These two products, with combined worldwide sales of 100 million pounds in 1990, are intermediates in the production of polyamides for several specialty applications. The syntheses of both the lactam and the dicarboxylic acid begin with hydrogenation of the triene to cyclododecane over a heterogeneous catalyst. The cyclododecane is oxidized with air to a mixture of cyclododecanol and cyclododecanone as described here.

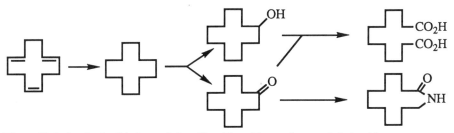

Figure 10.5 Synthesis of dodecanedioic acid and lauryl lactam from cyclododecatriene.

This mixture is oxidized to dodecanedioic acid by a nitric acid oxidation like that used to make adipic acid. As in adipic acid synthesis, the nitric acid oxidation can be catalyzed by $Cu(NO_3)_2$ and NH_4VO_3 [72].

The air oxidation of cyclododecane to a mixture of alcohol and ketone is even more complex than the comparable oxidation of cyclohexane [70]. Almost certainly, the initial oxidation of the cyclododecane forms cyclododecyl hydroperoxide by a mechanism like that for cyclohexane oxidation. The cobalt-catalyzed decomposition of cyclododecyl hydroperoxide yields cyclododecanol by a cyclic process analogous to that of Figure 10.1.

$$C_{12}H_{24} \xrightarrow{O_2} C_{12}H_{23}OOH \xrightarrow{Co^{II}} C_{12}H_{23}OH$$

As in the cyclohexane oxidation, the cyclododecanol is oxidized by air to give cyclododecanone together with a variety of undesirable byproducts. Formation of the byproducts can be minimized by removing the cyclododecanol from the reaction mixture. This removal can be achieved either by low conversion operation or by complexation with boric acid. If the reaction liquids are separated and distilled after about 23% conversion, a 90% yield of useful products is obtained. This separation is complicated by the high boiling points of the products and the fact that cyclododecane is a solid (m.p. 60°C). Because of these problems, the oxidation is often carried out commercially with addition of boric acid.

In a process described by Chemische Werke Hüls [73], moist *meta*-boric acid is added continuously to the reaction mixture to esterify cyclododecanol as it forms. Cyclododecane which contains a trace of a cobalt(II) carboxylate is oxidized with air at 160-180° and 1-3 atmospheres pressure. As the hydrocarbon is converted to the alcohol, the latter reacts with the boric acid to form a borate ester:

$$3\ C_{12}H_{23}OH + 3\ H_3BO_3 \longrightarrow [C_{12}H_{23}OBO]_3 + 6\ H_2O$$

This product is fairly resistant to further oxidation and is easily separated from the reaction mixture. Hydrolysis of the borate ester gives free cyclododecanol. The aqueous boric acid solution is concentrated and recycled to the oxidation reactor. Cyclododecanol and cyclododecanone (5:1) are formed in 80-82% yield at 33% conversion. The oxidation is completed using nitric acid, with reactions similar to those descibed above for adipic acid.

10.5   ACETIC ACID SYNTHESIS

Acetic acid is a major intermediate in the organic chemical industry [74,75]. It is used extensively as a solvent in xylene oxidation (Section 10.6) and as a starting material for synthesis of vinyl acetate, which is used in making polyvinyl acetate and polyvinyl alcohol. Acetic anhydride is an acetylating agent for production of cellulose acetate and aspirin. While all recently constructed plants for manufacture of acetic acid have been based on carbonylation of methanol (Section 5.2), some synthetic acetic acid is still made by oxidation processes. Oxidation of butane and other alkanes is the largest scale process. Another major process is the oxidation of acetaldehyde produced from ethylene by the Wacker process (Section 6.1).

The butane and acetaldehyde oxidations are discussed separately below. Both processes employ cobalt(II) and other transition metal salts as catalysts, but the detailed chemistry is substantially different.

Butane Oxidation

In the United States, butane from natural gas is relatively cheap and abundant. It became a major feedstock for acetic acid production by oxidation according to the equation:

$$C_4H_{10} + \text{}^5\!/_2\, O_2 \longrightarrow 2CH_3CO_2H + H_2O$$

In Europe, where natural gas is less abundant, light naphtha fractions from petroleum refining are used. With either feedstock, the yield of acetic acid is only moderate (30-60%) although the yield from butane is higher than that from naphtha. The yield is strongly dependent on operating conditions and conversion.

The oxidation of alkanes to acetic acid involves C-C bond cleavage in addition to C-H oxygenation. The reaction pathway becomes very complex in the oxidation of petroleum naphtha in which a mixture of aliphatic hydrocarbons is degraded to C_2 fragments. The success of these oxidations reflects the relative stability of acetic acid toward radical processes. Apart from CO_2, it is the most stable product and tends to accumulate under a variety of reaction conditions. This same resistance to oxidation is the reason that acetic acid is employed as a solvent for oxidation reactions.

Typically, the oxidation is carried out by sparging air through an acetic acid solution of butane and the catalyst at 160-200°C and 60-80 atmospheres pressure [76]. A portion of the reaction mixture is continually withdrawn and distilled to recover unconverted hydrocarbon. The products - acetic, propionic, and butyric acids and some 2-butanone - are separated and the residual catalyst solution is recycled to the reactor. An acetic acid yield of 45% at 30% conversion is typical. Economics depend heavily on sale of the other products; this reaction has been a commercially significant source of methylethylketone (2-butanone), and there has even been an effort to improve the yield of MEK [77].

Modifications of the reaction which allow it to be run under milder reaction conditions give higher yields and conversions. Ordinarily, high reaction temperatures are necessary to sustain a radical chain reaction because butane is relatively inert. Continuous recycle of the 2-butanone byproduct to the oxidation reactor [78] sustains the radical chain sequence under mild conditions (110-130°C) because it is very susceptible to radical attack at the α-methylene group. The 2-butanone also contributes to the acetic acid yield [79].

$$C_2H_5COCH_3 + \sqrt[3]{_2} O_2 \longrightarrow 2CH_3CO_2H$$

The overall yield of acetic acid from butane and 2-butanone is about 75% at 85% butanone conversion.

An additional beneficial effect of 2-butanone addition is assignable to an increase in the level of cobalt(III) ions in the reaction mixture [79]. [Oxidation of cobalt(II) to cobalt(III) appears to be the slow step in the process.] The oxidation of the ketone proceeds via an α-hydroperoxide **13**, which can oxidize cobalt(II) by a mechanism like that of Figure 10.1 [80].

$$\underset{\underset{O}{\underset{\|}{}}}{CH_3C} - \underset{\underset{OOH}{\underset{|}{}}}{CHCH_3}$$
$$\mathbf{13}$$

Another approach to operation under mild conditions is to preoxidize part of the cobalt(II) salt with ozone [81]. This treatment eliminates the induction period in batch oxidations of butane and permits continuous operation at 110-130°C and 35 atmospheres pressure. The oxidation converts cobalt(II) acetate to a μ_3-oxo bridged trimer $Co_3O(OAc)_6(AcOH)_3$, which is a good initiator for alkane oxidations [82].

The use of "promoters" such as ketones, cobalt(III), or bromide ion point to the importance of the initiation step in the overall mechanism of oxidation. In the batch oxidation of butane or in the start-up of a continuous process, it is necessary to supply a free-radical initiator that is sufficiently energetic to abstract a hydrogen atom from secondary C-H bonds. The RO· and ROO· radicals fulfill this requirement at least as equilibrium reactions, and are generally present in substantial quantities in oxidation mixtures at 150°C. There is some question as to whether or not cobalt(III) ions attack alkanes or cycloalkanes directly [6].

Once initiation has occurred, a radical chain reaction like that described for cyclohexane oxidation converts butane to 2-butyl hydroperoxide [80]:

$$n\text{-}C_4H_{10} + O_2 \longrightarrow \underset{\underset{OOH}{\underset{|}{}}}{CH_3CHC_2H_5}$$

The selectivity for attack at the secondary carbons is moderate and substantial amounts of propionic and butyric acids are formed by attack at the terminal carbons of butane.

The 2-butylhydroperoxide that forms initially yields the ultimate oxidation products by a variety of routes. Simple decomposition in the absence of catalyst is reported to give largely C_4 products [13] via peroxide dimerization [35]:

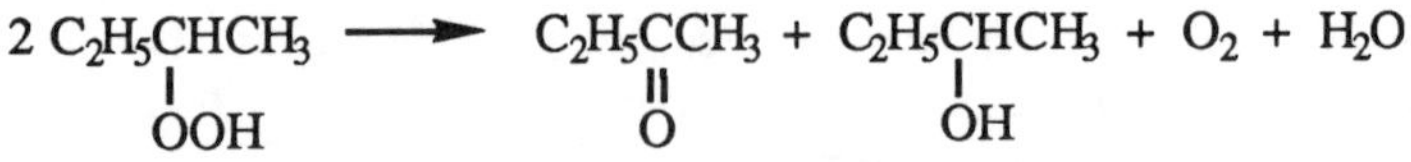

$$2\ C_2H_5CHCH_3 \longrightarrow C_2H_5CCH_3 + C_2H_5CHCH_3 + O_2 + H_2O$$

$$\underset{OOH}{|} \qquad \underset{O}{\overset{||}{}} \qquad \underset{OH}{|}$$

In practice, however, catalytic decomposition by cobalt salts gives $C_4H_9O\cdot$ and $C_4H_9OO\cdot$ radicals that sustain the cyclic oxidation. They replace radicals lost by chain termination processes such as radical dimerization and disproportionation. In addition, they provide mechanisms for fragmentation of the C_4 chain into the desired C_2 products, as well as the C_1 and C_3 byproducts always observed. A β-cleavage of the 2-butoxy radical provides one pathway for C-C fission:

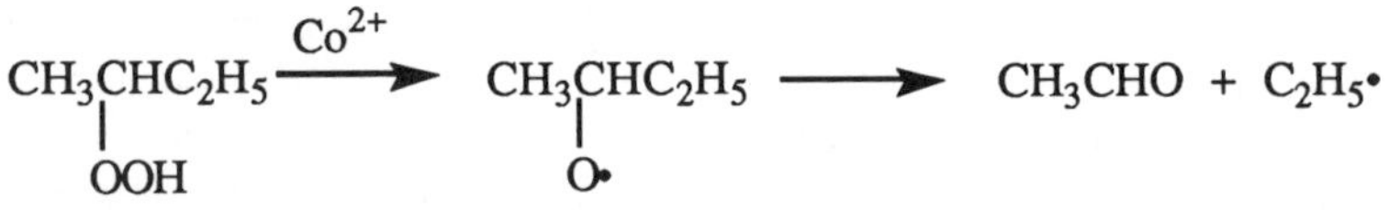

$$CH_3CHC_2H_5 \xrightarrow{Co^{2+}} CH_3CHC_2H_5 \longrightarrow CH_3CHO + C_2H_5\cdot$$

$$\underset{OOH}{|} \qquad \qquad \underset{O\cdot}{|}$$

The CH_3CHO intermediate is easily oxidized to acetic acid as discussed in the next section. The ethyl radicals are converted to ethanol, which is observed under the reaction conditions [83].

Another C-C cleavage mechanism has been proposed to involve intramolecular reactions of the hydroperoxy radical [13] though it is more likely that this is an intermolecular reaction involving the butoxy radical:

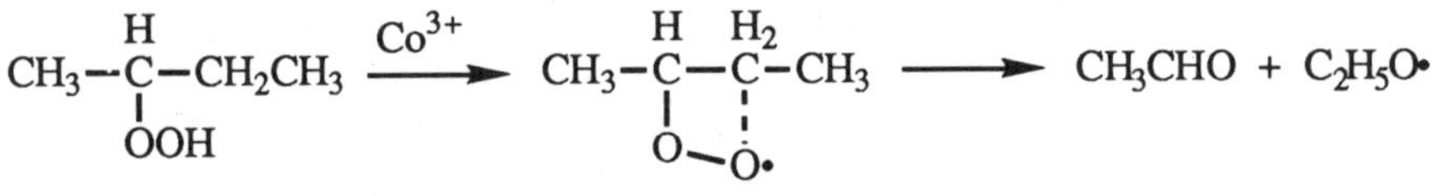

$$CH_3-\underset{OOH}{\overset{H}{C}}-CH_2CH_3 \xrightarrow{Co^{3+}} CH_3-\underset{O-O\cdot}{\overset{H\ H_2}{C-C}}-CH_3 \longrightarrow CH_3CHO + C_2H_5O\cdot$$

A path which may be more important involves 2-butanone, a product of both thermal and catalytic decomposition of 2-butylhydroperoxide. The α-methylene C-H bonds of the ketone are especially susceptible to oxidation and form oxy radicals that produce acetic acid precursors by C-C fission:

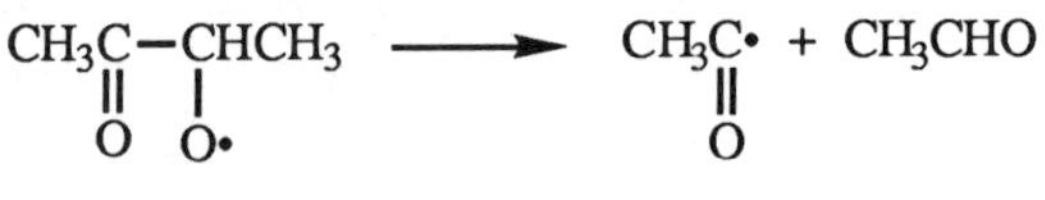

$$CH_3C-CHCH_3 \longrightarrow CH_3C\cdot + CH_3CHO$$

$$\underset{O\ \ O\cdot}{\overset{||\ \ |}{}} \qquad \qquad \underset{O}{\overset{||}{}}$$

The overall process gives moderate yields of acetic acid. These yields contrast sharply with the high yields of acetic acid obtained by lower temperature acetaldehyde oxidation, as discussed below.

Acetaldehyde Oxidation

Grain-based ethanol and acetylene-based acetaldehyde have been major feedstocks for acetic acid production. Both starting materials were oxidized by air with cobalt(II) acetate as the catalyst [84]. The invention of the Wacker process for production of acetaldehyde from ethylene (Section 6.1) provided another industrial route to acetic acid. An especially attractive aspect was that the technology for oxidation of acetaldehyde was well developed.

Acetaldehyde is much easier to oxidize than butane and gives much higher yields. The oxidations are typically run at 60-80°C and 3-10 atmospheres pressure in contrast to the 150-225°C and 60 atmospheres used in butane oxidation [75]. Acetaldehyde diluted with approximately 20% acetic acid is fed to the reactor along with N_2-diluted oxygen. Cobalt(II) or manganese(II) acetates are used as catalysts; salts of nickel [85], chromium [84], or copper [86] are often added to control the product distribution. With nickel addition, the yield of acetic acid exceeds 90% at 92-97% conversion of acetaldehyde; the purity of the final product is >99%. Acetic anhydride can be made the major product when a mixture of cobalt and copper salts is used as the catalyst [87]. Rapid separation of the products by flash evaporation avoids hydrolysis of the anhydride.

The oxidation of acetaldehyde resembles alkane oxidation in that a hydroperoxide is a major intermediate. In this instance, the hydroperoxide is peracetic acid **14**:

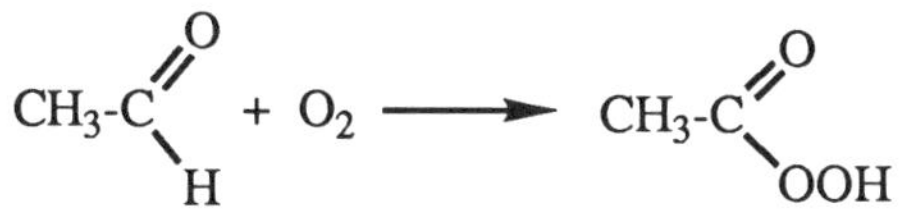

The initial oxidation of acetaldehyde to peracetic acid occurs by a radical chain mechanism (Figure 10.6) analogous to that of alkane oxidation. The aldehyde C-H is readily attacked by RO•, ROO•, or Co(III) to form an acetyl radical that propagates the cyclic oxidation process. A dioxygen adduct of a cobalt porphyrin complex can also abstract the aldehyde hydrogen even at 10°C [88]. The acetyl radical combines with O_2 to give the acetylperoxy radical. It selectively abstracts an aldehyde hydrogen to form peracetic acid **14** and an acetyl radical, thus repeating the cycle.

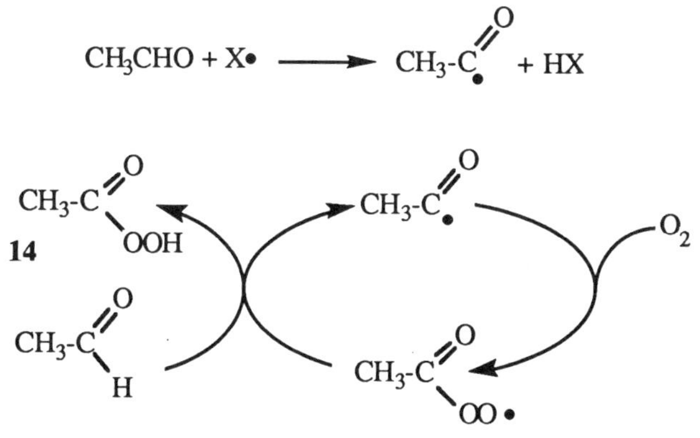

Figure 10.6   Cycle for oxidation of acetaldehyde to peracetic acid.

The yield of peracetic acid is almost quantitative at low temperature and low conversion. As noted below, the peracetic acid is present largely as its acetaldehyde adduct. The oxidation can be carried out in such a way as to make peracetic acid a major product [87]. The main chain-terminating reaction is bimolecular combination of the acetylperoxy radicals to give a tetroxide [89] which can decompose to give the observed byproducts, such as methanol, carbon dioxide, formaldehyde, and formic acid. The peracetic acid reacts with additional acetaldehyde at low temperatures to form an addition product, acetaldehyde monoperacetate **15**:

$$CH_3-C\begin{smallmatrix}O\\ \\OOH\end{smallmatrix} + \begin{smallmatrix}O\\ \\C-CH_3\\H\end{smallmatrix} \rightleftharpoons CH_3-C\begin{smallmatrix}O\ HO\ H\\ \\O-O\end{smallmatrix}C-CH_3$$

15

The adduct can be decomposed thermally to form a 30% solution of peracetic acid in acetic acid. Such solutions are used extensively as epoxidation reagents in pharmaceutical syntheses. They are also used in the production of glycerol from allyl alcohol. The in situ generation of peracetic acid from acetaldehyde is utilized in a process for oxidation of propylene to propylene oxide now being developed commercially (Section 6.5).

In acetic acid synthesis, the aldehyde monoperacetate **15** undergoes a hydride shift in a Baeyer-Villiger reaction. In a side reaction, a methyl rather than a hydride shift occurs, producing methyl formate [90].

$$H_3C-C\begin{smallmatrix}O\\ \\O^-\end{smallmatrix} + H_3C-C^+\begin{smallmatrix}OH\\ \\OH\end{smallmatrix} \longrightarrow 2\ H_3C-C\begin{smallmatrix}O\\ \\OH\end{smallmatrix}$$

$$H_3C-C\begin{smallmatrix}O\ H_3C\\ \\O-O\end{smallmatrix}C\begin{smallmatrix}OH\\ \\H\end{smallmatrix}$$

15

$$H_3C-C\begin{smallmatrix}O\\ \\O^-\end{smallmatrix} + H-C^+\begin{smallmatrix}OCH_3\\ \\OH\end{smallmatrix} - - - \rightarrow H_3C-C\begin{smallmatrix}O\\ \\OH\end{smallmatrix} + H-C\begin{smallmatrix}OCH_3\\ \\O\end{smallmatrix}$$

This type of oxidation is also employed for higher aldehydes for which the alkyl migration becomes more pronounced. Branching at the α-position further enhances alkyl migration.

The metal ion catalysts affect the ultimate product distribution by selective decomposition of peracetic acid and its aldehyde adduct [87,88]. Cobalt or manganese ions (the preferred catalyst) appear to form acetic acid by a redox cycle somewhat analogous to that of Figure 10.1. One major role of manganese is to suppress the concentration of acetylperoxy radicals through reduction [91]:

$$AcOO\bullet + Mn^{2+} \longrightarrow AcOO^- + Mn^{3+}$$

$$Mn^{3+} + AcH \longrightarrow Mn^{2+} + H^+ + Ac\bullet$$

The reduction of cobalt(III) or manganese(III) ions in the second part of the cycle occurs by oxidation of the aldehyde rather than peracetic acid as is the case in most hydroperoxide decomposition cycles. This reaction, which is first order in manganese, peracid, and aldehyde, increases the rate of formation of acetic acid, apparently without going through the peroxide-aldehyde adduct [86]. Copper(II) ion can divert some of the acetyl radical into anhydride formation by oxidation to the acetylium ion [92]:

$$Ac\bullet \xrightarrow[- Cu^+]{+ Cu^{2+}} Ac^+ \xrightarrow{AcOH} Ac\text{-}O\text{-}Ac + H^+$$

It does not seem able to bring about decomposition of peracetic acid solutions at 30°C in the absence of cobalt or manganese [87]. Hence, a combination of the metal ions is synergistic for acetic anhydride production [93].

Celanese and others use technology similar to that for acetaldehyde oxidation to produce some fatty acids, butyric acid, nonanoic acid, 2-ethylhexanoic acid, and other alkyl acids. In each case, an aldehyde obtained by hydroformylation (Chapter 5) is oxidized to the carboxylic acid. The mechanisms of these oxidations closely resemble that described above [94], but greater care must be exercised to avoid the alkyl transfer reaction mentioned.

10.6   OXIDATION OF METHYLBENZENES

The largest industrial scale homogeneous catalytic reaction is the oxidation of *p*-xylene to terephthalic acid and its esters. On a smaller but still substantial scale, *m*-xylene is oxidized to isophthalic acid and toluene is oxidized to benzaldehyde and benzoic acid. Benzoic acid is primarily an intermediate in the production of other compounds such as phenol (Section 7.3) and terephthalic acid. Some of the latter is produced by a disproportionation of potassium benzoate catalyzed by carbonate salts [95,96]:

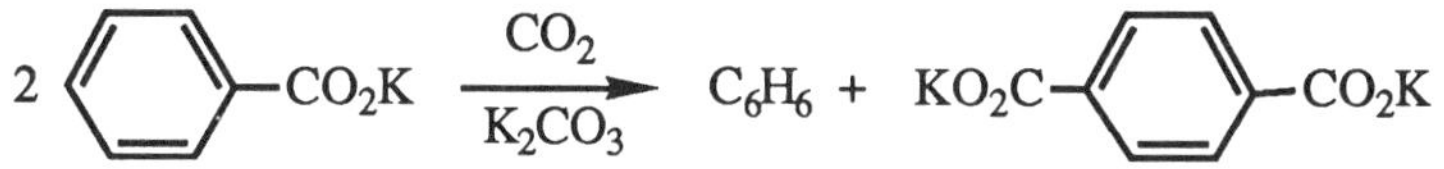

The oxidation of methylbenzenes resembles the oxidation of alkanes and cycloalkanes in some respects but gives higher yields at higher conversions [6]. Benzylic C-H bonds are more susceptible to free-radical attack than are alkyl C-H bonds. Even mildly energetic radicals such as bromine atoms can attack benzylic C-H bonds directly. These reactions probably initiate the complex sequence by which a methyl group is converted to a carboxyl function:

$$ArCH_3 \longrightarrow ArCH_2OOH \longrightarrow ArCH_2OH + ArCHO$$
$$\longrightarrow ArC(O)OOH \longrightarrow ArCOOH$$

Even though methylbenzenes as a class are relatively easy to oxidize, individual differences strongly influence the conditions required. Generally, the ease of oxidation parallels the electron density of the aromatic ring. Methylnaphthalene and p-methoxytoluene are more easily oxidized than toluene and can be oxidized by a different mechanism [6]. The oxidation of p-methoxytoluene to p-methoxybenzoic acid has been considered as a route to hydroxybenzoic acid, but carboxylation of potassium phenoxide is still the preferred route. Toluene and the xylenes are comparable in their initial susceptibility to oxidation. However, once one methyl group of a xylene is oxidized, the remaining methyl group is deactivated by the electron-withdrawing effect of the carboxyl group. As a result, the oxidation of p-xylene is sometimes carried out stepwise:

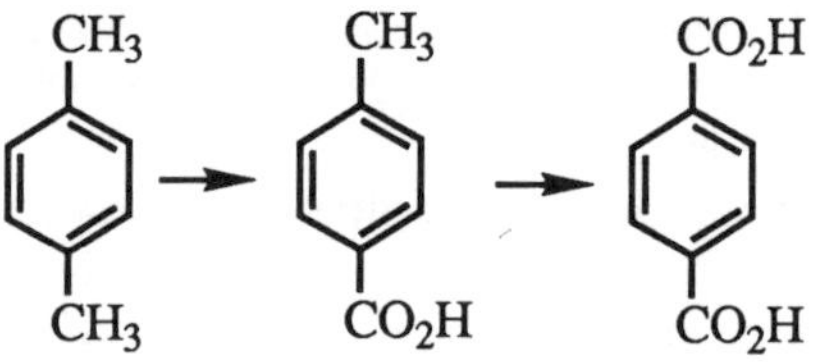

More vigorous conditions are required for the second step than for the first.

Toluene Oxidation

The oxidation of toluene differs from many xylene oxidation processes in that it is usually carried out in the absence of an added solvent [97-99]. Typically, a toluene solution of cobalt(II) 2-ethylhexanoate is reacted with air at 140-190°C and up to 10 atmospheres pressure [98]. The reaction is a free-radical chain process [100]. Liquid reaction mixture is withdrawn at a rate such that toluene conversion is 40-65%. Distillation or crystallization are used to separate toluene for recycle from the crude benzoic acid. Yields of benzoic acid are about 80% after purification by redistillation or recrystallization. A mixture of nickel and manganese salts can give improved results [99]. When toluene oxidation is carried out at lower conversions, benzaldehyde constitutes a major byproduct which is easily recovered by distillation. Because benzoic acid is a large-scale, captive intermediate for the preparation of phenol and caprolactam, byproduct benzaldehyde has largely displaced on-purpose manufacture of benzaldehyde by the hydrolysis of benzal chloride.

Xylene Oxidation

Many processes have been developed for oxidation of p-xylene to terephthalic acid or dimethyl terephthalate. Generally, these processes use air as the oxidant and cobalt and manganese salts as the catalyst and give high yields. Two major processes are discussed below. The Mid-Century process, which was commercialized by Amoco Chemicals [101], produces terephthalic acid by a one step oxidation of p-xylene in acetic acid. The Dynamit Nobel/Hercules process yields dimethyl terephthalate in a multistep process that avoids the use of a solvent. Each process has particular advantages. As noted in Chapter 11, many polymeric terephthalate esters have been made from dimethyl terephthalate by ester exchange, largely for reasons of purity of starting materials. High-quality terephthalic acid is used increasingly as a starting material for polymer production.

The Mid-Century/Amoco process may be used for oxidation of both m- and p-xylenes. Typically, the oxidation is carried out in acetic acid at about 225°C and 15 atmospheres pressure with a mixture of cobalt(II) and manganese(II) acetates and bromides as the catalyst [102]. Bromide is unique among the halogens in its catalytic activity [103]. During a 90-minute residence time in the reactor, most of the xylene is converted to terephthalic acid which crystallizes in about 99.95% purity. The acetic acid solvent keeps the intermediates and byproducts in solution; the terephthalic acid is virtually insoluble in the acetic acid as well as in other organic solvents. The slurry of product in acetic acid is withdrawn continuously, the crude terephthalic acid is separated, and the acetic acid is recycled. The terephthalic acid is recrystallized from aqueous acetic acid under pressure to attain a temperature at which the terephthalic acid has significant solubility. A product of about 99.96% purity is reported [102].

The Dynamit Nobel/Hercules process [104] is based on a series of sequential oxidations and esterifications:

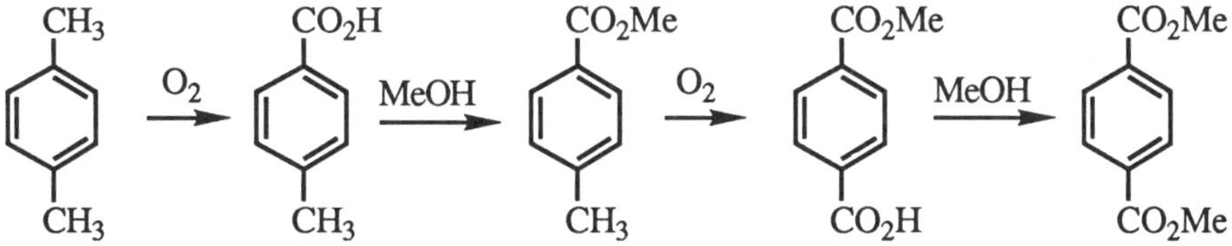

Although it is more complex than the Mid-Century/Amoco process in a chemical sense, some engineering problems may be minimized because the reaction mixtures are less corrosive. The interesting aspect of this oxidation/esterification process is that it uses the easy xylene **16** oxidation to promote the more difficult oxidation of a toluic acid derivative **18**. Neat p-xylene can be oxidized under the conditions described for toluene oxidation, but the major product is toluic acid **17**. However, when the toluic acid is esterified and co-oxidized with p-xylene, the major product is monomethyl terephthalate **19**. Evidently, the oxy radical intermediates formed in xylene oxidation assist in the attack on the methyl group of the the methyl p-toluate **18** in a manner similar to bromine atoms in the Mid-Century/Amoco process.

In the Dynamit Nobel process [104], the oxidation of p-xylene is initiated with air at 6-8 atmospheres sparged through a xylene solution of cobalt(II) 2-ethylhexanoate at 140-170°C, optionally accompanied by manganese(II) 2-ethylhexanoate as a cocatalyst. Once this easier oxidation is started, recycled methyl p-toluate **18** is introduced. The oxidation is carried out continuously [105] and part of the liquid reaction mixture is withdrawn. The heat of reaction is removed by vaporization of both the water produced and some excess xylene which is recycled [106]. The p-toluic acid and monomethyl terephthalate products taken out in the liquid phase are esterified with methanol without a catalyst at 250-280°C and 20-25 atmospheres to maintain a liquid phase. The esters are separated by distillation. Methyl p-toluate is recycled to the oxidation reactor. The dimethyl terephthalate is crystallized to obtain the purity required for polymer production.

Many variations of these processes are practiced commercially. Several are conceptually related to the Dynamit Nobel process in that they use other, more easily oxidized substrates as a source of radicals to promote the relatively sluggish oxidation of p-toluic acid. Mobil used the cooxidation of 2-butanone [107], Eastman may still use acetaldehyde [108], and Toray has used paraldehyde [109].

Mechanism

Methylbenzene oxidations are much like those of butane and cyclohexane in that it is necessary to generate benzylic radicals that can couple with O_2 in a radical chain process like that described for cyclohexane oxidation. A major difference is that the initiation step is much easier than for aliphatic hydrocarbons. Two initiation mechanisms can be distinguished [6] (Figure 10.7):

1. Electron transfer from the arene to a cobalt(III) ion to give an arene radical cation which, in turn, forms a benzyl radical by proton loss. This mechanism is not available with alkanes.

2. Abstraction of benzylic hydrogen by Br·, R·, RO·, and ROO· radicals and possibly even dioxygen complexes. This abstraction reaction is much easier for methylbenzenes than for alkanes because the benzylic radicals are relatively stable.

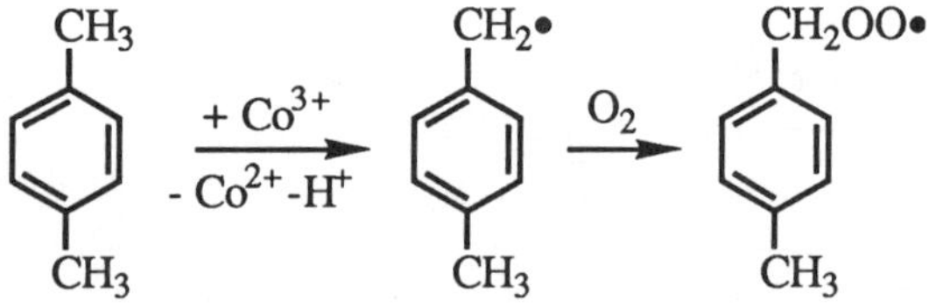

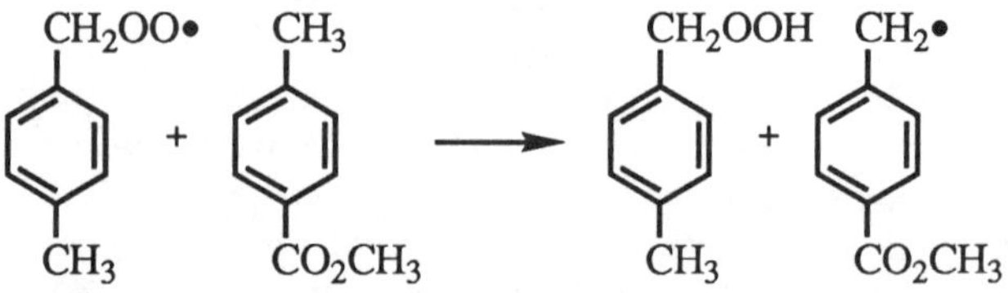

Figure 10.7 Combination of electron transfer and hydrogen abstraction mechanisms in co-oxidation of p-xylene and methyl p-toluate.

The toluene oxidation process and the first step in the Dynamit Nobel/Hercules process probably involve Mechanism **1** to a considerable extent. Mechanism **2** operates in all oxidations of methylbenzenes but seems especially important in the Mid-Century/Amoco process and the second step in the Dynamit Nobel process.

Catalysis of toluene and xylene oxidations by cobalt salts is characterized by an induction period in which cobalt(II) ions are oxidized to cobalt(III) [110]. Monomeric cobalt(III) ion is a powerful oxidizing agent ($E_0 = 1.82$) when it is surrounded only by O-donor ligands such as water, OH^- ions, or $RCOO^-$ ions. It is made even stronger by coordination to a diethylphosphate ligand [111]. Cobalt(III) oxidizes toluene and the xylenes to radical cations by electron transfer [110,112,113]. The cations yield benzyl radicals by proton loss. In the oxidation of toluene, a major pathway is

This sequence opens the way to oxygenation of the methyl group to form benzyl alcohol and benzaldehyde. These compounds are relatively easy to oxidize to benzoic acid by mechanisms like those discussed for acetaldehyde oxidation.

The cobalt(III) initiation pathway (Mechanism **1**, Figure 10.7) is effective in many oxidations, but it has severe limitations. It is strongly inhibited by cobalt(II) ions, which seem to form dimers with cobalt(III) [110]. The dimers are not sufficiently potent oxidizing agents to oxidize toluene and the xylenes directly. As a result, the rate of oxidation is inversely dependent on cobalt(II) concentration. This phenomenon puts a practical limit on total catalyst concentration and the rate that can be attained by cobalt(III) initiation. Even more seriously, the cobalt(III) ion does not seem very effective in the oxidation of *p*-toluic acid, the intermediate in *p*-xylene oxidation. Evidently, the electron-withdrawing carboxyl group raises the oxidation potential of *p*-toluic acid to the point that it is inert to cobalt(III).

The hydrogen abstraction (Mechanism **2**, Figure 10.7) is less sensitive to arene π-electron density than is the electron transfer mechanism. Odd electron species such as bromine atoms and R·, RO·, or ROO· radicals abstract hydrogen from methylbenzenes. Although some of these reagents can be quite discriminating in hydrogen abstraction, all are capable of hydrogen removal from the methyl group of *p*-toluic acid [114]. As noted above, both the Dynamit-Nobel/Hercules and Mid-Century/Amoco processes apparently use this approach to initiate the difficult second step in xylene oxidation.

The Dynamit Nobel process uses a combination of Mechanisms **1** and **2** illustrated in Figure 10.7. The electron transfer Mechanism **1** provides a supply of *p*-methylbenzyl radicals which react with O_2 in a now familiar pattern to form alkylperoxy radicals. These radicals can attack the *p*-methyl group of methyl *p*-toluate to abstract hydrogen and generate a new radical. The *p*-methylbenzyl hydroperoxide can form *p*-methylbenzyl alcohol and *p*-methylbenzaldehyde by standard mechanisms.

Similarly, the new benzylic radical can be oxygenated and initiate new reaction cycles. In the course of doing so, it is converted to terephthalate precursors.

The Mid-Century/Amoco process catalyst, a mixture of manganese, cobalt, and bromide salts [95], functions largely by hydrogen abstraction (Mechanism **2**). The manganese(III) ion is not a sufficiently potent oxidizing agent to abstract an electron from *p*-xylene [115]. The cobalt and manganese ions perform the usual function of hydroperoxide decomposition to produce RO• and ROO• radicals. However, these ions also play another major role in the oxidation of bromide ions to bromine atoms [114, 116]. This electron transfer is rapid and provides a constant supply of bromine atoms. The latter are extremely effective in the abstraction of hydrogen atoms from benzylic methyl groups. At low temperatures [116], the major first step for methyl group oxidation is hydrogen abstraction by bromine atoms. Under commercial conditions, R•, RO•, and ROO• may also play significant roles.

The bromide-bromine cycle probably involves both free and metal-complexed species as indicated in Figure 10.8 [114,116]. Coordination of bromide ion to cobalt(II) may facilitate electron transfer to oxygen or peroxy species to form a cobalt(III) complex. Cobalt(III) ion is a powerful oxidizing agent that can abstract an electron from a bromide ligand to yield a bromine atom. The free or complexed bromine atom can then abstract a hydrogen atom from a methyl group to complete the cycle [114].

Another significant species in oxidations in acetic acid is the •CH$_2$COOH radical. Manganese(III) acetate reportedly decomposes to give this radical and Mn(OAc)$_2$ [115]. This process is significant in several ways. The carboxymethyl radical can abstract hydrogen from methylbenzenes to initiate the desired oxidation process. However, the •CH$_2$COOH radical can also add to the aromatic ring or couple with benzylic radicals to form unwanted byproducts. Even more seriously, this radical provides a pathway for the oxidation of acetic acid to carbon dioxide and water. This loss of the acetic acid solvent is an economic handicap and the major area for improvement in this otherwise very efficient reaction.

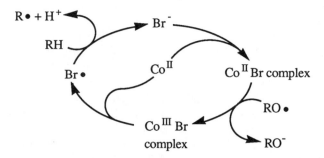

Figure 10.8 The bromine cycle in a bromide-promoted oxidation of a hydrocarbon.

The oxidation of methylbenzenes, like the oxidation of alkanes and cycloalkanes, is obviously an extremely complex process. The nature of the major chain-carrying steps is not totally understood, nor are the relative rates of the dozens of reactions that occur simultaneously. The picture is further complicated by a very strong coupling between the chemical aspects of the reaction and physical effects such as mass transfer. As a result, these reactions can exhibit oscillatory or chaotic behavior under conditions suitable for commercial production [117]. Despite the commercial importance of this reaction, a complete kinetic description of the processes is only slowly emerging and will require considerably more effort. The oxidation of the intermediate species, benzaldehyde, has been studied in more detail [118] and complex models are emerging [119,120]. There remains disagreement over questions as fundamental as the order of the reaction of Co^{II} with peroxyacids [119-121]. It is clear that this will continue to be a fertile area of research for those particularly adept at complex physical studies.

10.7 SPECIALTY OXIDATIONS

Halcon and Arco formed a joint venture, Oxirane, to commercialize a process for propylene oxide based upon the intermediacy of t-butylhydroperoxide. The second step of this process is discussed in more detail in Section 6.5. The first step is based upon the air oxidation of isobutane to t-butylhydroperoxide:

$$Me_3CH + O_2 \longrightarrow Me_3COOH$$

$$Me_3COOH + MeCH=CH_2 \xrightarrow{\text{Cat.}} MeHC\overset{\diagdown\diagup}{\underset{O}{-}}CH_2 + Me_3COH$$

Many aspects of the oxidation are similar in nature to those discussed for cyclohexane [122,123]. One major difference is that the C-H bond of the tertiary carbon is more susceptible to oxidation, so the reaction is reasonably facile and selective. Another is that the reaction is run uncatalyzed to maximize the yield of hydroperoxide. Despite the absence of catalyst, some t-butanol is formed in the oxidation step. The hydroperoxide is then utilized in a homogeneous catalytic reaction to epoxidize the propene [124,125]. Shell operates a related process based upon the oxidation of ethylbenzene and a heterogeneously catalyzed epoxidation [126-129].

In addition to toluene and the xylenes, many other oxidations of methylbenzenes have been proposed or operated on a modest scale. Pyromellitic dianhydride was produced by Du Pont by nitric acid oxidation of 1,2,4,5-tetramethylbenzene (durene) as an intermediate for synthesis of high-performance polyimide resins and films. This route has been replaced by oxidation of 2,4,5-trimethylbenzaldehyde. A Du Pont Japan - Mitsubishi Gas Chemical joint venture carbonylates pseudocumene to 2,4,5-trimethylbenzaldehyde using HF/BF_3 [130] and completes the transformation by an aqueous air oxidation catalyzed with bromide, manganese, and possibly cobalt [131]. Most other specialty acids and anhydrides, such as 3,4,3',4'-benzophenone-tetracarboxylic dianhydride or 2,2-bis(3,4-dicarboxy-phenyl)hexafluoropropane, are prepared by stoichiometric oxidations of the methyl precursors using nitric or chromic acids. Cresols are prepared by selective oxidation of xylenes to methylbenzoic acids followed by decarbonylation. An alternative route is uncatalyzed oxidation of cymenes (methyl-isopropylbenzenes) to the cymene hydroperoxides, followed by acid cleavage to the cresols.

Trimellitic acid is produced by Amoco by air oxidation of pseudocumene in acetic acid [132]; this process competes very effectively with the nitric acid oxidation process of Saarbergwerke which requires the recovery of large quantities of NO_2, NO, and N_2O. The main uses of trimellitic acid are for polyimide-amide resins and plasticizers for PVC used inside automobiles. While most phthalic anhydride is prepared by heterogeneous, fixed-bed oxidation of naphthalene or o-xylene, some is prepared at Rhone-Progil by liquid phase oxidation of o-xylene. Some phthalic anhydride is converted to the dipotassium salt and rearranged to the terephthalic salt [133,134].

The 2,6-naphthalenedicarboxylic acid, obtained by oxidation of 2,6-dimethyl or 2,6-diisopropyl naphthalene, has been commercialized as an intermediate for high-performance polyester resins and films [135]. Because 2,6-dimethylnaphthalene is available in limited quantities and is difficult to obtain in high purity, other routes to the diacid have been developed; each relies on a final air oxidation. Mitsubishi Gas and others are investigating the synthesis of 2,6-naphthalenedicarboxylic acid by acylation of 2-methylnaphthalene 21 in the 6-position with acetyl fluoride/HF from acetic anhydride followed by oxidation to the methylnaphthalenecarboxylate 22 and then to the dicarboxylate 23 [135-139]. The oxidation may be carried out in two sequential steps [140].

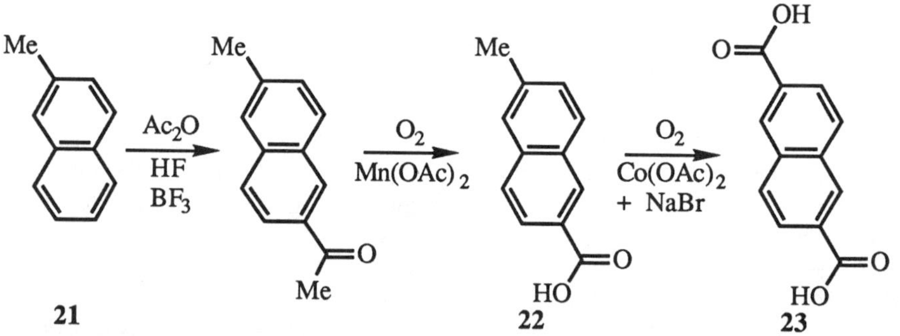

Naphthalene is available in large quantity and is an attractive starting material. Dialkylation with propylene using $AlCl_3$ yields 2,6-diisopropylnaphthalene [141,142] which is then oxidized using standard cobalt-manganese-bromide catalyst systems [143]. The isomeric 1,8-naphthalenedicarboxylic acid or its anhydride is prepared by the metal-catalyzed air oxidation of acenaphthene [144,145].

Lilial, the fragrance of lily of the valley, is produced commercially from two components, both of which are synthesized using homogeneous catalysis. Selective oxidation of p-t-butyltoluene 24 to p-t-butylbenzaldehyde 25 is accomplished in a mixed water-acetic acid medium with the cobalt/bromide system [146]. The water helps stop the reaction at the aldehyde stage.

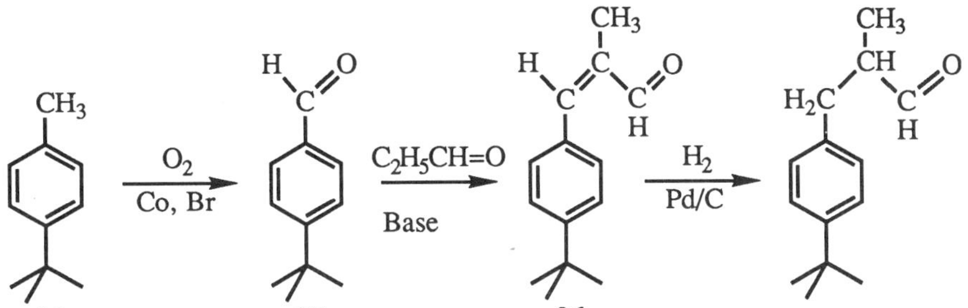

CH$_3$ $\xrightarrow[\text{Co, Br}]{O_2}$ (25) $\xrightarrow[\text{Base}]{C_2H_5CH=O}$ (26) $\xrightarrow[\text{Pd/C}]{H_2}$

24   **25**   **26**

An aldol condensation with propionaldehyde, made by homogeneous hydroformylation of ethylene (Chapter 5) yields the methylcinnamaldehyde **26**. Currently, the synthesis is completed by hydrogenation using heterogeneous Pd/C [147], but a homogeneous Co$_2$(CO)$_8$/pyridine hydrogenation (Section 3.1) has also been explored [148].

Union Carbide manufactures high-purity 1-naphthol for manufacture of the insecticide Sevin by an interesting route going through tetralin **27** [149]. Naphthalene is partially hydrogenated to tetralin. This "benzocyclohexane" is oxidized selectively to the 1-hydroperoxide; the benzylic nature of the 1-position provides a strong directional influence on the reaction. Decomposition of the hydroperoxide yields a mixture of the alcohol and ketone.

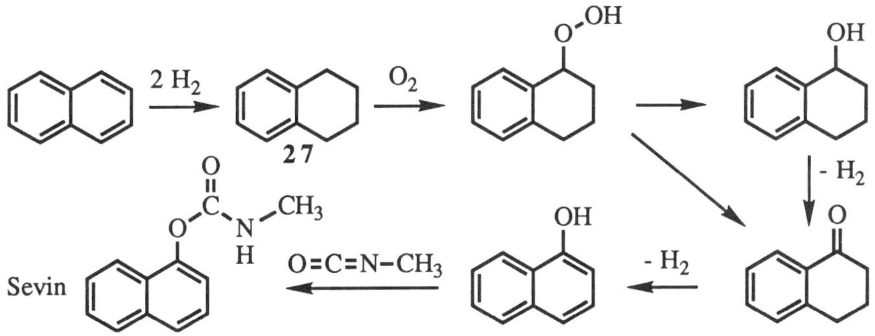

$\xrightarrow{2\,H_2}$ **27** $\xrightarrow{O_2}$ ... $-H_2$ ... $O=C=N-CH_3$ ... Sevin

The alcohol is dehydrogenated to 1-tetralone before it is aromatized to 1-naphthol over a heterogeneous catalyst. A final reaction with methyl isocyanate yields the desired 1-naphthyl methylcarbamate. This process was commercialized in the United States and in Bhopal, India, necessitating the storage of methyl isocyanate.

The isomeric 2-naphthol is prepared by alkylation of naphthalene to the 2-isopropylnaphthalene and air oxidation to the 2-naphthyl-2-propyl hydroperoxide in the presence of some alkali and peroxide and possibly a transition metal catalyst. Cleavage of the product with sulfuric acid yields 2-naphthol and acetone, both in greater than 90% yield [150,151]. Though not commercial, oxidation of phenylcyclohexane yields 1-phenylcyclohexylhydroperoxide which can be decomposed cleanly to phenol and cyclohexanone [152]. The phenylcyclohexane is prepared in good yield by hydrogenation of benzene over a tungsten-modified nickel on aluminosilicate.

GENERAL REFERENCES

T. Dumas, and W. Bulani, *Oxidation of Petrochemicals: Chemistry and Technology*, Halstead Press, New York, 1974.

N. M. Emanuel and D. Gal, *Modelling of Oxidation Processes*, Akademiai Kiado, 1986.

N. M. Emanuel, E. T. Denisov, and Z. K. Maizus, *Liquid Phase Oxidation of Hydrocarbons*, Plenum Press, New York, 1967.

E. T. Denisov, N. I. Mitskevich, and V. E. Agabekov, *Liquid Phase Oxidation of Oxygen-Containing Compounds*, Consultants Bureau, 1977.

J. K. Kochi, *Organometallic Mechanisms and Catalysis*, Academic Press, New York, 1978.

R. A. Sheldon and J. K. Kochi, *Metal Catalyzed Oxidations of Organic Compounds*, Academic Press, New York, 1981.

A. E. Shilov, *Activation of Saturated Hydrocarbons by Transition Metal Complexes*, D. Reidel, Dordrecht, 1984.

C. L. Hill, Ed., *Activation and Functionalization of Alkanes*, Wiley Interscience, New York, 1989.

SPECIFIC REFERENCES

1. C. Moureu and C. Dufraisse, *Chem. Rev.*, **3**, 113 (1926); early review of organic oxidation.
2. N. A. Shilov, *O Sopryazhennykh Reaktsijakh Okislenija (On the Coupled Oxidation Reactions)*, 1905.
3. E. Davey, *Philos. Trans. R. Soc. London*, 108 (1820).
4. J. S. Valentine, *Chem. Rev.*, **73**, 235 (1973); review of oxygen complexes.
5. *Oxygen Complexes and Oxygen Activation by Transition Metals*, A. E. Martell and D. T. Sawyer, Eds., Plenum, New York, 1988.
6. R. A. Sheldon and J. K. Kochi, *Adv. Catal.*, **25**, 272 (1976).
7. R. G. Wilkins, "Uptake of Oxygen by Cobalt(II) Complexes in Solution," in R. Dessy, J. Dillard, and L. Taylor, Eds., *Bioinorganic Chemistry, Advances in Chemistry Series*, Vol. 100, American Chemical Society, Washington, DC, 1971, pp. 111-134.
8. A. E. Martell and M. Calvin, *Chemistry of the Metal Chelate Compounds*, Prentice-Hall, New Jersey, 1952, pp. 336-350.
9. A. J. Adduci, 107th National Meeting, American Chemical Society, Chicago, IL, 1975.
10. G. Henrici-Olive and S. Olive, *J. Organomet. Chem.*, **52**, C49 (1973).
11. F. A. Cotton and G. Wilkinson, *Advanced Inorganic Chemistry*, Wiley-Interscience, New York, 1980, pp. 156.

12. L. Vaska, *Acc. Chem. Res.*, **9**, 175 (1976).

13. N. M. Emanuel, Z. K. Maizus, and I. P. Skibida, *Angew. Chem. Int. Ed.*, **8**, 97 (1969).

14. I. I. Chuev, V. A. Shushunov, M. K. Shchennikova, and G. A. Abakunov, *Kinet. Katal.*, **10**, 75 (1969).

15. I. I. Chuev, V. A. Shushunov, M. K. Shchennikova, and G. A. Abakunov, *Kinet. Katal.*, **11**, 426 (1970).

16. F. Haber and J. Weiss, *Naturwissenshaften*, **20**, 948 (1932).

17. Y. Kamiya and K. Ingold, *Can. J. Chem.*, **42**, 2424 (1964).

18. J. F. Black, *J. Am. Chem. Soc.*, **100**, 527 (1978).

19. V. D. Luedecke, "Adipic Acid," in J. J. McKetta and W. A. Cunningham, Eds., *Encyclopedia of Chemical Processing and Design,* Vol. 2, Marcel Dekker, New York, 1977, pp. 128-146.

20. I. V. Berezin, E. T. Denisov, and N. M. Emanuel, *The Oxidation of Cyclohexane,* Pergamon, New York, 1966.

21. H. J. Boonstra and P. Zwietering, *Chem. Ind.*, 2039 (1966).

22. S. A. Miller, *Chem. Proc. Eng.*, **50**, 63 (June 1969).

23. M. T. Musser, "Cyclohexanol and Cyclohexanone," in W. Gerhartz, Eds., *Ullmann's Encyclopedia of Industrial Chemistry,* Vol. A8, VCH Verlagsgesellschaft, 1987, pp. 217-226.

24. M. Costanti, N. Crenne, M. Jouffret, and J. Nouvel, U.S. Patent 3,923,895 (1975).

25. J. C. Brunie and N. Crenne, U.S. Patent 3,925,316; U.S. Patent 3,927,105 (1975).

26. J. L. Russell, U.S. Patent 3,665,028 (1972).

27. J. L. Russell, U.S. Patent 3,932,513 (1976).

28. *Eur. Chem. News*, 15, 22 (March 29, 1969).

29. J. Alagy, L. Asselineau, C. Busson, and B. Cha, French Patent 1,573,834 (1969).

30. R. L. Marcell, U.S. Patent 3,317,614 (1967).

31. A. N. Bashkirov, *Dokl. Akad. Nauk. SSSR*, **118**, 149 (1958).

32. P. L. Won, V. A. Itskovich, and V. M. Potekhin, *Neftekhimiya*, **20**, 4, 568-72 (1980).

33. A. V. Nikitin, V. V. Kasparov, V. L. Rubailo, and A. B. Gagarina, *Izv. Akad. Nauk SSSR, Ser. Khim.*, 2, 281-6 (1987).

34. G. F. Pustarnakova, V. M. Solianikov, and E. T. Denisov, *Izv. Akad. Nauk SSSR, Ser. Khim.*, 547 (1975).

35. G. A. Russell, *J. Am. Chem. Soc.*, **79**, 3871 (1957).

36. R. E. Kellogg, *J. Am. Chem. Soc.*, **91**, 5433 (1969).

37. R. F. Vassilev, *Prog. React. Kinet.*, **4**, 305 (1967).

38. C. A. Tolman, J. D. Druliner, M. J. Nappa, and N. Herron, "Alkane Oxidation Studies in Du Pont's Central Research Department," in C. L. Hill, Ed., *Activation and Functionalization of Alkanes,* Wiley-Interscience, New York, 1989, pp. 303-360.

39. J. D. Druliner, L. D. Greller, and E. Wasserman, *J. Phys. Chem.*, **95**, 4, 1519-1521 (1991).

40. J. P. M. Bonnart, U.S. Patent 3,510,526 (1970).

41. N. Rieber, R. Platz, W. Fuchs, J. Stabenow, G. Herrmann, H. J. Wilfinger, and H. Hellbach, European Patent 31,113 (1981).

42. H. A. Sorgenti and S. N. Rudnick, U.S. Patent 3,949,004 (1976).

43. C. A. Tolman, J. D. Druliner, P. J. Krusic, M. J. Nappa, W. C. Seidel, I. D. Williams, and S. D. Ittel, *J. Mol. Catal.*, **48**, 129 (1988).

44. A. Yashima, T. Matsuda, T. Sato, K. Sakai, and M. Takahashi, Eur. Patent 164,463 (1985).

45. P. E. Ellis, J. E. Lyons, and H. K. Meyers, U.S. Patents 4,895,682 and 4,895,680 (1990).

46. T. Adams, U.S. Patent 3,404,185 (1968).

47. W. J. Barnett, D. L. Schmitt, and J. O. White, U.S. Patent 3,987,100 (1976).

48. A. M. Syroezhko and V. A. Proskuryakov, "Mechanism of Catalytic Decomposition of Cyclohexyl Hydroperoxide," in A. I. Rakhimov, Ed., *Org. Peroksidy Homoliticheskie Reakts. Ikh. Uchastiem,* 1989, pp. 51.

49. R. V. Kucher, A. P. Pokutsa, V. I. Timokhin, and T. B. Chernyak, *Dolk. Akad Nauk. SSSR,* **313**, 2, 373-6 (1990).

50. M. S. Furman, I. L. Arest-Yakubovich, V. V. Lipes, F. A. Geberger, L. E. Mitauer, and G. Z. Lipkina, *Zh. Khim.,* **15B**, 1175 (1975).

51. W. O. Bryan, U.S. Patent 4,238,415 (1980).

52. W. W. Croch and J. C. Hillyer, U.S. Patent 2,931,834 (1960).

53. J. P. M. Bonnart, Y. Bonnet, and P. P. M. Rey, U.S. Patent 3,557,215 (1971).

54. C. G. M. van der Moesdijk and A. M. J. Thomas, U.S. Patent 3,927,108 (1975).

55. A. M. J. Thomas and C. G. M. Van de Moesdijk, U.S. Patent 3,927,108 (1975).

56. J. Nouvel, U.S. Patent 3,694,511 (1972).

57. J. D. Druliner, *J. Org. Chem.,* **43**, 2069 (1978).

58. K. Tanaka, *Chem. Tech.,* 555 (1974).

59. T. Alagy, P. Trambouze, and H. van Landeghem, *Ind. Eng. Chem. Proc. Des. Dev.,* **13**, 317 (1974).

60. A. K. Suresh, T. Sridhar, and O. E. Potter, *AIChE Journal,* **34**, 1, 55, 69, 81 (1989).

61. A. K. Suresh, T. Sridhar, and O. E. Potter, *AIChE Journal,* **36**, 1, 137-140 (1990).

62. D. D. Davis, "Adipic Acid," in *Ulmann's Encyclopedia of Industrial Chemistry,* 5th Ed., Vol. A1, VCH Verlagsgesellshaft, 1985, pp. 269-278.

63. A. F. Lindsay, *Chem. Eng. Sci.,* **3** (Special Supplement), 78-93 (1954).

64. W. J. van Asselt and D. W. van Krevelen, *Chem. Eng. Sci.,* **18**, 471-483 (1963).

65. W. J. van Asselt and D. W. van Krevelen, *Recl. Trav. Chim.,* **82**, 51-56, 429-437, 438-449 (1963).

66. H. Godt and J. Quinn, *J. Am. Chem. Soc.,* **78**, 1461-1464 (1956).

67. I. Y. Lubyanitskii, R. Minati, and M. Furman, *Russ. J. Phys. Chem.,* **32**, 294-297 (1962).

68. Ya. I. Lubyanitskii, *Zh. Prikl. Khim. (Leningrad),* **36**, 819-823 (1963).

69. M. H. Thiemens and W. C. Trogler, *Science (4996),* **251**, 932-4 (1991).

70. O. S. Shchatinskaya, S. M. Sedova, T. L. Vesel'chakova, and A. M. Gol'dman, *Sov. Chem. Ind.,* **8**, 502 (1976).

71. W. Griehl and D. Ruestem, *Ind. Eng. Chem.,* **62**, 16 (Mar. 1970).

72. J. O. White and D. D. Davis, U.S. Patent 3,637,832 (1972).

73. F. Broich and H. Grasemann, *Erdöl Kohle Erdgas Petrochemie,* **18**, 360, (1965); British Patent 1,110,396 (1968).

74. K. S. McMahon, "Acetic Acid," in J. J. McKetta and W. A. Cunningham, Eds., *Encyclopedia of Chemical Processing and Design,* Vol. 1, Marcel Dekker, New York, 1976, pp. 216, 240, 258.

75. R. P. Lowry and A. Aguilo, *Hydrocarbon Proc.,* 103 (Nov. 1974).

76. F. Broich, H. Höfermann, W. Hunsmann, and H. Simmrock, *Erdöl Kohle-Erdgas Petrochemie,* **16**, 284 (1963).

77. N. Cox, U.S. Patent 3,196,182 (1965) to Union Carbide.

78. J. G. D. Schulz and R. Seekircher, U.S. Patent 4,032,570 (1977).

79. A. Onopchenko and J. G. D. Schulz, *J. Org. Chem.,* **38**, 909 (1973).

80. F. Broich, *Chem. Ing. Tech.,* **36**, 417 (1964).

81.   J. S. Bartlett, B. Hudson, and J. Pennington, U.S. Patent 4,086,267 (1978).

82.   T. Szymanska-Buzar and J. J. Ziolkowski, *Koord. Khim.*, **2**, 1172 (1976).

83.   C. C. Hobbs, T. Horlenko, H. R. Gerberich, and F. G. Mesich, *Ind. Eng. Chem. Proc. Des. Dev.*, **11**, 59 (1972).

84.   D. C. Hull, U.S. Patent 2,578,306 (1951).

85.   G. Roscher, H. Schaum, and H. Schmitz, British Patent 1,483,724 (1977) to Hoechst.

86.   G. C. Allen and A. Aguilo, "Metal-Ion Catalyzed Oxidation of Acetaldehyde," in F. R. Mayo, Ed., *Oxidation of Organic Compounds Vol. 2, Advances in Chemistry Series* **76**, American Chemical Society, Washington, DC, 1968, pp. 363-381.

87.   G. H. Twigg, *Chem. Ind.*, 476 (1966).

88.   M. Tezuka, O. Sekiguchi, Y. Ohkatsu, and T. Osa, *Bull. Chem. Soc. Jpn.*, **49**, 2765 (1976).

89.   N. A. Clinton, R. A. Kenley, and T. G. Traylor, *J. Am. Chem. Soc.*, **97**, 3746, 3752, 3757 (1975).

90.   J. Royer and M. Beugelmans-Verrier, *C. R. Hebd. Seances Acad. Sci. Ser. C*, **272**, 1818 (1971).

91.   W. J. de Klein and E. C. Kooyman, *J. Catal.*, **4**, 626 (1965).

92.   R. A. Sheldon and J. K. Kochi, *Metal Catalyzed Oxidation of Organic Compounds*, Academic Press, New York, 1981.

93.   European Patent 2,696 (1981).

94.   A. P. Litovka, V. V. Baluev, and S. A. Rakhimzhanova, *Kinet. Katal.*, **19**, 567 (1978).

95.   Y. Ichikawa and Y. Takeuchi, *Hydrocarbon Proc.*, **51**, 103 (Nov. 1972).

96.   Japanese Patent 4,724,549 (1972).

97.   R. W. Ingwalson and G. D. Kyker, "Benzoic Acid," in J. J. McKetta and W. A. Cunningham, Eds., *Encyclopedia of Chemial Processing and Design*, Vol. 4, Marcel Dekker, New York, 1977, pp. 296-308.

98.   C. H. Bell, U.S. Patent 3,631,204 (1971).

99.   K. Namie, T. Harada, and T. Fuji, U.S. Patent 3,903,148 (1975).

100.   N. Ohta and T. Tezuka, *Kogyo Kagaku Zasshi*, **59**, 71 (1956).

101.   R. Landau and A. Saffer, *Chem. Eng. Prog.*, **64**, 20 (Oct. 1968); P. H. Towle and R. H. Baldwin, *Hydrocarbon Proc.*, **43**, 149 (Nov. 1964).

102.   C. M. Park and D. G. Micklewright, U.S. Patent 4,053,506 (1977).

103.   D. E. Burney, G. H. Weisemann, and N. Fragen, *Petrol. Refiner*, **38**, 186 (1959) ; A. Saffer and R. S. Barker, U.S. Patent 2,833,816 (1958).

104.   G. Hoffman, K. Irlweck, and R. Cordes, British Patent 1,344,383 (1974).

105.   *Hydrocarbon Proc.*, **56**, 147, Nov. 1977.

106.   *Hydrocarbon Proc.*, **58**, 154, 1979.

107.   British Patent 1,129,386 (1968).

108.   P. H. Towle and R. H. Baldwin, *Hydrocarbon Proc.*, **56**, 11, 149 (1977).

109.   *Hydrocarbon Proc.*, **56**, 11, 230 (1977).

110.   E. J. Y. Scott and A. W. Chester, *J. Phys. Chem.*, **76**, 1520 (1972).

111.   J. Hanotier and M. Hanotier-Bridoux, *J. Mol. Catal.*, **12**, 133 (1981).

112.   E. I. Heiba, R. M. Dessau, and W. J. Koehl, *J. Am. Chem. Soc.*, **91**, 6830 (1969).

113.   K. Sakota, Y. Kamiya, and N. Ohta, *Can. J. Chem.*, **47**, 387 (1969).

114.   D. A. S. Ravens, *Trans. Faraday. Soc.*, **55**, 1768 (1959).

115.   E. I. Heiba, R. M. Dessau, and W. J. Koehl, *J. Am. Chem. Soc.*, **91**, 138 (1969).

116. Y. Kamiya, *J. Catal.*, **33**, 480 (1974).

117. *Chem. Eng. News*, 25, January 21, 1991.

118. J. H. Jensen, *J. Am. Chem. Soc.*, **105**, 2639-2641 (1983).

119. M. G. Roelofs, E. Wasserman, and J. H. Jensen, *J. Am. Chem. Soc.*, **109**, 4207-4217 (1987).

120. A. J. Colussi, E. Ghibaudi, Z. Yuan, and R. M. Noyes, *J. Am. Chem. Soc.*, **112**, 8660-8670 (1990); E. Boga, G. Peintler, and I. Nagypal, *ibid.*, 151-153 (1990).

121. G. H. Jones, *J. Chem. Res. Miniprint*, 2801-2868 (1981).

122. J. C. Jubin, U.S. Patent 4,128,587 (1978).

123. H. R. Grane, U.S. Patent 3,445,523 (1975).

124. J. Kollar, U.S. Patent 3,352,972 (1968).

125. J. Kollar, U.S. Patent 3,507,809 (1970).

126. J. Poloczek and J. Bobinski, *Int. Chem. Eng.*, **11**, 1, 87 (1971).

127. H. P. Wulff and F. Wattlemenn, U.S. Patent 3,634,464 (1972).

128. H. P. Wulff, U.S. Patent 3,829,392 (1974).

129. S. E. Heary and S. F. Newmann, U.S. Patent 3,632,482 (1972).

130. S. Fujiyama, *Jpn. Chem. Assoc. Monthly*, **36**, 2, 11 (1983).

131. M. Komatsu, T. Tanaka, and H. Tooru, Japanese patent JP 56,026,839 (1981).

132. W. Partenheimer, U.S. Patent 4,992,580 (1991) to Amoco.

133. British Patent 975,113 (1964).

134. Japanese Patent 4,220,525 (1967).

135. A. Wood, *Chemicalweek*, March 6, 7 (1991).

136. *Plastics Focus*, **21**, 1 (June 5, 1989).

137. T. Tanaka and M. Inari, European Patent 324,342 (1989) to Mitsubishi Gas Chemicals.

138. S. Fujiama, S. Matsumoto, and T. Yanagawa, European Patent 215,351 (1987) to Mitsubishi Gas Chemicals.

139. T. Maki and Y. Asahi, Japan Kokai JP62-61946 and JP62-61947 (1987) to Mitsubishi Gas Chemicals.

140. M. Feld, U.S. Patent 4,764,638 (1988).

141. Y. Nagao, R. Minami, and Y. Idegami, Japanese Kokai JP01-61436 (1989) to Sumikin.

142. M. Kojima, British Patent 2,207,614 (1989).

143. I. Hirose, Japan Kokai JP62-120343 (1987), 63-66150, 63-104943, and 63-104944 (1988).

144. R. Hasegawa , U.S. Patent 4,212,813 (1980) to Nipon Kayaku.

145. H. Okushima, T. Komae, Y. Issei, and K. Rikizo, German Patent DE2,520,094 (1975) to Mitsubishi Chemicals.

146. P. A. Verbrugge, P. A. Kramer, R. Van Helden, and J. J. K. Boulton, German Patent DE2,804,115 (1978).

147. M. Dunkel, D. J. Eckhardt, and A. Stern, U.S. Patent 3,520,934 (1970).

148. K. Kogami and J. Kumanotani, *Bull. Chem. Soc. Jpn.*, **46**, 3562-3565 (1973).

149. V. I. Trofimov, Y. L. Levkov, and A. M. Yakubson, *Sov. Chem. Ind.*, **3**, 168-171 (1973).

150. H. Hosaka, K. Tanimoto, and H. Yamachika, U.S. Patent 4,049,720 (1977).

151. M. I. Farberov, B. N. Bychkov, and G. D. Mantyukov, USSR Patent 593,729 (1978).

152. G. N. Koshel, M. I. Farberov, M. M. Makarov, and B. N. Bychkov, *Neftekhimiya*, **17**, 705-9 (1977).

11 | ESTERIFICATION, POLYCONDENSATION, AND RELATED PROCESSES

Many of the industrial homogeneous catalytic processes discussed earlier in this book produce intermediates for the synthesis of condensation polymers such as polyesters and polyamides. Indeed, some of the largest applications of homogeneous catalysis are in the production of intermediates having the high purity required for polymer formation [1,2]. The selectivity of a homogeneous catalyst is a major advantage when the final product must be better than 99.9% pure.

In addition to the synthesis of polymer intermediates, homogeneous catalysis is used in some polymerization processes. Soluble metal compounds are used as catalysts in most processes for the manufacture of polyesters. Similar catalysts are often used for the production of polyurethanes and polycarbonates. Polyamide synthesis is not usually catalyzed by metal complexes, but the topic is considered in this chapter as a result of the important role of homogeneous catalysis in the production of nylon intermediates.

11.1 POLYESTER SYNTHESIS

The most widely used synthetic fiber is polyester - poly(ethylene terephthalate) or PET - which is sold under such familiar trade names such as Du Pont's Dacron®, ICI's Terylene®, or Eastman's Kodel®. PET is also commonly encountered in the familiar plastic bottles for soft drinks, and in films such as Du Pont's Mylar® which are used for food or liquid packaging, magnetic tapes, novelty balloons, and a host of other applications. While the applications of PET overwhelm all other polyesters combined, world production is well into the leveling portion of a typical "S"-curve (Figure 11.1) of a maturing, commodity product. Other specialized applications represent the more rapidly growing segment of the industry. Fully aromatic polyesters, often referred to as polyarylates, are used for higher-performance applications because the relatively rigid, highly extended polymer chains give better

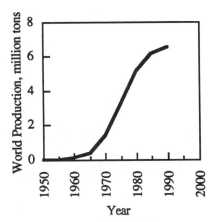

Figure 11.1 World production of polyester for the last half century.

strength and high-temperature properties. Lower molecular weight aliphatic polyester diols find applications as intermediates in automotive finishes and polyurethanes.

When production of polyester began about 40 years ago, the major process was based on dimethyl terephthalate, DMT [3,4]. This approach, illustrated in Figure 11.2, involves two steps, both of which are catalyzed by soluble metal compounds. Transesterification of the dimethyl ester with ethylene glycol gives methanol and bis(hydroxyethyl)terephthalate, BHET [3]. The hydroxyethyl ester is then heated under vacuum to drive out one equivalent of ethylene glycol forming the polyester [5]. For fiber applications, it is common to add a low percentage of the sodium salt of the 5-sulfo-bis(hydroxyethyl)-isophthalate as a site for dying the fiber [6,7].

Today, the availability of high-purity terephthalic acid from processes such as the Mid Century-Amoco process discussed in Section 10.6 makes the acid more attractive as a starting material for polyester synthesis. The acid is generally the same price by weight as dimethyl terephthalate but is cheaper on a molar basis. In addition, there is no need to recycle methanol when the acid is the starting material. Direct esterification of terephthalic acid often employs a different sort of catalyst from that used in transesterification. However, the conversion of BHET to polymer ("polycondensation") may use the same catalyst whether ester or acid is the ultimate starting material.

Transesterification

The usual preparation of poly(ethylene terephthalate) uses two different catalyst systems [8]. The first step in the process of Figure 11.2, the *transesterification* of dimethyl terephthalate with ethylene glycol, is catalyzed by a variety of metal ions. The relative activities of a number of metals [9,10] are given by

$$Zn^{2+}>Pb^{2+},Cd^{2+}>Mn^{2+}>Co^{2+}>Ca^{2+}>Ba^{2+}>Na^{+}>Li^{+}>Hg^{2+}=Fe^{3+}>Ni^{2+}>Cu^{2+}$$

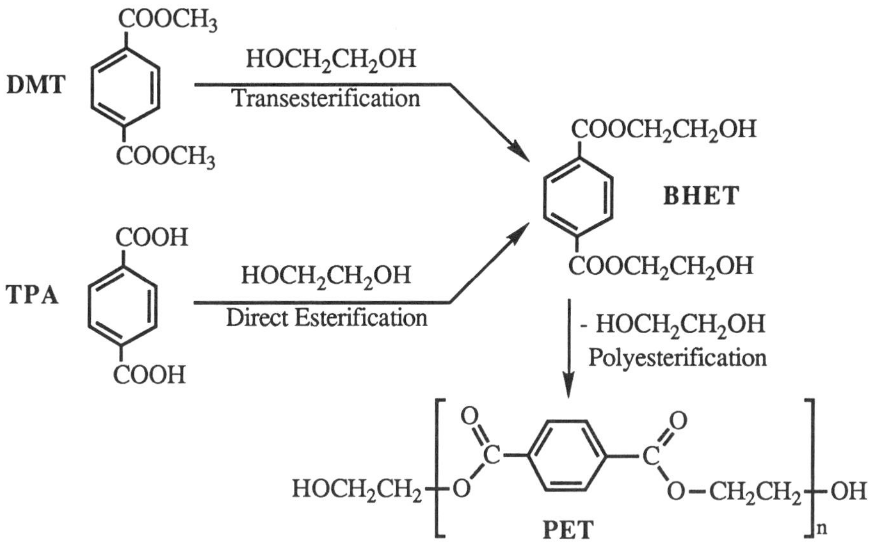

Figure 11.2 Routes to polyester [poly(ethylene terephthalate)] from dimethyl terephthalate and terephthalic acid.

Acetate salts of Zn and Mn are used commercially [3]. Although these salts are very effective in the first step, which is carried out at 150-200°C, they cause undesirable side reactions at the higher temperatures necessary for the second step. The *polycondensation* of bis(hydroxyethyl) terephthalate is carried out at 250-300°C under vacuum in order to remove ethylene glycol as it is produced. Antimony compounds almost always are used to catalyze this reaction, which is also a transesterification of sorts. Even though antimony and germanium do not catalyze transesterification at low temperatures, they cleanly polymerize the hydroxyethyl ester at about 275°C. Presumably they catalyze transesterification by a mechanism different from that of the simple metal ions. They work poorly in the presence of free alcohols.

Polyester synthesis is simple from a chemist's viewpoint and can be carried out in the laboratory in a reaction sequence that simulates commercial practice. In a laboratory preparation [11], the transesterification catalyst, $Ca(OAc)_2 \cdot 2H_2O$, and the polycondensation catalyst, Sb_2O_3, are mixed with dimethyl terephthalate and a slight excess of ethylene glycol. The mixture is heated at 197°C and is bubbled with N_2 to expel methanol as it is produced by transesterification. After three hours at 197°C, the solution of bis(hydroxyethyl) terephthalate and low oligomers is heated to 222°C to distill excess ethylene glycol. The polycondensation step is effected by heating the mixture at 283°C under vacuum. Ethylene glycol from the transesterification of two molecules of the hydroxyethyl ester distills out of the mixture as viscosity builds. Heating for three hours produces high molecular weight polymer that can be melt-spun to form fibers or hot-pressed to produce films.

Industrial polymerization is carried out very similarly in either a batch or a continuous process [3]. The transesterification catalyst is an acetate of cobalt, zinc, or manganese, all of which produce rapid reaction of ethylene glycol with the terephthalate ester. Unfortunately, they also catalyze the formation of diethylene glycol at the high temperatures used in polycondensation. The net reaction is given by

$$2 \; HOCH_2CH_2OH \longrightarrow (HOCH_2CH_2)_2O + H_2O$$

It is unlikely that this is a direct dehydration of two molecules of ethylene glycol to form the ether. Because the benzoyl group would be a better leaving group, it is more likely that the reaction proceeds through the intermediacy of the ester [12]:

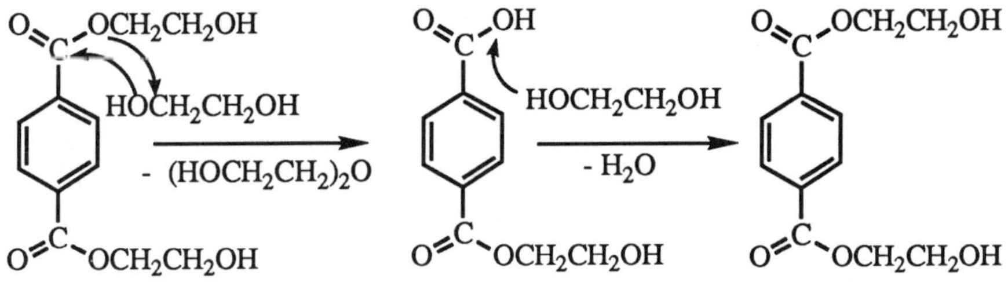

The diethylene glycol formed in this way is incorporated into the final polymer and changes the chemical and physical properties. In fiber applications, the diethylene glycol can act as a dye site, so it may even be added to the polymerization in an effort to keep its concentration constant, thereby improving dye uniformity. In film applications, its presence is generally minimized. It is common practice to add an "inhibitor" such as phosphoric acid or triphenyl phosphate [13] before the polycondensation step. The inhibitor converts the metal salt to a complex phosphate that has little catalytic activity and does not interfere with the antimony-based polycondensation catalyst.

Another difficulty encountered under the high temperatures required to complete polymerization is the dehydration of ethylene glycol or decomposition of hydroxyethyl esters to acetaldehyde [14]. Antimony catalysts are observed to contribute less to this reaction than are catalysts typically used for the transesterification step [15]. These reactions represent a loss of ethylene glycol, but more importantly, they lead to very undesirable color formation. The acetaldehyde undergoes a variety of condensation reactions, and CO is observed as an off-gas from the polymerization.

In a commercial polymerization [16], ethylene glycol, dimethyl terephthalate, and the metal acetate catalyst may be added continuously to a transesterification reactor that is heated to 150-200°C. Methanol and some glycol distill out of the mixture. The molten bis(hydroxyethyl) terephthalate that results is added to a polycondensation reactor along with the inhibitor and antimony(III) oxide which has been predissolved in glycol. As in the laboratory experiment, the mixture is gradually heated to 270-280°C under vacuum to distill glycol as it forms. Mass transfer often becomes a limiting feature of the polymerization, especially in the presence of highly active catalysts. While the "catalysts" are added for chemical reasons, there is a possibility

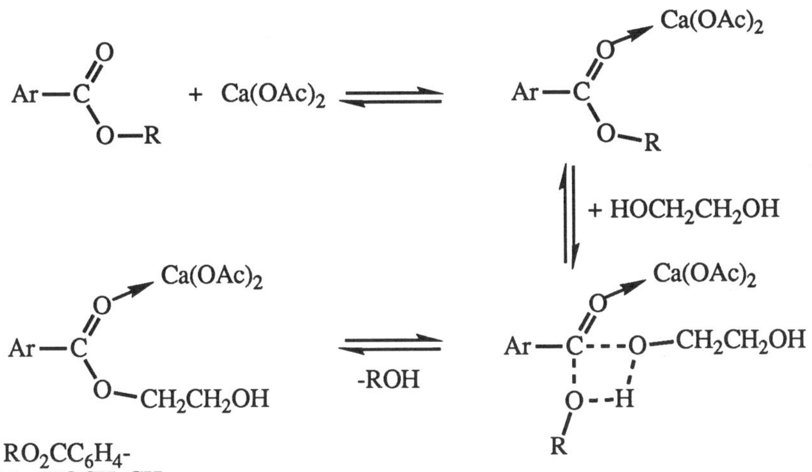

Ar = RO$_2$CC$_6$H$_4$-
R = Me, HOCH$_2$CH$_2$-

Figure 11.3 Calcium-catalyzed transesterification of aromatic esters.

that they also catalyze mass transfer by forming fine solids which act as nucleation sites for glycol vapor formation - essentially acting as "boiling chips." At the high viscosities achieved toward the end of polymerizations, mass transfer is a critical problem not only in commercial reactors, but also in laboratory-scale physical measurements. Thus, experiments are sometimes carried out in thin molten films [17].

Divalent metal ions are said to act as Lewis acids in the catalysis of transesterification [18]. Coordination of the carbonyl group of the ester to the cation is believed to activate it to nucleophilic attack by the oxygen of an alcohol. The complexation and activation of dimethyl terephthalate by calcium acetate are depicted in Figure 11.3. When the carbonyl group coordinates to a calcium(2+) ion, electron density is transferred to the metal ion. The electron depletion of the carbonyl is most pronounced at the carbon atom. This effect makes the carbon especially susceptible to nucleophilic attack. The attack by a hydroxyl group is often written with a four-center transition state as shown. Regardless of the timing of events, the overall effect is breaking and making of C-O and O-H bonds to produce a glycol ester and free methanol. Dissociation of the metal ion from the carbonyl group completes the catalytic cycle.

The same kinds of reactions occur during the first stages in growth of the polymer chain. The aryl group can be a terephthalate unit at the carboxyl end of a growing polymer chain. Similarly, the ethylene glycol in this scheme can be replaced by a glycol-terminated chain:

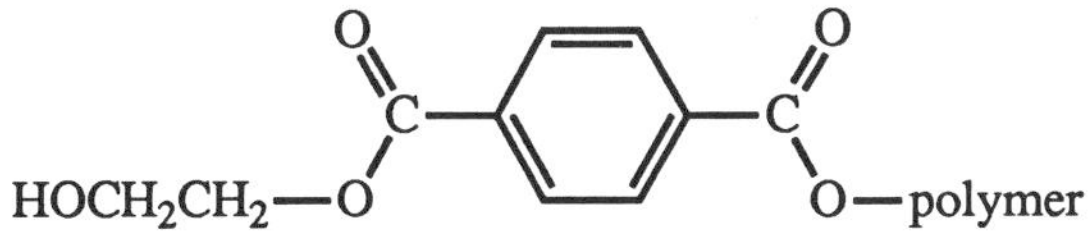

Under these conditions, a substantial increase in molecular weight can occur. However, since all the reactions are reversible, residual methanol or glycol in the system can lead to chain cleavage. For this reason, the transesterification and polycondensation reactions are conducted with continuous removal of the alcohols.

The mechanism of Figure 11.3 seems quite plausible for the transesterification catalyzed by divalent metal ions but is not universally accepted. It was proposed earlier [19] that the metal ion reacts with ethylene glycol to form an alkoxide, for example, [Ca-OCH$_2$CH$_2$OH]. The nucleophilic alkoxide ion then attacks the carbonyl group of a coordinated ester. This proposal was supported by the kinetics of the transesterification catalyzed by M^{2+} ions. The reported kinetics (first order each in glycol, ester, and catalyst) [19-21] would also be consistent with the Lewis acid mechanism if the three reactants were assembled in a complex like

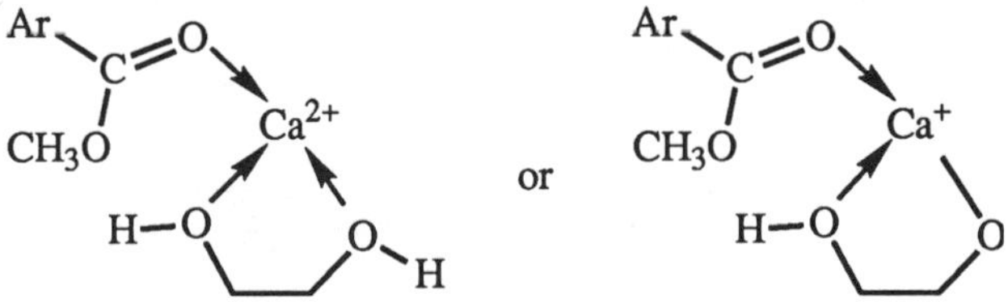

or

Kinetics for alkali-metal-catalyzed reactions are reported to be zero order in ester supporting a reaction in which the metal ion serves to generate carboxylate anion which acts as a nucleophile reacting with an ester carbonyl [21].

The mechanism of the polycondensation reaction is also unresolved. In commercial practice, this final portion of the polymerization is antimony-catalyzed. It has been suggested [21,22] that hydroxyethyl end groups of two growing polymer chains are joined to an antimony (or titanium) center as alkoxides in the intermediate:

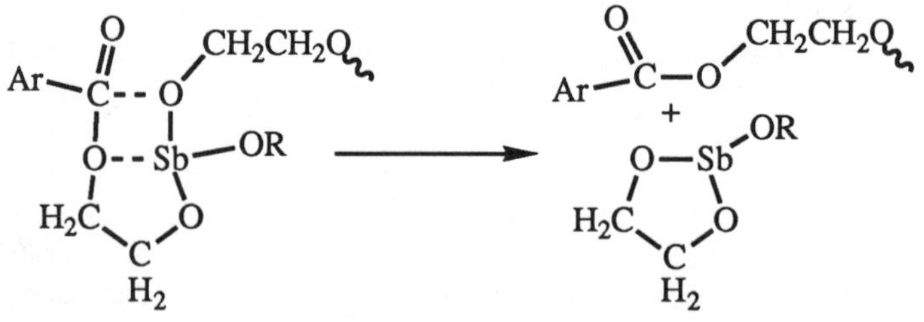

Ligand exchange within the coordination sphere of antimony forms a new ester link which couples the two growing polymer chains. The other product is an antimony salt of ethylene glycol which can release ethylene glycol by reaction with other hydroxyethyl groups:

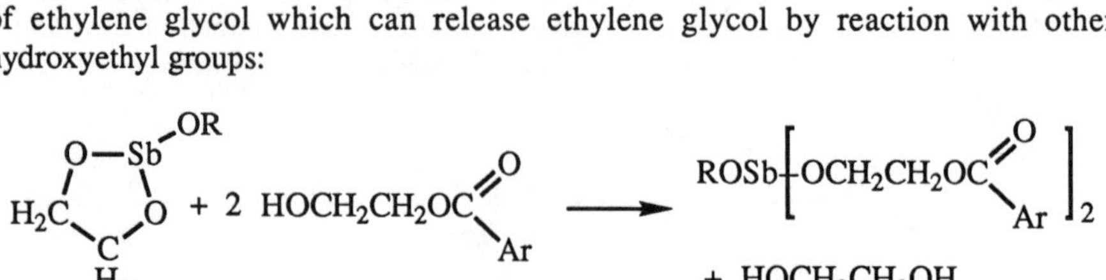

This mechanism is speculative, but it seems consistent with the kinetic data [23] available. The polycondensation reaction is also catalyzed by other metals whose activity is reported [24] to be in the order

$$Ti>Sn>Sb>Mn>Zn>Pb$$

The influence of temperature varies for each of the catalysts, indicating differing mechanisms of activity and deactivation. Catalysts such as Mn, Zn, and Pb typically used for the transesterification step are very active in media having both a high and a low hydroxyl content, but they are easily poisoned by very small amounts of carboxylic end groups.   Antimony trioxide, which is typically used in the polycondensation step, is insensitive to the presence of acidic carboxylic end groups, but its activity is inversely proportional to the hydroxyl group concentration [8]. Thus ethylene glycol must be removed rapidly [12].

The importance of the chelation shown in the intermediates sketched above is highlighted by studies of the hydrolysis of pyridylmethyl hydrogen phthalates [25]. Divalent metal ions such as Ni^{2+}, Co^{2+}, and Zn^{2+} are very good catalysts for the reaction. This is in spite of the fact that they bind to the substrate very weakly; saturation of activity is not observed even at a 100-fold excess of metal over ester.

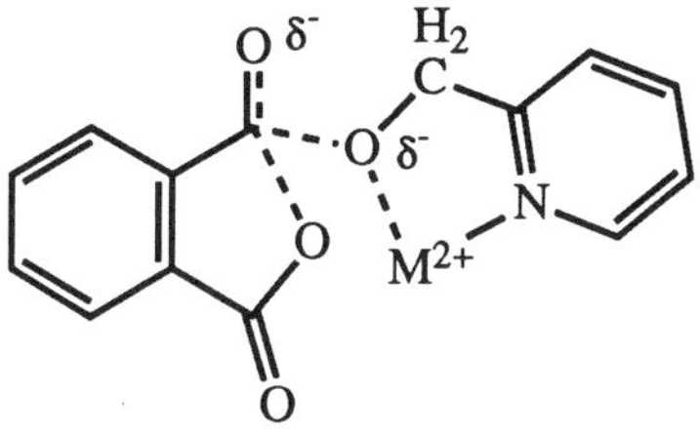

Catalysis is dependent upon the chelation afforded by the presence of the pyridine nitrogen in the 2-position; metal ions have no effect on hydrolysis of the 4-pyridyl analog. Since breakdown of a tetrahedral intermediate must be rate-determining in the carboxyl nucleophilic reactions of esters with poor leaving groups such as the pyridylmethyl, the metal ions must exert their effect in the transition state by stabilization of the leaving group, or in the case of transesterification reactions, the incoming group. In contrast, the metal ion effects observed in hydrolysis of 2-pyridylmethyl benzoate represent metal-ion-promoted hydroxyl catalysis.

Direct Esterification

The change from dimethyl terephthalate to terephthalic acid as a starting material for polyester production requires a change in catalyst. The divalent metal salts usually used as *transesterification* catalysts have little activity in the *esterification* of a carboxylic acid with an alcohol [26]. Extensive screening of potential catalysts for esterification has led to the ordering of activities

$$Ti^{4+}>Sn^{2+}>Sn^{4+}>Bi^{3+}>Zn^{2+}>Pb^{2+},Sb^{3+}>Al^{3+},Mn^{2+}>Co^{2+},Cd^{2+}>Mg^{2+}$$

for reaction of ethylene glycol with benzoic acid to form hydroxyethyl benzoate [26], and

$$Sn^{2+}>Ti^{4+}>>Zn^{2+}>Pb^{2+},Co^{2+},Cd^{2+}$$

for reaction of ethylene glycol with terephthalic acid to form bis(hydroxyethyl) terephthalate [27]. This work has focused attention on compounds of tin and titanium. The tin compounds reported to have good activity in direct esterification include tin(II) oxide and oxalate and organotin(IV) compounds such as dialkyltin(IV) oxides and carboxylates. The titanium catalysts are typically titanium(IV) alkoxides.

The industrial application of titanium alkoxide catalysts is typified by a laboratory or semiworks experiment that may simulate plant operation [28]. A slurry of terephthalic acid in excess ethylene glycol is fed continuously to a reactor along with titanium(IV) tetraisopropoxide. The hot (250-300°C) reaction mixture which contains bis(hydroxyethyl) terephthalate and low molecular weight polymer dissolves the terephthalic acid and permits esterification to proceed in a relatively homogeneous medium. Continuous distillation removes some excess glycol and the water as it is formed. The hot esterified material is fed to a polycondensation reactor in which it is heated at 260-300°C under reduced pressure to remove ethylene glycol. The titanium compound serves as a polycondensation catalyst but is supplemented with a conventional catalyst such as Sb_2O_3. As in the transesterification process, formation of ether linkages by condensation of hydroxyethyl groups is an unwanted side reaction.

The insolubility of terephthalic acid has impeded mechanistic studies on polyester synthesis by direct esterification. As an alternative, kinetic studies have been done on the reactions of benzoic acid, isophthalic acid, and substituted isophthalic acids with ethylene glycol [7,26,27]. Once benzoic acid is added to one end of the ethylene glycol, the remaining OH group is more active toward esterification [12]. Tin(II) and titanium(IV) show similar kinetic dependencies, although the order of the reaction varies with the acid that is being esterified. Since the rate depends on the concentrations of acid, glycol, and catalyst, it appears that all three components must be assembled in a reactive complex such as $(ArCO_2)_xM(OCH_2CH_2OH)_y$ [26].

The role of such a complex in esterification of benzoic acids is illustrated in the catalytic cycle of Figure 11.4. Dimethyltin oxide is normally an oligomer, but much of its chemistry is that expected from the hypothetical monomer, $Me_2Sn=O$. The initial reaction within the catalytic cycle is the addition of the alcohol O-H to form a hydroxy-alkoxy derivative. Esterification of this compound with the arenecarboxylic acid generates the proposed $(ArCOO)M(OCH_2CH_2OH)$ intermediate. Intramolecular nucleophilic attack on the carbonyl group by the alkoxide ligand produces ester and Me_2SnO to complete the cycle. This mechanism seems quite plausible for the direct esterification, but much additional work is required to establish it conclusively.

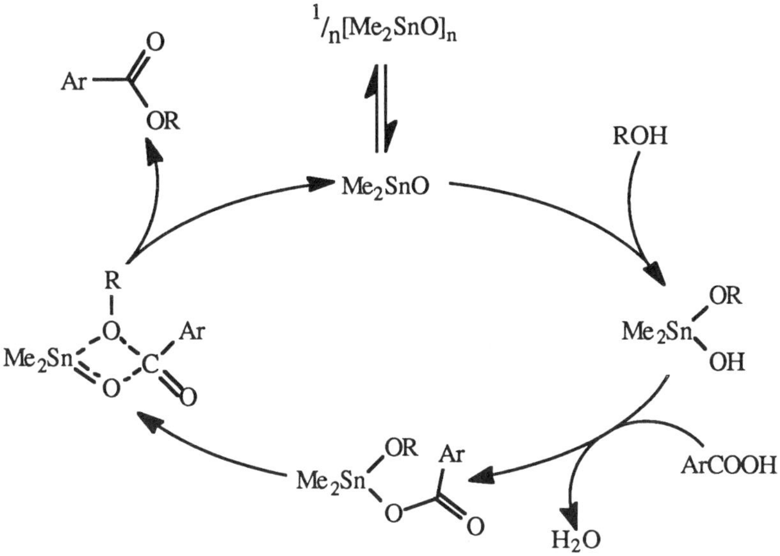

$^1/_n[Me_2SnO]_n$

$Ar-C(=O)-OR$

ROH

Me_2SnO

$Me_2Sn\begin{smallmatrix}OR\\OH\end{smallmatrix}$

Me_2Sn (four-membered ring with O, C, Ar, O=C) R—O

$Me_2Sn\begin{smallmatrix}OR & Ar\\O-C=O\end{smallmatrix}$

ArCOOH

H_2O

Figure 11.4 Tin-catalyzed esterification of benzoic acid. Dimethyl tin oxide is polymeric; the mononuclear species shown in this diagram are not meant to imply the nuclearity of the catalytically active species.

Catalysts based upon ladder-tetramer $Sn_4R_8O_2X_2Y2$ (combinations of $X = Cl$, NCS; $Y = Cl$, OH, NCS) have been employed quite successfully in esterification and transesterification reactions [29].

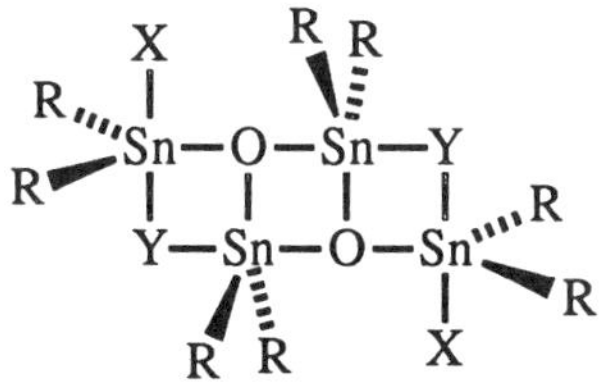

$$\begin{array}{c} X \quad R\ R \\ R_{\cdots}\ |\ \ \|\ \\ R^{\diagup}Sn-O-Sn-Y \\ |\ \ |\ \ |\ \ |\ _{\cdots R} \\ Y-Sn-O-Sn_{\cdots} \\ \diagup\ \ |\ \ \searrow R \\ R\ R \quad X \end{array}$$

It is felt that the tetramer stays intact during catalysis and two metal sites are involved during the catalytic step. These better-defined species might allow a more detailed description of the mechanism of tin-catalytic esterification.

Other Esterification Reactions

The transesterification and direct esterification processes discussed above serve to illustrate the general principles involved in most metal-catalyzed esterification and transesterification reactions. These catalysts are also applied to a number of lower-volume polymers and chemicals.

Du Pont and Teijin have introduced a polyester film based upon ethylene glycol and 2,6-naphthalenedicarboxylate for food packaging, magnetic tapes, and high-performance electrical applications. Advantages include higher transparency and better resistance to permeation of oxygen and moisture.

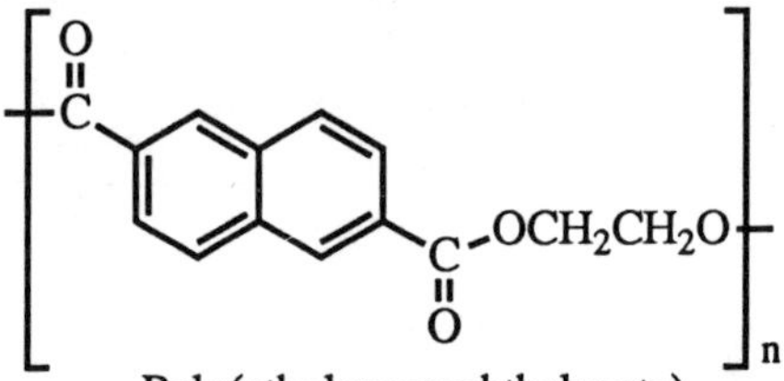

Poly(ethylene naphthalenate)

While most of the polymers discussed to this point involve ethylene glycol, other glycols or diols are also employed in polymer synthesis. Terephthalate polyesters are synthesized from 1,4-butanediol (BDO) [30], diethyleneglycol (DEG), or cyclohexanedimethanol (CHDM) [31].

HO(CH$_2$)$_4$OH   (HOCH$_2$CH$_2$)$_2$O   HOCH$_2$—⬡—CH$_2$OH

BDO   DEG   CHDM (cis and trans)

The butanediol product finds wide use as a high-performance elastomer in applications such as tires on small vehicles and protective joint "boots" for front-wheel drive cars. Terephthalate esters of DEG and CHDM find use in more limited specialty applications.

The reaction of butanediol with dimethyl terephthalate is generally catalyzed using titanate esters. A model for this reaction is beginning to appear [21,32,33], but it is clear that the ligands on titanium and therefore the molecularity of the starting titanate ester can have an effect on the catalytic reaction [34]. Aggregation of the titanates was found to depend on the nature of the alkoxide ligand, with the complexes displaying both bridging and nonbridging alkoxide groups. In solutions of tetrabutyltitanate, there is a rapid equilibrium between monomeric and dimeric species with the catalytically less-active dimer favored at the higher concentrations [21,34,35]. Exchange of alcohol with the metal center occurs primarily through the nonbridging alkoxide ligands in a reaction which is rapid on the NMR time scale. Alkoxide is transferred to the ester within the coordination sphere of the metal. Catalysis has been ascribed to the coordination of the titanium atom to the carbonyl oxygen [32], though coordination is not observed in 4-hydroxybutyl benzoate. A more likely explanation is that carbonyl coordination of ester is a nonproductive side reaction, and that productive catalysis takes place through intermediates such as (where the two additional alkoxide ligands around titanium are not shown for simplicity). This intermediate would be more in keeping with the chelation effects mentioned above [25] and would explain the lack of C=O coordination.

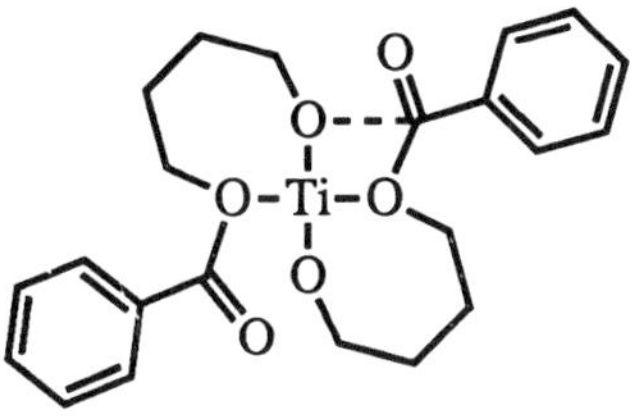

Analogous chemistry takes place in the synthesis of esters of adipic and methylglutaric acid, which are used as solvents and synthetic lubricants, and polyesters of ethylene glycol, BDO, DEG, and CHDM which are used as macrodiols for the polyurethanes described later in this chapter. These oligomers are often synthesized from mixtures of diols or diacids to minimize crystallinity in the resulting products; a good example is poly(ethylene-butylene adipate).

Tributyltin(IV) catalysts are effective for the ring-opening polymerization of e-caprolactone in solution or bulk [36]. The polymers can be taken to high molecular weight, but there are also commercial applications for lower molecular weight species. The related species, poly(glycolide), poly(lactide), and their copolymers [37], have been commercialized for resorbable sutures [38] and bone screws and other high-value medical applications [39] where controlled degradations of the polymers to innocuous fragments is required.

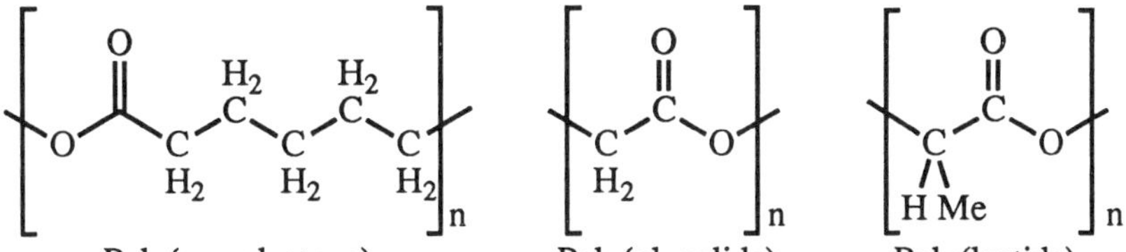

Poly(caprolactone)   Poly(glycolide)   Poly(lactide)

Polymer properties, especially crystallinity, can be varied over a wide range through copolymerization of stereoisomers or monomers. Broader application of these polymers in recyclable or biodegradable consumer applications awaits lower-cost monomer syntheses.

Polyglycolides and polylactides are synthesized by ring-opening of their cyclic dimer esters using tin catalysts in solution or in the melt [40,41], but many other metal-based catalysts are also effective [42,43]. The choice of metal for the catalyst can have a pronounced effect upon the rate of polymerization and also the subsequent thermal stability of the polymer. Because of their biological origin, lactide dimers can be obtained as the enantiomerically pure *d,d*-, *l,l*-, or racemic *d,l*-species, thus providing stereochemically "labeled" starting materials. Formally, the polymerization is an addition reaction and each polymerization event should add two monomers to the growing chain. It is possible to use nmr to differentiate between the addition of these dimer units and "transesterification," which cleaves both of the linkages in the initial dimer and has the apparent effect of adding monomers. $ZnCl_2$ causes extensive transesterification while $Zn(acac)_2$ and $Al(O\text{-}i\text{-}Pr)_3$ cause intermediate transesterification and $Al(acac)_3$ effects pure dimer incorporation [44]. Thus both the metal and its ancillary ligands guide the course of the polymerization.

There is a substantial industry based upon the esterification and transesterification of organics. Phthalic acid esters of long-chain alcohols derived from oxo synthesis (Chapter 5) or ethylene oligomerization (Chapter 4) are used as plasticizers. They are often synthesized from the convenient anhydride. The mechanism of the titanium-catalyzed reaction has been explored [45], and the proximity of the two carboxylate groups has an influence on the course of the reaction. The degree of hydrolysis of titanium catalysts can have a pronounced effect on the catalyst activity [46,47].

Esters are widely used as fragrances, and there is a major industry built upon their synthesis. Esters are also employed in medicinals. Long-chain esters of aliphatic diacids are used as high-performance synthetic lubricants in demanding applications such as turbojet engines. While conventional strong-acid catalysis is most commonly used in the synthesis of these low-volume, high-value products, titanate catalysis has been explored. Chiral titanium catalysts can even be used to carry out stereoselective esterifications on a-arylcarboxylic acid derivatives [48] and intermediates for insecticides [49].

Methyl methacrylate is synthesized on a commodity basis; other longer-chain esters used in homopolymerizations and copolymerizations with methyl methacrylate are synthesized from it by transesterification [50,51]. Alcohols used for exchange can contain functional groups designed for specific medical [52] or for post-polymerization crosslinking or adhesion [53,54]. While many of these transesterifications have traditionally been carried out using classical acid catalysts, most are now synthesized using titanate or tin catalysts.

11.2   POLYURETHANE SYNTHESIS

The diversity of applications of polyurethanes (a urethane is an N- and O-substituted derivative of carbamic acid) is very impressive. They range from upholstery foams to rigid structural resins, from durable flooring and automotive finishes to designer fibers like Du Pont's Lycra® spandex, and from rubbers to quick-bonding adhesives. They are prepared by reaction of diisocyanates with dihydroxy compounds. A simple example is the synthesis of Perlon-U, the first commercial polyurethane resin [55]:

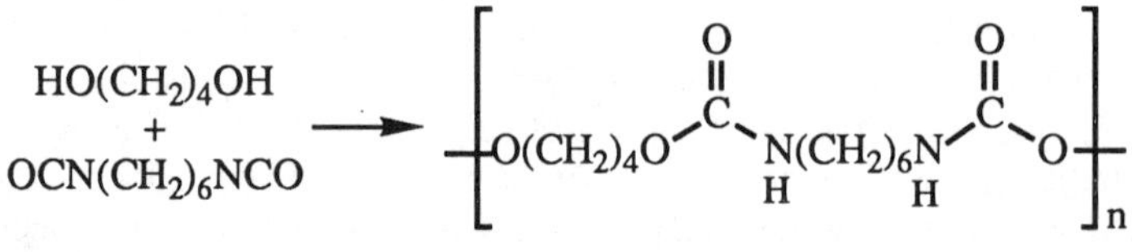

This is a good example of an aliphatic polyurethane, but most modern polyurethanes are prepared from aromatic diisocyanates such as MDI or TDI in combination with diols such as ethylene or propylene glycol, polyether diols, aliphatic polyesterdiols, or siliconediols. As in other condensation polymers, high-purity monomers are required to achieve the molecular weights required for desirable physical properties [56]. Polyester diols mentioned above also find application in specialty applications. Thus the final polymers are often segmented block copolymers consisting of polyester or polyether soft segments between polyurethane hard segments.

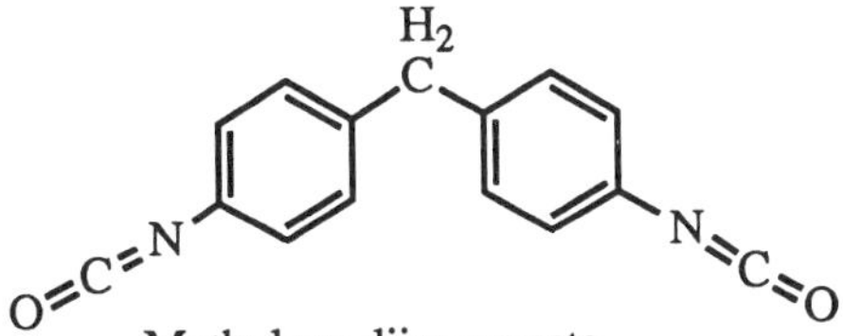

Methylene diisocyanate,
MDI

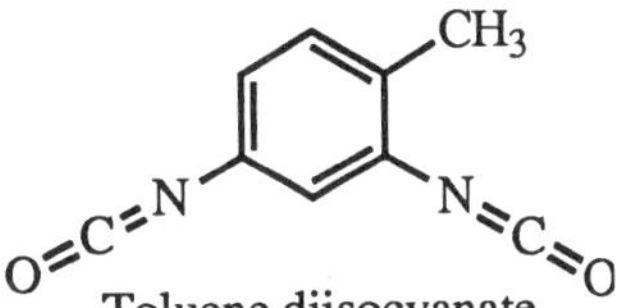

Toluene diisocyanate,
TDI

Polyurethanes can also be prepared from simple carbamic esters by catalytic transesterification [57]. This polymerization process is much less common than the isocyanate-based process, but it finds some use in the synthesis of plasticizers.

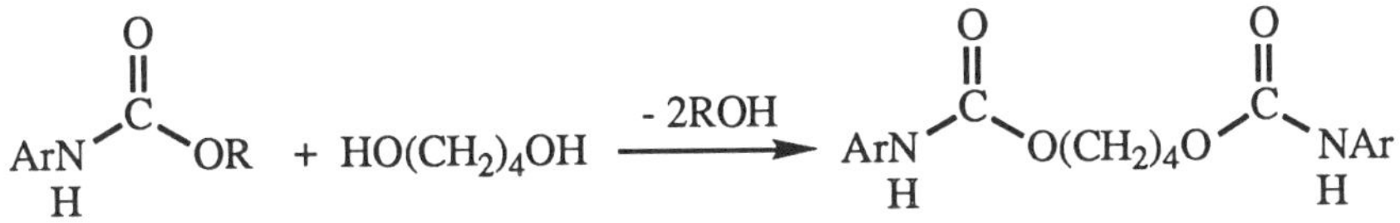

It is a true condensation polymerization like polyester production because a mole of alcohol is eliminated in the "condensation" reaction that forms the urethane group. The carbamate ester process may be used more frequently when these starting materials become available from the reaction of nitrobenzenes with CO and alcohols.

The reaction of an aromatic diisocyanate with a glycol occurs readily and, for laboratory-scale operations, no catalyst is required [58]. However, catalysts are used to accelerate polyurethane formation in most commercial processes. For aliphatic isocyanates, catalysts are required. The most common catalysts are tertiary amines, but metal complexes are also widely used. The most commonly used metal catalysts are tin derivatives, although titanium tetraalkoxides are very effective with both isocyanates and carbamic esters. Polyurethane formation is also catalyzed by many of the other metal compounds that are used for polyester production [59-64]. This similarity in catalysts is not surprising because the urethane group is similar to an ester in structure and chemistry.

Polyurethanes are used in many different ways - fibers, coatings, adhesives - but one of the largest uses is in the preparation of foams. The polymerization is carried out in the presence of a "blowing agent" which releases a gas such as N_2 or CO_2 or a fluorocarbon. The gas forms bubbles in the viscous prepolymer. The foam structure is retained after the polymer attains high molecular weight and becomes solid. Trifunctional reactants such as triols provide crosslinks for rigidity. The process is illustrated by a patent [65] in which a nonvolatile amine and tin(II) octoate are used as cocatalysts to produce both flexible and rigid foams. A mixture of the polyalcohol (a propylene oxide adduct of glycerin), the catalysts, the blowing agent, and a trace of

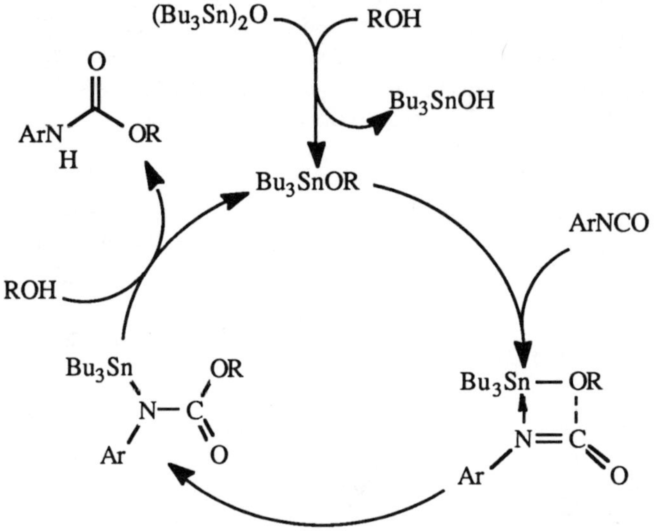

Figure 11.5 Tin-catalyzed addition of alcohols to isocyanates to yield carbamates.

water are vigorously stirred while 2,4-toluenediisocyanate is added. The foaming mixture is poured into a heated box and cured at 100-105°C for one hour in order to produce a cube of polyurethane foam. The rigidity of the foam depends on the number of polymer crosslinks provided by the alcohol component.

The mechanism of catalytic polyurethane production may be analogous to that of polyester production. Lewis acidic M^{2+} salts can activate the isocyanate function by coordination to oxygen [59] as in transesterification. Organotin compounds reportedly [66] act through alkoxide and carbamate intermediates as shown in Figure 11.5. Coordination of the isocyanate to the tin activates the C=N group to intramolecular nucleophilic attack by alkoxide to form a tin carbamate. Alcoholysis of the Sn-N bond liberates the carbamate-containing polymer and regenerates the tin alkoxide for another cycle of reaction. For reactions with glycols, intramolecular hydrogen bonding may play a role in the reaction [67].

11.3 POLYCARBONATE SYNTHESIS

The original synthesis of polycarbonates [68] was by transesterification of aromatic diesters of carbonic acid with dihydroxydiaryls such as bisphenol-A in the presence of basic catalysts. The reaction is also catalyzed by $Ti(OBu)_4$ [69], much like the polyesterifications discussed above.

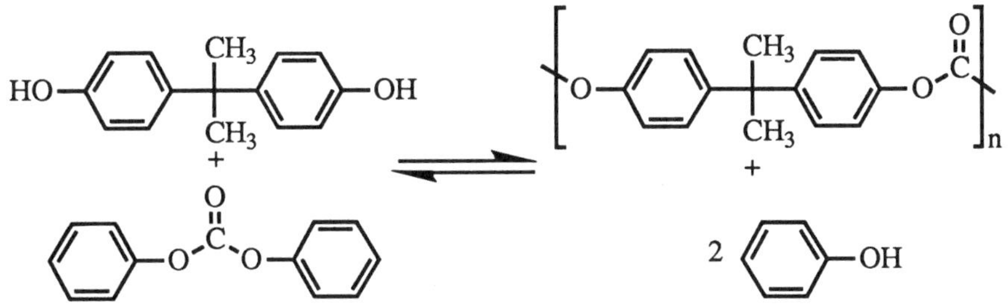

To achieve the high molecular weights necessary for desired properties, the equilibrium is shifted by distilling off the phenol at high temperature. This process was largely displaced by reaction of the sodium salt of bisphenol-A directly with phosgene [70].

General Electric [71] has developed a synthesis based upon macrocyclic monomers [72-74] which are polymerized by ring-opening polymerization using basic catalysts such as phenyl lithium, lithium phenoxide, tetrabutylammonium tetraphenylborate, and other strong anions [75-77]. The polymerization can also be accomplished using β-diketonate complexes of a variety of metals including Al, Fe, Ce [78], and Ti [79]. Mechanistic details of this reaction are still emerging, but it is thought that the action of the metal catalysts is very similar to that described in the polyesterification reactions.

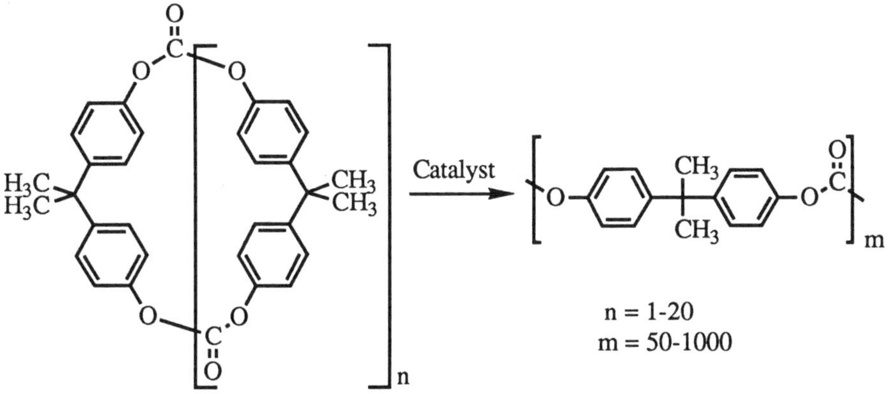

$n = 1-20$
$m = 50-1000$

There are many advantages to this new synthesis. The reaction is almost thermoneutral and is driven almost entirely by entropy. Thus heat removal is not a problem. At equilibrium, almost no cyclics remain. Since no byproducts are formed in the reaction, the final molecular weight is determined solely by monomer purity and the level of initiator used. The molecular weights of 2-400,000 typically achieved are significantly higher than those obtained in conventional polymerizations. Because the melt viscosities of the cyclics are much lower than the polymers, composites, crosslinking, copolymerizations, and extruder polymerizations can now be considered.

Metal catalysis plays an important if originally unintended role in the blending of polyesters [in particular poly(ethylene)- and poly(butylene)terephthalates] and polycarbonates. Under melt conditions, residual catalyst from the polyester synthesis catalyzes the exchange of carbonate and ester linkages, yielding block polyester-polycarbonate copolymers [80,81]; titanium residues, normally found in poly(butylene)terephthalate, are more active than the antimony normally found in poly(ethylene)terephthalates [82]. This covalent linking of the two types of groups stabilizes the resulting blend.

11.4 POLYAMIDE SYNTHESIS (NYLON)

The most widely used polyamide in the United States is poly(hexamethyleneadipamide), usually known as nylon-6,6 [83]. The numerical designation indicates that it is prepared from a six-carbon diamine and a six-carbon dicarboxylic acid. Second behind nylon-6,6 and equal on a worldwide scale is nylon-6, which is made by the ring-opening polymerization of ε-caprolactam. The largest outlet for both polymers is in the production of fibers for carpets, hosiery, and other applications. The polymers are also used as engineering plastics. There is a growing use of nylon-6,12 and nylon-12,12 for more demanding applications because their changes in physical dimensions upon exposure to humidity are reduced because of the higher aliphatic content. Nylon-12,T, where T stands for terephthalic acid, is also under investigation. Nylon-46 is produced in Europe and Japan for its improved properties at higher temperatures.

The usual preparation of nylons from diamines and diacids begins with the reaction of the two components to form a salt - thus guaranteeing balanced stoichiometry. In the polymerization step, the salt is heated under vacuum to expel water and form the amide bonds [84]:

$$H_2N(CH_2)_6NH_2 + HO_2C(CH_2)_4CO_2H \longrightarrow [H_3N(CH_2)_6NH_3][O_2C(CH_2)_4CO_2]$$

$$\downarrow -H_2O$$

$$H-[HN(CH_2)_6NHOC(CH_2)_4CO]_n-OH$$

Although the amidation reaction ordinarily does not require a catalyst, the commercial manufacture of nylon intermediates is heavily dependent on the homogeneously catalyzed reactions mentioned in earlier chapters. The oxidation of cyclohexane to adipic acid (Section 10.3) employs soluble catalysts in two steps. In a similar process, cyclododecane is oxidized to dodecanedioic acid, a component of nylon-6,12 and other specialty polyamides (Section 10.4).

In contrast to adipic acid, which is made by similar processes by most producers, 1,6-hexanediamine is manufactured by several different processes though each goes through adiponitrile as an intermediate. The dinitrile is hydrogenated smoothly to the diamine with a heterogeneous catalyst. The oldest route is reaction of adipic acid with ammonia to give ammonium adipate that dehydrates to form adiponitrile when it is heated strongly. Electrolytic reductive dimerization of acrylonitrile to adiponitrile is widely practiced and is especially competitive on a smaller scale. Attempts to dimerize acrylonitrile catalytically have had only moderate success. The addition of 2 moles of HCN to butadiene regioselectively to form adiponitrile is one of the most sophisticated commercial applications of homogeneous catalysis and is covered in Section 3.6. Routes through chlorination of butadiene are obsolete, but because this chemistry is widely used for the production of chloroprene, a monomer for oil-resistant synthetic rubbers [85], it is discussed in Section 12.1.

The synthesis of nylon-6 from ε-caprolactam does not require the salt-forming step because the acid and amine functionality are combined into a single cyclic molecule, thereby assuring balance of the two functionalities. However, the polymerization is an equilibrium between polymer and monomer. The monomer is only slightly less stable than polymer, so there is an appreciable equilibrium concentration. The uncatalyzed polymerization of pure monomer proceeds very slowly at temperatures as high as 250°C, and monomer becomes more favored at these higher temperatures. Fortunately, the reaction is polymerized by nylon-66 salt or small concentrations of free acid which arise from hydrolysis [86]. Although manganese salts of phosphorous acid and other phosphorus or boronic species will catalyze the polymerization, they are not commonly added to commercial polymerizations. In the synthesis of nylon-4 from the more stable pyrrolidone ring, high polymer is disfavored at higher temperatures [87]; the polymerization must be carried out at very low temperatures using alkaline catalysts [88,89].

11.5   POLYSILOXANE (SILICONE) SYNTHESIS

The silicon-oxygen bond is one of the most stable bonds in nature and, as a result, it confers a great deal of stability to the siloxane $(-SiR_2-O-)_x$ backbone of silicones. The physical properties of silicones, which range from mobile fluids to elastomers and highly crosslinked resins, are retained over a wide range of temperatures and conditions. They also have unique surface properties and low toxicity. These features make them useful in applications such as protective coatings, sealants, medical implants, hydraulic oils, and water-repellants. Originally, silicones were synthesized by the addition of Grignard reagents to $SiCl_4$ in a difficult and costly process. Development of a direct process brought silicones into common use. Today silicone chemistry is dominated by the methyl silicones.

The process for making methyl silicones starts with the addition of two equivalents of methyl chloride to silicon metal. The resulting dimethyl silicon dichloride is hydrolyzed with one equivalent of water yielding the desired Si-O-Si linkages and two equivalents of HCl.

$$n\ Si\ +\ 2n\ MeCl \xrightarrow{\text{Catalyst}} n\ Me_2SiCl_2 \xrightarrow{+\ n\ H_2O} 2n\ HCl\ + \left[\begin{array}{c} Me \\ | \\ -Si-O- \\ | \\ Me \end{array} \right]_n$$

$$+\ 2n\ MeOH$$
$$-\ 2n\ H_2O$$

The HCl is recycled by reaction with methanol to give methyl chloride; thus the overall reaction is methanol plus silicon to give silicone and water. Cyclic oligomers are the simplest of the useful products obtained. Linear polymers are obtained by incorporating Me_3SiCl or other monochlorides into the hydrolytic reaction, thereby controlling end-group functionality and molecular weight. The $RSiCl_3$ molecules provide branching points in the polymer to further adjust the desired properties.

The cyclic oligomers are polymerized to high polymer by a number of catalysts. Under some conditions, nucleophile-catalyzed polymerization of the cyclic trimer (n = 3), which is ring-strained, is enthalpically driven and can go straight to linear high polymers without redistribution or formation of larger cyclics [90]. On the other hand, polymerizations with other catalysts or polymerizations of larger, unstrained cyclics are entropically driven equilibrium polymerizations in which cyclics appear in concert with the higher linear species. The ratio of cyclic and linear products is dependent on concentration [91].

$$\text{LiR} + m \left[\text{(Si-O)}_n \right]_{Me_2} \rightleftharpoons R\left[\text{(Si-O)}_n \right]_{Me_2} \Big]_m \text{Li}$$

$$\left[\text{(Si-O)}_{nm} \right]_{Me_2}$$

The molecular weight can be controlled by the concentration of initial base or by the use of siloxane dimers which act as endcaps. Platinum and palladium compounds will also catalyze the polymerization [92], but only the acidic and basic catalysts are used commercially.

In the 1960s, silicone elastomers that could undergo room-temperature vulcanization (RTV), that is, cure or crosslink in situ without application of heat, were introduced. These products quickly became an appreciable fraction of the entire industry. Single-component RTV elastomers based upon an acetate cure are most familiar to consumers as bathtub caulks. Of the four commercially important curing processes to form silicone rubbers, three - the one- and two-component condensation cures and the silane/olefin-addition cure - utilize metallic catalysts. The fourth is a peroxide-induced, free-radical cure and is not covered here.

The condensation cure is used for most RTV silicone rubbers, as well as most solvent-borne silicone resins. The reaction is much like esterification or transesterification if silicon methoxides are considered to be esters of silicic acid. In one form, a trialkoxy silicon alkyl is reacted with three silanol-terminated polysiloxanes to yield a branch or crosslink in the $(-SiR_2-O-)_x$ backbone.

$$R'\!-\!\underset{\underset{OCH_3}{|}}{\overset{\overset{OCH_3}{|}}{Si}}\!-\!OCH_3 + 3\ H\!-\!O\!\left[\underset{CH_3}{\overset{CH_3}{|}}{Si}\!-\!O\right]_n\!\!R \xrightarrow[\text{Catalyst}]{-3\ CH_3OH} R'\!-\!Si\!\left(O\!\left[\underset{CH_3}{\overset{CH_3}{|}}{Si}\!-\!O\right]_n\!\!R\right)_3$$

The catalyst is kept in a separate component and mixing the two components initiates the reaction. As shown, the resulting polymer would be three times the molecular weight with one branch. In practice, the group, R, on the other end of the chain may be a hydrogen, thereby providing hydroxyl groups on both ends of the polymer and developing a highly crosslinked network. The evolution of methanol drives the reaction to completion. The original catalysts for these reactions were amines, including aminopropylsilane derivatives. These have been displaced by carboxylic acid salts of tin and zinc, with tin(II) octoates and laurates and dibutyl tin(IV) compounds finding greatest utility. The mechanism of the reaction is analogous to transesterification of carboxylic acids.

In the single-component version of this reaction, a silanol-terminated silicone is functionalized with methylsilicon triacetate or a related molecule. This leaves one or more acetate groups on the polymer ends. Exposure to moisture hydrolyzes the acetoxy groups to hydroxyl groups which are then able to chain-extend or crosslink the polymer by formation of new Si-O-Si linkages. Loss of acetic acid, which gives these polymers their characteristic smell during application, drives the reaction

$$2 \ CH_3CO_2-Si\!\!\!< \ + \ 2H_2O \ \xrightarrow{\ Sn^{2+}\ } \ >\!\!\!Si-O-Si\!\!\!< \ + \ 2 \ CH_3CO_2H$$

to high conversion. The cure is catalyzed by tin soaps which both accelerate and control the reaction [93]. It is believed that the tin forms an active complex with the silanol and then reacts with the crosslinking agent [94,95]. Because the curing process depends upon ambient moisture, curing is from the outside in. Surfaces can be tack-free in 15-30 minutes, but depending upon conditions, strength can develop slowly for about three weeks. These formulations adhere well to ceramic tiles and grouting because of hydroxyls present on the surfaces [96].

Crosslinking by hydrosilation of pendant olefin groups catalyzed by transition metals has become important in the last decade (see Section 3.5 for a discussion of hydrosilylation). Commercial catalysts include complexes of platinum, rhodium, and ruthenium, with chloroplatinic acid or platinum olefin catalysts among the most efficient.

$$R-O\!\!\left[\!\!\begin{array}{c} CH_3 \\ | \\ Si-O \\ | \\ CH_3 \end{array}\!\!\right]_{n}\!\!\!CH=CH_2 \ + \ \begin{array}{c} -Si- \\ | \\ O \\ | \\ H-Si-CH_3 \\ | \\ O \\ | \\ -Si- \\ | \end{array} \ \longrightarrow \ R-O\!\!\left[\!\!\begin{array}{c} CH_3 \\ | \\ Si-O \\ | \\ CH_3 \end{array}\!\!\right]_{n}\!\!\!\begin{array}{c} H_2 \ H_2 \\ C-C \end{array}\!\!\!\begin{array}{c} -Si- \\ | \\ O \\ | \\ Si-CH_3 \\ | \\ O \\ | \\ -Si- \\ | \end{array}$$

Concentrations of 1-2 ppm of platinum are often sufficient to effect the cure [97]. The cure usually occurs at room temperature over several days or can be accelerated by heating to temperatures from 50 to 100°C. The fact that no coproduct is evolved makes this type of cure particularly advantageous in certain applications. Because the catalyst can be poisoned by a number of compounds, some care must be taken with the environment in which this cure is used. Water and alcohols will react with the Si-H in the polymer, liberating hydrogen, which can be used to foam the polymer if desired.

11.6 RING OPENING OF CYCLIC ETHERS

The ring-opening polymerization of cyclic ethers is one of the oldest examples of macromolecule formation [98], dating to 1863. Today polymers and oligomers of ethylene and propylene oxide enjoy extensive use as adhesives, coatings, plasticizers, surfactants, and intermediates in the polyurethanes discussed above. Polymers of tetrahydrofuran terminated with hydroxyl groups are also used in polyurethanes. Polymers of epichlorohydrin and 3,3-bis(chloromethyl)oxetane also find wide use.

When propylene oxide is polymerized using basic catalysts such as potassium hydroxide, the methyl groups are distributed randomly on either side of the polymer chain. When the polymerization is catalyzed with metal systems such as $ZnEt_2/H_2O$, or $AlEt_3/H_2O$/acetylacetone, the polymer is isotactic, having sequences in which consecutive asymmetric carbon atoms have the same handedness [99,100]. Because the starting propylene oxide is racemic, this isotacticity must result from selective incorporation of a single antipode of the monomer into a given chain. While there is some evidence for chain-end control in polymerization of sterically hindered epoxides [101], it is generally accepted that the polymerization is controlled by an asymmetric catalytic site [102]. This view is supported by the use of chiral catalyst systems such as $ZnEt_2$/(+)-borneol [19] or $ZnEt_2$/3,3-dimethyl-1,2-butanediol [103] for the optical resolution of racemic propylene oxide by stereoselective polymerization of a single enantiomer.

Propylene oxide has been polymerized with the $ZnEt_2$/pyrogallol system in dioxane [104]. Each polymer chain is terminated by the catalyst group, as evidenced by increasing the molecular weight of the polymer with increasing conversion of the monomer. An idealized view of the catalyst and a proposed mechanism for polymerization is shown in Figure 11.6. Unfortunately, the system is not this simple, because of aggregation of the catalyst. This aggregation results in catalytic sites of different activity and a resulting broad molecular weight distribution of the polymer.

Polymers of controlled molecular weight have been prepared with a truly homogeneous catalyst system based upon aluminum complexes of tetraphenyl-porphyrin (AlTPP) [105,106].

AlTPP(X)
X = Cl, OMe

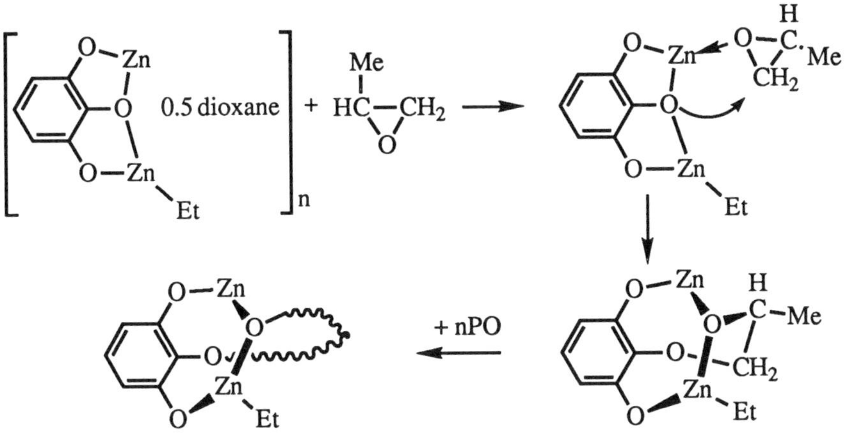

Figure 11.6 Zinc-catalyzed polymerization of propylene oxide.

The resulting polymers have the very narrow molecular weight distribution expected of a living polymerization. That is, each polymer chain is capped by an active catalyst species and the number of polymer chains formed is equal to the number of catalyst molecules put into the reaction. Thus, the molecular weight (g/mole) is determined solely by the grams of monomer divided by the moles of catalyst. If an additional aliquot of monomer is added to the system, the molecular weight of the polymer begins to increase again [107]. It is also possible to make block copolymers by sequential addition of different 1,2-epoxides. Propylene oxide will even copolymerize with carbon dioxide [108]. It is clear from detailed analyses of initial oligomeric products that the polymerization proceeds by a mechanism in which each chain is capped by the aluminum porphyrin at one end and the starting chloride at the other.

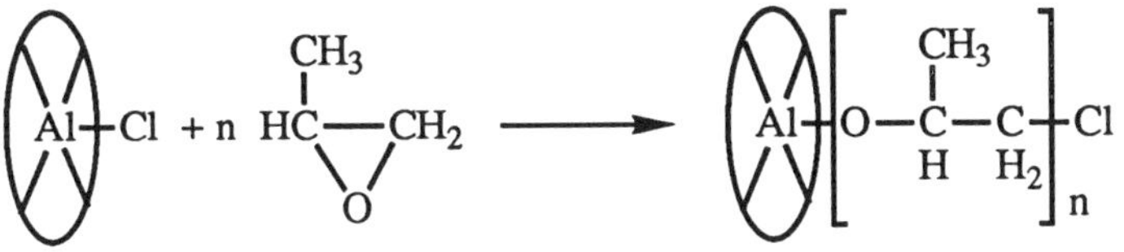

11.7 EPOXY RESINS

Epoxy resins are characterized by their good adhesion to a variety of substrates, high stability to chemicals, temperature, and corrosion, and good electrical and mechanical properties. The most important epoxy resins are based upon the product of the reaction of epichlorohydrin with bisphenol-A.

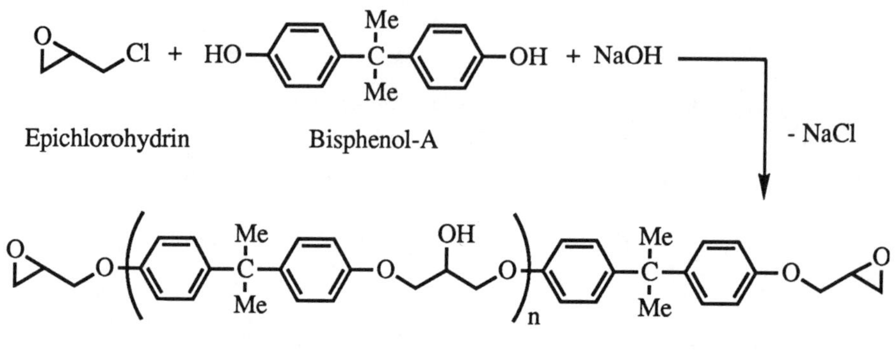

Epichlorohydrin   Bisphenol-A   - NaCl

DGEBPA (Diglycidylether-bisphenol-A)   n ≈ 0.2 for liquids n = 3-20 for solids

The resulting liquid is a reactive bis-epoxide which is cured to solid polymer by a variety of ring-opening reactions, the choice of which depends upon the application. These reactions include reactions with anhydrides, primary and secondary aliphatic polyamines, and catalytic curing agents.

Anhydride-cured systems usually rely upon a direct, uncatalyzed reaction, but the desire to increase the rate of cure while leaving the storage-life of the resin unaffected has led to the discovery of "latent accelerators" for the reaction. The complex $BF_3 \cdot NH_2Et$ is effective for this application, but has an adverse effect on electrical properties [109]. Transition metal β-diketonates catalyze the reaction and have no effect on the electrical or mechanical properties [110-112]. The $Co(acac)_3$-catalyzed curing reaction follows first-order kinetics, but surprisingly, it is 0.25 order in catalyst. A variety of chromium compounds exhibited excellent latent catalytic properties in the curing of bisphenol-A epoxy resins with 1-methyltetrahydrophthalic anhydride [113]. Catalyst residues in a copper-catalyzed isomethyltetrahydrophthalic anhydride system catalyze the oxidative thermal degradation of the resulting epoxy polymer [114].

Matrices formed by crosslinking of bisphenol-A epoxy resins with metal carboxylates chelated with polyamine ligands, for example, $Zn(en)_2(MeCO_2)_2$, $Co(dien)(HOC_6H_4CO_2)_2$, or $Fe(trien)(HOC_6H_4CO_2)_3$, show significantly higher strength and lower heat deformation than nonmodified epoxy resins. The higher energy-absorbing capacity of the composites in comparison with nonmodified resins [115] is dependent on the nature of the metal, with the metals falling in the order

Zn>Cu>Co>Ni>Fe>Mn

The effect is thought to arise by induction of polar interactions or ionomeric crosslinks between the polymer chains by the electropositive metals. It is not clear whether some of this effect could also arise from a more selective curing of the epoxy groups through the metal-mediated process.

11.8   GROUP TRANSFER POLYMERIZATION

Free-radical polymerization has been the standard method of preparation of poly(methyl methacrylate) polymers. It is a relatively inexpensive and simple procedure to get high molecular weight homopolymer. The method is equally effective for random copolymerization of mixed methacrylates. If one desires homopolymers having a low or narrow molecular weight or copolymers with customized sequences of the comonomers, one must use anionic or group transfer polymerization (GTP). Generally, anionic polymerization uses hindered lithium alkyl compounds as initiators and the polymerizations are carried out at -60°C [116]. While satisfactory for the laboratory, this technique does not lend itself to commercial preparations. Group transfer polymerization [117,118] overcomes a number of these difficulties and, as a result, is finding an increasing number of commercial applications, primarily in automotive finishes where improved resins are required to meet today's stricter volatile organic emission codes.

Group transfer polymerization is initiated by the Michael addition of a silylketene acetal to an a,b-unsaturated ketone. In the illustration for methylmethacrylate, the addition results in a new silyl ketene acetal which can undergo further additions. The polymerization is living, the molecular weight distribution approaches one, all of the

initiating groups are consumed, and initiation is fast relative to propagation. The polymerization requires a catalyst to function. The original catalyst was TAS bifluoride [tris(dimethylamino)sulfonium bifluoride], but tetrabutylammonium fluoride trihydrate is also effective [119]. The waters of hydration do not need to be removed. Though less effective, tetrabutylammonium carboxylates or bicarboxylates are the preferred catalysts because of availability and reproducibility [120,121]. All of these catalysts operate by coordinating to and activating the silicon group for transfer. Lewis acids such as $ZnBr_2$ or HgI_2 are also effective catalysts, though they operate by a completely different mechanism [122]. Rather than activate silicon, most of the Lewis acids activate the incoming methacrylate by coordination to the carbonyl oxygen atom. Because there is a large quantity of monomer compared to silicon, relatively large amounts of Lewis acid are required, thereby complicating isolation of the polymer. HgI_2 is unusual in that much less is required, and it is thought that it activates the silyl ketene acetal which is present in much lower quantities.

Group transfer polymerization allows the custom design of many interesting polymers. Dienoates and a trienoate are polymerized more readily than methyl methacrylate itself. Chain ends of the polymer can be functionalized after the polymerization.   Because of the large differences in reactivities, acrylates, methacrylates, and acrylonitrile cannot be randomly copolymerized, but all monomers within a given class (such as methyl and butyl acrylate) can be randomly copolymerized. AB and ABA blocks are readily prepared be sequential addition of the monomers.   Graft, comb, star, and ladder polymers can be prepared; the star copolymers are particularly useful in automotive finish applications.

A form of GTP in which the transferring group is an aluminum tetraphenylporphyrin results in a polymerization which has been termed "immortal" because of its robust nature [123,124].   Polymerization of methyl methacrylate initiated with AlTPP(SMe) (see description of TPP in Cyclic Ethers, Section 11.6) gives a polymer

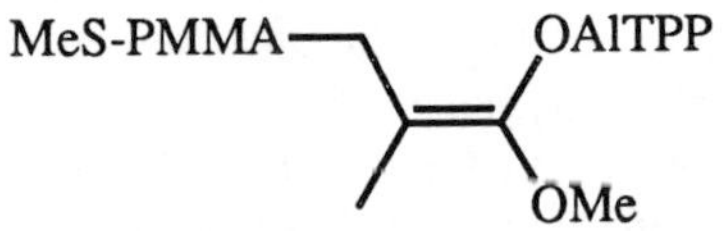

$$\text{MeS-PMMA} \diagdown \overset{\text{OAlTPP}}{\underset{\text{OMe}}{\diagup}}$$

which is terminated on one end with a methyl sulfide group and on the other with an aluminum porphyrin. This system has been used to make a variety of interesting copolymers.

A living polymerization of methacrylates can also be initiated using tri(isopropoxy)titanium enolates [125]:

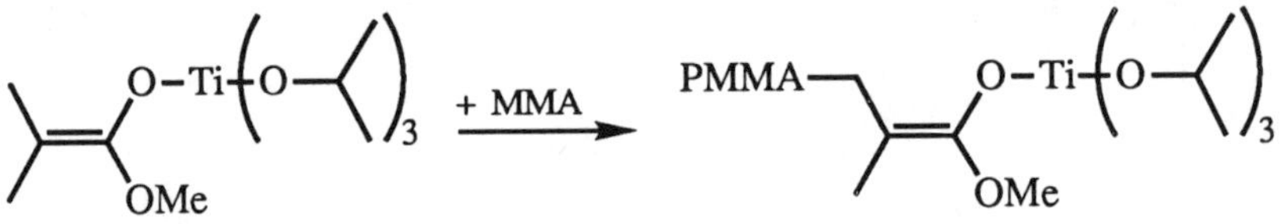

$$\diagup\diagdown\underset{\text{OMe}}{\overset{\text{O--Ti}(\text{O--CH(CH}_3)_2)_3}{}} \;+\; \text{MMA} \;\longrightarrow\; \text{PMMA}\diagup\diagdown\underset{\text{OMe}}{\overset{\text{O--Ti}(\text{O--CH(CH}_3)_2)_3}{}}$$

The molecular weight distribution is narrow, indicating a living polymerization, but few other details were given.

GENERAL REFERENCES

M. Lewin and E. M. Pierce, Eds., *Handbook of Fiber Science and Technology: Volume IV, Fiber Chemistry, International Fiber Science and Technology Series,* Marcel Dekker, New York, 1985.

C. E. Schildknecht and I. Skeist, Eds., *Polymerization Processes*, Wiley-Interscience, New York, 1977.

G. Woods, Ed., *The ICI Polyurethane Handbook*, 2nd Ed., Wiley, New York, 1990.

K. J. Ivin and T. Saegusa, *Ring-Opening Polymerization*, Elsevier, New York, 1984.

H. F. Mark, N. M. Bikales, C. Overberger, and G. Menges, Eds., *Encyclopedia of Polymer Science and Engineering*, 2nd Ed., Wiley-Interscience, New York.

SPECIFIC REFERENCES

1. J. K. Stille and T. W. Campbell, Eds., *Condensation Monomers*, Wiley-Interscience, New York, 1972.
2. G. W. Parshall, *J. Mol. Catal.*, **4**, 243 (1978).
3. H. Ludewig, *Polyester Fibres*, Wiley-Interscience, New York, 1971.
4. M. Katz, "Preparations of Linear Saturated Polyesters," in C. E. Schildknecht and I. Skeist, Eds., *Polymerization Processes*, Wiley-Interscience, New York, 1977, pp. 468-496.
5. R. E. Wilfong, *J. Polym. Sci.*, **54**, 385 (1961).
6. British Patent 826,248 (1959) to Du Pont.
7. J. Vejrosta, E. Zelena, and J. Malek, *Coll. Czech. Chem. Commun.*, **43**(2), 424-33 (1978).
8. S. G. Hovenkamp, *J. Polym. Sci.*, Part A-1, **9**, 3617 (1971).
9. K. Yoda, K. Kimoto, and T. Toda, *J. Chem. Soc. Jpn., Ind. Chem.*, **67**, 909-914 (1964).
10. M-D. Lee and R-J. Ho, *Hua Hsueh*, **3**, 93-102 (1974).
11. F. B. Cramer, *Macromol. Syn.*, **1**, 17 (1963).
12. J. Otton and S. Ratton, *J. Polym. Sci.*, Part A, **29**(3), 377-91 (1991).
13. H. Zimmerman, *Faserforsch Textiltech*, **19**, 372 (1968).
14. K. Dimov, E. Terlemezyan, A. Grosheva, and S. Hubawii, *Angew. Makromol. Chem.*, **28**, 1-11 (1973).
15. F. Kobayashi, K. Matsukura, K. Suga, and H. Obara, *Kogyo Kagaku Zasshi*, **74**(6), 1244-7 (1971).
16. I. Goodman, "Polyesters," in H. F. Mark, N. M. Bikales, C. Overberger, and G. Menges, Eds., *Encyclopedia of Polymer Science and Engineering*, 2nd Ed., Vol. 12, Wiley-Interscience, New York, 1988, pp. 1-75.
17. G. Rafler, F. Tesch, and D. Kunath, *Acta Polym.*, **39**(6), 315-20 (1988).
18. K-H. Wolf, B. Kuster, H. Herlinger, C-J. Tschang, and E. Schrollmeyer, *Angew. Makromol. Chem.*, **68**, 23 (1978).
19. C. M. Fontana, *J. Polym. Sci.*, **6**, 2343 (1968).
20. K. Yoda, K. Kimoto, and T. Toda, *J. Chem. Soc. Jpn., Ind. Chem. Sect.*, **67**, 909 (1964).
21. J. Otton, S. Ratton, V. A. Vasnev, G. D. Markova, K. M. Nametov, V. I. Bakhmutov, L. I. Komarova, S. V. Vinogradova, and V. V. Korshak, *J. Polym. Sci.*, Part A, **26**(8), 2199-224 (1988).
22. S. B. Maerov, *J. Polym. Sci., Polym. Chem*, **17**, 4033-40 (1979).
23. R. Lasarova and K. Dimov, *Angew. Makromol. Chem.*, **55**, 1 (1976).
24. T. H. Shah, J. I. Bhatty, G. A. Gamlen, and D. Dollimore, *Polymer*, **25**(9), 1333-6 (1984).
25. T. H. Fife and T. J. Przystas, *J. Am. Chem. Soc.*, **102**(24), 7297-300 (1980).
26. O. M. O. Habib and J. Malek, *Coll. Czech. Chem. Commun.*, **41**, 2724-36 (1976).
27. L. Nondek and J. Malek, *Makromol. Chem.*, **178**, 2211-21 (1977).
28. Research Disclosure 16710 (Mar. 1978).
29. J. Otera, N. Dan-oh, and H. Nozaki, *J. Org. Chem.*, **56**, 5307-5311 (1991).
30. S. G. Hovenkamp, *J. Polym. Sci.*, Part A-1, **9**, 3617 (1971).
31. F. B. Cramer, *Macromol. Synth.*, **1**, 17 (1963).
32. F. Pilati, P. Manaresi, B. Fortunato, A. Munari, and P. Monari, *Polymer*, **24**(11), 1479-83 (1983).
33. A. Fradet and E. Marechal, *J. Macromol. Sci., Chem.*, **A17**(5), 881-91 (1982).

34. M. Marsi and D. C. Roe, Abstract INOR-320, 196th National Meeting of the American Chemical Society, Los Angeles, CA, September, 1988.

35. B. Fortunato, P. Manaresi, A. Munari, and F. Pilati, *Polym. Commun.*, **27**(1), 29-31 (1986).

36. H. R. Kricheldorf, M. V. Sumbel, and I. Kreiser-Saunders, *Macromolecules*, **24**(8), 1944-9 (1991).

37. H. R. Kricheldorf and I. Kreiser, *Makromol. Chem.*, **188**(8), 1861-73 (1987).

38. E. E. Schmitt and R. A. Polistina, U.S. Patents 3,297,733 (1967) and 3,463,158 (1969) to American Cyanamide.

39. D. L. Wise, T. D. Fellmann, J. E. Sonderson, and R. L. Wentworth, in G. Gregoraides, Ed. *Drug Carriers in Biology and Medicine*, Academic Press, New York, 1979, pp. 237-270.

40. H. R. Kricheldorf, M. Berl, and N. Scharnagl, *Macromolecules*, **21**(2), 286-93 (1988).

41. H. R. Kricheldorf and M. Sumbel, *Eur. Polym. J.*, **25**(6), 585-91 (1989).

42. H. R. Kricheldorf and M. Sumbel, *Makromol. Chem.*, **189**(2), 317-31 (1988).

43. S. J. McLain and N. E. Drysdale, U. S. Patent 5,028,667 (1991) to Du Pont.

44. M. Bero, J. Kasperczyk, and Z. J. Jedlinski, *Makromol. Chem.*, **191**, 2287-2296 (1990).

45. S. R. Bhutada and V. G. Pangarkar, *J. Chem. Technol. Biotechnol.*, **36**(2), 61-6 (1986).

46. E. A. Khrustaleva, Y. G. Yatluk, and A. L. Suvorov, *Izv. Akad. Nauk SSSR, Ser. Khim.*, (6), 1247-50 (1989).

47. F. Pilati, A. Munari, P. Manaresi, and V. Bonora, *Conv. Ital. Sci. Macromol.*, *[Atti], 6th*, Vol. 2, Assoc. Ital. Sci. Tecnol. Macromol., Genoa, Italy, 1983, pp. 393-60.

48. K. Narasaka, F. Kanai, M. Okudo, and N. Miyoshi, *Chem. Lett.*, (7), 1187-90 (1989).

49. K. Narasaka, T. Myoshi, F. Kanai, and M. Okudo, Japanese Patent 3,099,037 to Sumitomo Chemical Co., Ltd. (1991).

50. M. A. Korshunov and F. N. Bodnaryuk, *Zh. Org. Khim.*, **4**(7), 1204-9 (1968).

51. C. Falize and A. Bouniot, German Patent 2,319,688 (1973) to Rhone-Progil.

52. N. J. Lewis and A. S. Wells, World Patent Application 91-09,005 (1991) to Smith Kline and French.

53. D. Farrar, U.S. Patent 4,609,755 (1986) to Allied Colloids.

54. E. Canivenc and M. Gay, European Patent 281,718 (1988) to Rhone-Poulenc.

55. J. H. Saunders and K. C. Frisch, *Polyurethanes: Chemistry and Technology*, Pt. 1, Wiley-Interscience, New York, 1962.

56. J. K. Backus, "Polyurethanes," in C. E. Schildknecht and I. Skeist, Eds., *Polymerization Processes*, Wiley-Interscience, New York, 1977, pp. 642-80.

57. L. Heiss, German Patent 2,655,741 (1978).

58. T. W. Brocks, C. Bledsoe, and J. Rodriguez, *Macromol. Synth., Coll. Vol. 1*, 381 (1977) .

59. J . W. Britain and P. G. Gemeinhardt, *J. Appl. Polym. Sci.*, **4**, 207 (1960).

60. E. P. Squiller and J. W. Rosthauser, *Mod. Paint Coat.*, **77**(6), 28-35 (1987).

61. E. P. Squiller and J. W. Rosthauser, *Proceedings of the 14th Water-Borne Higher-Solids Coatings Symposium*, 460-77 (1987).

62. E. P. Squiller and J. W. Rosthauser, *Polym. Mater. Sci. Eng.*, **55**, 640-7 (1986).

63. R. Sojecki, *Acta Polym.*, **41**(6), 315-18 (1990).

64. R. Sojecki, R., *Acta Polym.*, **40**(8), 487-92 (1989).

65. I. Cuscurida and G. P. Speranza, U.S. Patent 4,101,462 (1978).

66. A. J. Bloodworth and A. G. Davies, *J. Chem. Soc.*, 5238 (1965).

67.   P. A. Berlin, R. P. Tiger, Y. N. Chirkov, and S. G. Entelis, *Kinet. Katal.* (English translation), **28**(6), 1168-72 (1987).

68.   H. Schnell, *Polymer Reviews, Chemistry and Physics of Polycarbonates*, Vol. 9, Wiley-Interscience, New York, 1964.

69.   P. G. Odell, G. Baranyi, and L. Alexandru, U.S. Patent 4,921,940 (1990) to Xerox.

70.   D. Fox, *Tech. Pap. Reg., Technical Conference of the Plastics Engineers, Baltimore-Washington Section,* 1985.

71.   D. J. Brunelle, T. L. Evans, T. G. Shannon, E. P. Boden, K. R. Stewart, L. P. Fontana, and D. K. Bonauto, *Polym. Preprints*, **30**(2), 569 (1989).

72.   D. J. Brunelle, E. P. Boden and T. G. Shannon, *J. Am. Chem. Soc.*, **112**, 2399-2402 (1990).

73.   D. J. Brunelle and T. G. Shannon, U.S. Patent 4,638,077 (1987) to General Electric.

74.   D. J. Brunelle, T. L. Evans, and T. G. Shannon, U.S. Patent 4,644,053 (1987) to General Electric.

75.   K. R. Stewart, *Polym. Preprints*, **30**(2), 575 (1989).

76.   T. L. Evans, U.S. Patent 4,605,731 (1987) to General Electric.

77.   K. R. Stewart, U. S. Patent 4,778,875 to General Electric.

78.   K. R. Stewart, U. S. Patent 4,853,459 to General Electric.

79.   T. L. Evans, C. B. Berman, J. C. Carpenter, D. Y. Choi, and D. A. Williams, *Polym. Preprints*, **30**(2), 573 (1989).

80.   J. Devaux, P. Godard, and J. P. Mercier, *Polym. Eng. Sci.*, **22**(4), 217-21 (1982).

81.   P. Godard, J. M. Dekoninck, V. Devlesaver, and J. Devaux, *J. Polym. Sci.,* Part A, **24**(12), 3301-3313 and 3315-24 (1986).

82.   F. Pilati, E. Marianucci, and C. Berti, *J. Appl. Polym. Sci.*, **30**(3), 1267-75 (1985).

83.   D. B. Jacobs and J. Zimmerman, "Preparation of 6,6 Nylon and Related Polyamides," in C. E. Schildknecht and I. Skeist, Eds., *Polymerization Processes*, Wiley-Interscience, New York, 1977, pp. 424-467.

84.   P. E. Beck and E. E. Magat, *Macromol. Synth., Coll. Vol . 1*, 317 (1977) .

85.   C. E. Hollis, *Chem. Ind.*, 1030 (1969).

86.   P. Schlack, U.S. Patent 2,241,321 (1941) to I. G. Farbenindustrie AG.

87.   W. H. Carothers, *J. Am. Chem. Soc*, **52**, 5289 (1938).

88.   W. O. Ney, W. R. Nummy, and C. E. Barnes, U.S. Patent 2,638,463 (1953) to Arnold, Hoffman and Co.

89.   W. O. Ney and M. Crowther, U.S. Patent 2,739,959 (1956) to Arnold, Hoffman and Co.

90.   J. E. McGrath, "Ring Opening Polymerization," in J. E. McGrath, Ed., *ACS Symposium Series, 286*, American Chemical Society, Washington, DC, 1985, p. 1.

91.   J. A. Semlyen and P. V. Wright, *Polymer*, **10**, 543 (1969).

92.   I. S. Akhrem, N. M. Christovalova, and E. I. Mysov, *Izv. Akad. SSSR Ser. Khim.*, **7**, 1598 (1978).

93.   S. Nitzsche and M. Wick, U.S. Patent 3,082,527 (1963) to Wacker Chemie.

94.   P. Bajaj, G. N. Babu, D. N. Khanna, and S. K. Varshney, *J. Appl. Polym. Sci.*, **23** 3505 (1979).

95.   K. G. Mayhan, A. W. Hahn, S. W. Dartch, S. H. Wu, B. W. Pease, M. E. Biolsi, and G. L. Bertrand, *Int. J. Polym. Mat.*, **5**, 231 (1977).

96. *Silicone Rubber Adhesive Sealants for Industrial Applications, S-2,* General Electric Company, Waterford, NY.

97. R. P. Eckberg, U.S. Patent 4,256,870 (1981) to General Electric.

98. A. Wurtz, *Ann. Chim. Phys.,* **69,** 330, 334 (1863).

99. J. Furakawa, T. Tsuruta, R. Sakata, and T. Saegusa, *Makromol. Chem.,* **32,** 90 (1959).

100. E. J. Vandenberg, *J. Polym. Sci.,* **47,** 486 (1960).

101. A. Sato, T. Hirano, and T. Tsuruta, *Makromol. Chem.,* **176,** 1187 (1975).

102. T. Tsuruta, S. Inoue, N. Yohida, and Y. Yokoda, *Makromol. Chem.,* **81,** 191 (1965).

103. C. Coulon, N. Spassky, and P. Sigwalt, *Polymer,* **17,** 821 (1976).

104. W. Kuran, A. Rokicki, and J. Pienkowski, *J. Polym. Sci., Polym. Chem.,* **17,** 1238 (1979).

105. T. Aida, R. Mizuta, Y. Yoshida, and S. Inoue, *Makromol. Chem.,* **182,** 1073 (1981).

106. T. Aida and S. Inoue, *Macromolecules,* **14,** 1162 (1981).

107. T. Aida and S. Inoue, *Makromol. Chem., Rapid Commun.,* **1,** 677 (1980).

108. T. Aida and S. Inoue, *Macromolecules,* **15,** 682 (1982).

109. A. J. Landau, *Am. Chem. Soc., Div. Org. Coat. Plast. Chem. Prep.,* **24,** 299 (Sept. 1964).

110. D. M. Stoakley and J. St. Clair, *J. Appl. Polym. Chem.,* **31,** 225-236 (1986).

111. J. D. B. Smith and R. N. Kaufmann, *IEEE Trans. Elect. Insul.,* **EI-19,** No. 1 (1984).

112. P. V. Reddy, R. Thiagarajan, M. C. Ratra, and N. M. Nanje Gowda, *J. Appl. Polym. Sci.,* **41**(1-2), 319-28 (1990).

113. J. D. B. Smith, *Proceedings of the 18th Electrical and Electronic Insulation Conference,* 1987, pp. 262-8.

114. A. V. Kurnoskin and V. A. Lapitskii, *Zh. Prikl. Khim. (Leningrad),* **62**(10), 2397-9 (1989).

115. A. V. Kurnoskin and M. Z. Kanovich, *Plast. Massy,* (5), 54-5 (1991).

116. D. M. Wiles in T. Tsusuto and K. F. O'Driscoll, Eds., *Structure and Mechanism in Vinyl Polymerization,* Marcel Dekker, New York, 1960, p. 223.

117. O. W. Webster, W. R. Hertler, D. Y. Sogah, W. B. Farnham, and T. V. RajanBabu, *J. Am. Chem. Soc.,* **105,** 5706 (1983).

118. O. W. Webster and B. C. Anderson, "Group Transfer Polymerization," in W. J. Mijs, Ed., *New Methods of Polymer Synthesis,* Plenum, New York, 1991.

119. D. Y. Sogah, W. R. Hertler, O. W. Webster, and G. M. Cohen, *Macromolecules,* **21,** 1473-88 (1988).

120. I. B. Dicker, W. R. Hertler, G. M. Cohen, W. B. Farnham, E. D. Laganis, and D. Y. Sogah, *Polym. Preps. Am. Chem. Soc. Div. Polym. Chem.,* **28**(1) 106-107 (1987).

121. I. B. Dicker, W. B. Farnham, W. R. Hertler, E. D. Laganis, D. Y. Sogah, T. W. Del Pesco, and P. H. Fitzgerald, U.S. Patent 4,588,795 (1986) to Du Pont.

122. W. R. Hertler, D. Y. Sogah, O. W. Webster, and B. M. Trost, *Macromolecules,* **17,** 1415-1417 (1984).

123. S. Inoue, *Polym. Preps. Am. Chem. Soc. Div. Polym. Chem.,* **29**(2) 42-43 (1988).

124. M. Kuroki, S. Nashimoto, T. Aida, and S. Inoue, *Macromolecules,* **21,** 3114-3115 (1988).

125. M. T. Reetz, *Pure and Appl. Chem.,* **57,** 1781-1788 (1985).

12 | HOMOGENEOUS CATALYSIS IN HALOCARBON CHEMISTRY

Recent concerns about the impact of chlorinated organic compounds on the ozone layer have called attention to the enormous role that these compounds play in our lives. Chlorinated polymers such as PVC and neoprene are found widely in our homes and automobiles. Chlorinated ethanes and ethylenes are amongst the most widely used solvents, particularly in dry cleaning and degreasing. The chlorofluorocarbons (CFCs) have been the fluids of choice for refrigerants, foaming agents, and aerosol propellants. Replacing the CFCs with ozone-innocuous fluids has proven to be a difficult challenge because few other materials have such a desirable combination of nonflammability, low toxicity, and vapor pressure properties.

Homogeneous catalysis complements heterogeneous catalysis in the production of organic compounds containing chlorine and fluorine. In some multistep processes, the two catalytic technologies are used sequentially. This sequential use of homogeneous and heterogeneous catalysts is illustrated in the ICI process for making HFC-134a, the new ozone-innocuous refrigerant. The C_1 CFCs are largely produced through the use of soluble catalysts while the C_2 analogs make greater use of heterogeneous catalysts. For both the CFCs and their replacements, however, chlorination of a hydrocarbon is the initial step.

12.1 CHLORINATION OF OLEFINS AND DIENES

The chlorinations of ethylene and butadiene are major industrial processes. They provide intermediates for the production of vinyl chloride monomer and for chloroprene, the monomer for neoprene rubber. Chlorination of ethylene has also come to replace acetylene chlorination (Section 8.3) in the preparation of highly chlorinated C_2 compounds used extensively as solvents for dry cleaning and for industrial degreasing. Tri- and tetrachloroethylene are also important as intermediates

297

in the production of ozone-innocuous replacements for the soon to be banned chlorofluorocarbon refrigerants, solvents, and foam-blowing agents.

Ethylene Chlorination

The addition of Cl_2 to the C=C bond of ethylene is the initial step in manufacture of the many important industrial chemicals, as illustrated in Figure 12.1 [1,2]. Two major processes are used to accomplish this addition. The most direct is the liquid-phase chlorination in which iron(III) chloride is used as the catalyst. In one process mode, chlorine and ethylene are reacted at 40-70°C in a dichloroethane solution of $FeCl_3$. The yields and conversions approach 100% under these conditions, but heat removal from this highly exothermic reaction is a challenge to the engineer. Operation at approximately 100°C with constant distillation of 1,2-dichloroethane as it is formed gives slightly lower yields, but simplifies the operation of the process.

The other major process for making 1,2-dichloroethane is oxychlorination [1-4]. In this process, HCl, a byproduct from vinyl chloride production (Figure 12.1), is used as the chlorine source:

$$C_2H_4 + 2\ HCl + \tfrac{1}{2}\ O_2 \longrightarrow ClCH_2CH_2Cl + H_2O$$

DCE

Oxychlorination is carried out as a gas phase reaction in which ethylene, HCl, and air are mixed in the presence of a heterogeneous catalyst. Typically the catalyst is $CuCl_2$ supported on alumina and is used in a fluidized bed mode. Yields of DCE are 91-95% at nearly quantitative conversion. These yields are inferior to those from direct chlorination of ethylene, but the combination of oxychlorination with chlorination in

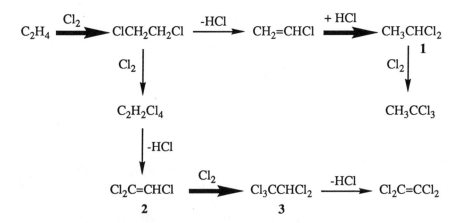

Figure 12.1 Ethylene chlorination products. The bold arrows indicate steps often carried out in the liquid phase with soluble, Lewis acidic catalysts.

an integrated vinyl chloride production facility .eads to efficient use of chlorine. A liquid-phase oxychlorination of ethylene in an aqueous HCl solution of $CuCl_2$ has been demonstrated [5], but appears not to be used industrially. The solutions are extremely corrosive and difficult to contain in normal materials of construction.

The largest economic incentive for DCE production is to make vinyl chloride for use in making "vinyl" plastics, but, as shown in Figure 12.1, vinyl chloride is also a valuable intermediate for other purposes. If the HCl eliminated from DCE in vinyl chloride production is allowed to react with vinyl chloride in a Lewis acidic solution, it adds back to the C=C bond to produce 1,1-dichloroethane **1** [6]. In effect, 1,2-dichloroethane is isomerized to the 1,1-isomer. This reaction is carried out on a large scale to manufacture **1**, which is chlorinated further to produce 1,1,1-trichloroethane, "methyl chloroform," an important solvent and intermediate. The addition of HCl to vinyl chloride is typically conducted in a solution of $FeCl_3$ in **1**. Using the desired product as the solvent simplifies the process. Reaction at approximately 60°C gives **1** in 95-98% yield [7]. A platinum complex, $PtCl_2(CO)(olefin)$, also catalyzes the addition of HCl to olefins [8], but seems to offer no practical advantages over the Lewis acid catalysts.

When the chlorination of ethylene is carried out under vigorous conditions with excess chlorine, the initially formed DCE reacts further to produce more highly chlorinated ethanes. Tetrachloroethanes produced in this way (or by chlorination of acetylene) are precursors to the major solvents tri- and tetrachloroethylenes. Thermal cracking of HCl from 1,1,2,2-tetrachloroethane gives "triclene" **2**. The chlorination of **2** to give pentachloroethane **3** is carried out somewhat like the chlorination of ethylene. The olefin **2** and Cl_2 react in the presence of a trace of $FeCl_3$ dissolved in **3**, the desired product. Reaction at 70-110°C gives a nearly quantitative yield [1]. Thermal cracking of HCl from **3** gives tetrachloroethylene ("perclene").

In the homogeneous catalytic chlorinations, it seems likely that $FeCl_3$ serves primarily as a Lewis acid to polarize and thereby activate the chlorine molecule:

$$Cl_2 \cdot FeCl_3 \quad \longleftrightarrow \quad Cl \overset{\delta+}{\cdot} - -Cl \cdot - -Fe \overset{\delta-}{} Cl_3$$

In contrast, most of the chlorinations of saturated hydrocarbons and hydrochlorocarbons are carried out under thermal or photochemical conditions that favor free radical-reactions. The inferior yields in the oxychlorination of ethylene are likely due to radical-mediated side reactions.

Butadiene Chlorination

The chlorination of butadiene is conducted on a large scale to produce the dichlorobutenes which are intermediates in making chloroprene [9,10], a monomer for oil-resistant rubbers. As shown in Figure 12.2, the dichlorobutenes are also intermediates in a now-obsolete process for 1,6-hexanediamine, a key nylon intermediate [11].

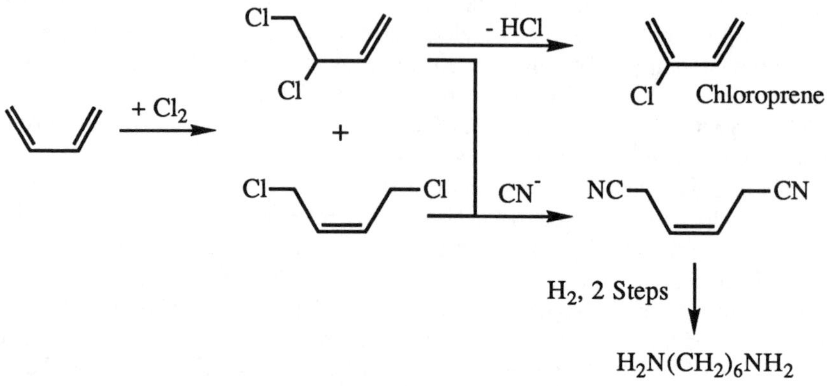

Figure 12.2 Butadiene-based routes to chloroprene and 1,6-hexanediamine.

The uncatalyzed vapor-phase chlorination of butadiene produces a mixture of about 33% 3,4-dichloro-1-butene and 66% *cis*- and *trans*-1,4 -dichloro-2-butene [11]. Liquid-phase chlorination with pyridine as a catalyst gives more of the 3,4-dichlorobutene [12]. For production of chloroprene [13], the 3,4-dichloro isomer is preferred, so the unwanted 1,4-dichloro-2-butenes are catalytically isomerized to the desired 3,4-dichloro compound. Both the isomerization and the reaction of dichlorobutenes with sodium cyanide are catalyzed by soluble copper(I) salts. The butadiene-based chloroprene synthesis replaces the expensive and hazardous acetylene-based process described in Section 8.2.

Dichlorobutene Isomerization

The vapor phase chlorination of butadiene produces a mixture of dichlorobutenes as mentioned above. The 3,4-dichloro-1-butene desired for chloroprene synthesis boils at 123°C and is easily separated from the 1,4-dichlorobutenes (b.p. 155°C) by distillation [14]. To avoid waste of the 1,4-isomers, they are isomerized to the 3,4-dichloro derivative. This isomerization is catalyzed by copper(I) complexes:

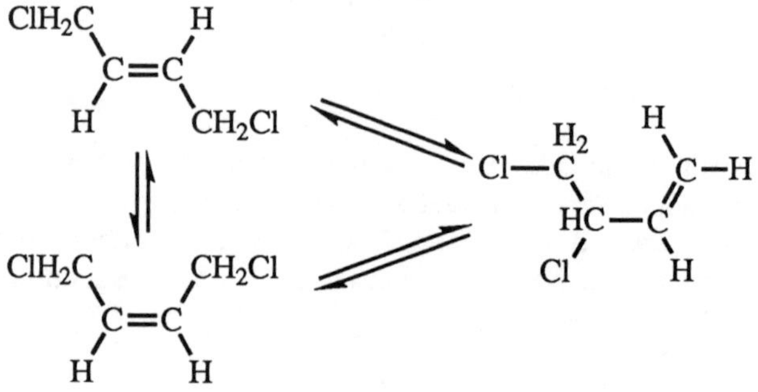

In one mode of operation [15], the crude dichlorobutenes are fed to a distillation column that contains the catalyst in the stillpot. The 3,4-dichloro-1-butene fed to the column distills continuously. The higher boiling isomers contact the catalyst and are converted to an equilibrium mixture from which the 3,4-dichloro compound distills. Eventually most of the chlorination product is converted to the desired 3,4 isomer. The same chemistry can be used in a different engineering context to produce the optimum mixture of dichlorobutenes for reaction with sodium cyanide to make 1,4-dicyanobutene.

The isomerization is catalyzed by many copper(I) compounds, but the patent literature cites advantages for nitrile complexes such as PhCN•CuCl [16] and chloride complexes like R_4N^+ $CuCl_2^-$ [17]. The mechanism of the copper-catalyzed isomerization has not been reported, but a kinetic study of dichlorobutene isomerization with a soluble iron catalyst led to a proposal based on π-allyl intermediates [18]. An analogous mechanism based on π-allylcopper intermediates is shown in Figure 12.3.

Presumably, the first step is coordination of a dichlorobutene molecule to copper(I) chloride as a simple olefin complex. Oxidative addition of a C-Cl bond to the Cu(I) ion generates a π-allyl complex in which the metal is formally trivalent. This sequence of events can occur with any of the dichlorobutene isomers. The thermodynamically disfavored *cis*-1,4-dichloro complex **4** forms the *anti*-π-allyl complex **5** in which the chloromethyl group is trans to the hydrogen on C-2. If the chloride migration from carbon to copper is reversed, but the chloride returns to C-3 rather than C-1 from which it came, the 3,4-dichloro-1-butene complex **6** forms. The *anti*-π-allyl complex **5** can isomerize to its syn isomer **7**. In this instance, return of a chloride ion from copper to C-1 generates the *trans*-1,4-dichloro-2-butene complex **8**.

The isomerization of dichlorobutenes via allylcopper intermediates seems plausible, but the evidence for such a mechanism is scant. Few organocopper(III) complexes have been characterized, even though they seem to be involved in many

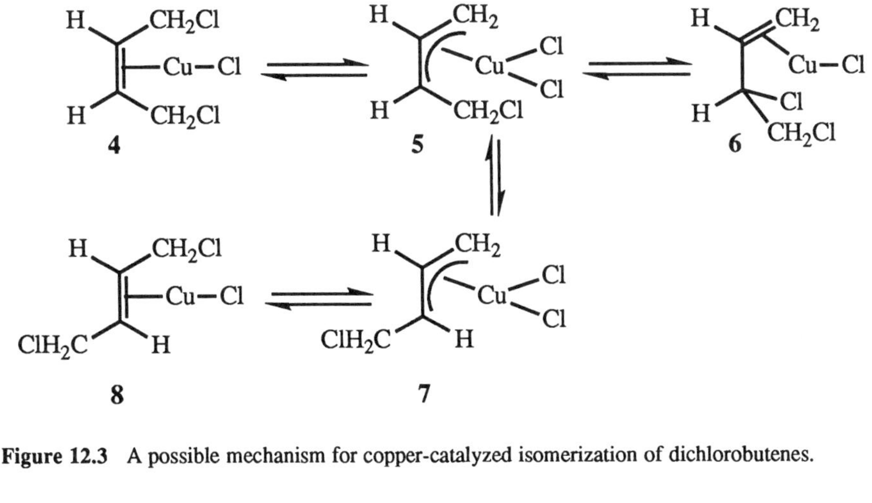

Figure 12.3 A possible mechanism for copper-catalyzed isomerization of dichlorobutenes.

reactions catalyzed by copper(I) salts. Allylcopper intermediates have been suggested in the CuCl-catalyzed hydrolysis of allyl chloride [19] and the reaction of allylic acetates with methylcuprate salts [20]. π-Allylcopper intermediates probably also occur in the copper-catalyzed cyanation of dichlorobutenes discussed next.

Dicyanobutene Synthesis

The cyanation of either 1,4-dichloro-2-butene or 3,4-dichloro-1-butene gives 1,4-dicyano-2-butene as a major product [11,21]:

$$
\begin{array}{c}
ClCH_2CH = CHCH_2Cl \\
\text{or} \\
H_2C = CHCHClCH_2Cl
\end{array}
\xrightarrow{\ CN^-\ }
NCCH_2CH = CHCH_2CN
$$

The reaction is catalyzed by copper salts. Detailed descriptions of the commercial process have not been published, but a large-scale laboratory experiment [21] illustrates operable conditions. A 40% solution of sodium cyanide in water was added to an acidic aqueous solution of 3,4-dichlorobutene-1 and $CuCl_2$ at 85-95°C. A rapid, exothermic reaction consumed all the cyanide. 1,4-Dicyano-2-butene and 4-chloropentenenitrile were formed in 95% yield. It seems likely that the copper(II) chloride was reduced to Cu(I) under these conditions. Most patent examples use copper(I) chloride or cyanide as the catalyst. In the presence of high concentrations of CN^-, the metal probably forms $[Cu(CN)_2]^-$ and $[Cu(CN)_3]^{2-}$ complexes.

Mechanistic information about this reaction is not available, but analogies suggest allylcopper intermediates like those proposed in Figure 12.3. Exchange of cyanide for chloride within the coordination sphere of copper in a complex such as **5-CN** provides a simple mechanism for cyanation:

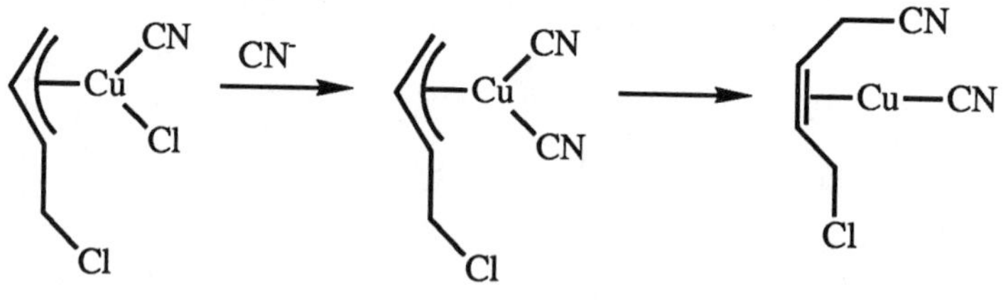

This exchange converts **5** to its cyanide analog (**9**). Migration of cyanide from the metal to C-1 produces **10**, the CuCN complex of 5-chloropentenenitrile. Since **5** is accessible from any of the dichlorobutenes, it seems quite reasonable that all can form **10** and, by further cyanation, the desired 1,4-dicyanobutene. This product was hydrogenated in two steps in a now-obsolete process for the manufacture of 1,6-hexanediamine, a major intermediate for nylon-6,6.

12.2   CFC PROCESSES

The most common procedure to introduce fluorine into a simple organic molecule is to replace Cl with F by simple exchange with HF [22,23]. For the widely produced chlorofluoromethanes, the major industrial processes have been based on antimony catalysts soluble in liquid hydrogen fluoride. On the other hand, C_2 products, such as the widely used cleaning agent CCl_2FCClF_2 (CFC-113), are usually manufactured by vapor-phase processes catalyzed by Cr-based heterogeneous catalysts. Both liquid- and vapor-phase processes are used to make the CFC replacements described in Section 12.3.

Chlorofluoromethanes

The reaction of carbon tetrachloride with liquid hydrogen fluoride is widely used to make the mono- and difluorochloromethanes [23-25].

$$CCl_4 \xrightarrow{\text{HF}} CCl_3F \xrightarrow{\text{HF}} CCl_2F_2 + HCl$$
$$\text{CFC-11} \qquad\qquad \text{CFC-12}$$

In common practice, CCl_4 and anhydrous HF are fed continuously to a reactor containing an antimony halide catalyst, typically $SbCl_5$. Reaction occurs at about 100°C under autogeneous pressure. Hydrogen chloride, which boils much lower than HF and the CFCs, is continuously removed to drive the reaction to the desired product ratio. Control is assisted by the fact that each successive C-Cl bond replacement is slower. If an Sb(III) catalyst is used, chlorine is added to the reactants to convert it to a catalytically active Sb(V) species. Yields in the process exceed 90% based on CCl_4.

A similar procedure is used for the manufacture of $CHClF_2$, a major hydrochlorofluorocarbon (HCFC-22). Chloroform is treated with liquid HF in the presence of an antimony catalyst under conditions similar to those used for making the CFCs [25,26].

$$CHCl_3 + 2\,HF \longrightarrow CHClF_2 + 2\,HCl$$
$$\text{HCFC-22}$$

In this process, as in the CCl_4 reaction, the product (b.p. -41°C) is distilled away from the antimony catalyst, which can be reused repeatedly if the operation is carried out batchwise.

While the technology for the safe handling of liquid HF on a large scale is complex, the processes sketched above have made possible the manufacture of the chlorofluoromethanes on a scale of hundreds of thousands of tons annually. All three compounds are used as refrigerants although each has found a particular niche use in automotive air conditioners, home refrigerators, or commercial coolers. CFC-11 has been widely used as a "blowing agent" to make polyurethane foams and CFC-12 for making polystyrene foams, although both compounds are being replaced by hydrocarbons or hydrofluorocarbons (HFCs). A major use of $CHClF_2$ is the manufacture of tetrafluoroethylene by high-temperature pyrolysis:

$$2\ CHClF_2 \longrightarrow F_2C{=}CF_2 + 2\ HCl$$

Antimony Catalysts

The discovery that antimony compounds catalyze Cl/F exchanges dates back 100 years. In 1892, Swartz [27] observed that SbF_3 underwent a stoichiometric Cl/F exchange with reactive CCl_3 groups. He later found that a trace of $SbCl_5$ catalyzed the exchange between SbF_3 and less reactive chlorocarbons such as CCl_4. This observation was put to use in 1928 when CCl_2F_2 was made for testing as a nonflammable refrigerant. It was found somewhat by chance that the stoichiometric exchange:

$$3\ CCl_4 + 2\ SbCl_2F_3 \longrightarrow 3\ CCl_2F_2 + 2\ SbCl_5$$

provides an efficient synthesis of CFC-12 [28]. Subsequently it was found that the $SbCl_5$ could be reconverted to an antimony fluoride in HF. This finding provided the basis for a catalytic HF exchange reaction [24,29] that has evolved into the manufacturing process described above.

One of the key understandings in this process evolution is that Sb(V) is the catalytically active oxidation state of antimony. As noted above, if an Sb(III) compound is used, it must be oxidized with Cl_2 to produce an active catalyst. This procedure is also used to regenerate an Sb(V) catalyst that has been deactivated by reduction by impurities.

Another important factor in the activity of the catalyst is the fluorine:antimony ratio [30]. Antimony pentafluoride is said to convert CCl_4 to CF_4 rapidly, even at low temperatures. Under commercial conditions with a high ratio of HF to $SbCl_5$, the major species is reported to be $SbCl_4F$ [26], a less active, but still efficient catalyst. Such a formulation is undoubtedly oversimplified because Sb(V) has a great affinity for fluoride ion. In the case of SbF_5, reaction with fluoride gives the $[SbF_6]^-$ ion. The reaction of $SbCl_5$ with a large excess of HF at room temperature gives a white solid characterized as $SbCl_5 \cdot 4HF$ [30]. This composition is probably better formulated as $[H_4F_3]^+[SbCl_5F]^-$, based on the strong hydrogen-bonding properties of liquid HF.

The addition of a Lewis acid such as $SbCl_5$ to liquid HF greatly increases the acidity of the solution, possibly to the "super acid" range [31]. This acidity appears to play a major role in the replacement of Cl by F in an aliphatic compound. It seems likely that protonation of the Cl probably leads into a four-center transition state in the exchange

$$
\begin{array}{c}
\mathrm{Cl} \cdot \text{-}\mathrm{H} \\
| \quad + \quad | \\
\mathrm{C} \qquad \mathrm{F}
\end{array}
\longrightarrow
\begin{array}{c}
\mathrm{Cl} \cdot \text{-}\mathrm{H} \\
\vdots \qquad \vdots \\
\mathrm{C} \text{-} \text{-} \text{-} \mathrm{F}
\end{array}
\longrightarrow
\begin{array}{c}
\mathrm{Cl} \text{---} \mathrm{H} \\
+ \\
\mathrm{C} \text{---} \mathrm{F}
\end{array}
$$

The stoichiometric reaction of SbF_5 or $SbCl_2F_3$ with a chloroalkane in an inert medium probably involves an analogous antimony-based exchange:

$$
\begin{array}{c}
\mathrm{Cl} \qquad \mathrm{Sb} \\
| \quad + \quad | \\
\mathrm{C} \qquad \mathrm{F}
\end{array}
\longrightarrow
\begin{array}{c}
\mathrm{Cl} \cdot \text{-}\mathrm{Sb} \\
\vdots \qquad \vdots \\
\mathrm{C} \text{-} \text{-} \text{-} \mathrm{F}
\end{array}
\longrightarrow
\begin{array}{c}
\mathrm{Cl} \text{---} \mathrm{Sb} \\
+ \\
\mathrm{C} \text{---} \mathrm{F}
\end{array}
$$

Alternatives to antimony(V) halides for catalysis of HF reactions include compounds such as TaF_5 [32], which generates even more acidic species in liquid HF.

12.3 CFC REPLACEMENTS

Two of the most challenging problems facing industrial chemists in recent years have been the identification of safe, ozone-innocuous replacements for the CFC's and development of safe, efficient, environmentally acceptable processes to make the replacements [33,34]. The identification of the replacements has been especially difficult because the CFCs were uniquely well suited to their applications by virtue of their nonflammability and low toxicity in addition to criteria based on vapor pressure, heat capacity, and other chemical and physical properties. The replacements should have all these desirable properties in addition to being unreactive with ozone and having little potential for global warming ("greenhouse effect"). The chemical industry has not yet discovered products that meet these criteria perfectly. As indicated in Table 12.1, some property deficiencies remain for all the potential replacements.

Table 12.1   Chlorofluorocarbons and Their Replacements

Application	CFC Product	Proposed Replacement	Comments on Replacement[a]
Aerosol propellant	CF_2Cl_2 + $CFCl_3$	Propane, butane	Highly flammable.
Refrigerant	CF_2Cl_2 CF_2HCl	CF_3CH_2F	Not suitable for existing equipment.
Foaming agent	$CFCl_3$	CH_3CFCl_2 CF_3CHCl_2	Toxicity? Tumorgenicity.
Cleaning	$CFCl_2CF_2Cl$	-	No single replacement identified.

[a]All the compounds proposed have significant global warming potential.

Perhaps the closest approach to the ideal replacement is the unsymmetrical tetrafluoroethane (CF_3CH_2F) proposed as a general purpose refrigerant. Its physical properties closely match that of CF_2Cl_2 and it is totally innocuous toward stratospheric ozone. Unfortunately, it has significant global warming potential and will be about five times as expensive as the CFC that it replaces. The global warming potential, which arises from the intense infrared absorption associated with C-F bonds, is mitigated by the limited lifetime of the compound in the atmosphere. Ultraviolet light leads to disruption of the molecule.

The strategy of including a C-H bond in the molecular structure of the CFC replacement is quite general as a way to ensure that the compound largely decomposes before rising to the ozone-rich troposphere. Although the tactic achieves the desired effect, it severely complicates the task of the process chemist seeking an efficient catalyst to manufacture the CFC replacement. The presence of C-H functions in the reaction mixtures shortens the lifetime of both the antimony catalysts used for liquid-phase HF exchange reactions and the chromium catalysts used for vapor-phase processes. In both situations, catalyst deactivation appears to occur by reduction of the catalyst. Catalyst lifetime has been extended in practice by a number of chemical and engineering expedients, but it remains a source of concern and expense.

Because of the rapid pace of development for the CFC replacements, it seems likely that the processes now being used are not optimal and that new processes will be introduced throughout the 1990s. Three processes illustrating the use of homogeneous catalysis are described below. In each instance, heterogeneous catalysts may be used to effect the same reaction.

unsym-Tetrafluoroethane

As noted above, 1,1,1,2-tetrafluoroethane (CF_3CH_2F) is the prime candidate for a general purpose refrigerant fluid for automotive air conditioners and home refrigerators. Many different processes for its manufacture have been studied [33,34]. One of simplest, conceptually, is the two-step process commercialized by ICI [35]:

$$Cl_2C = CHCl \xrightarrow{\text{HF(l)}} CF_3CH_2Cl \overset{\text{HF(g)}}{\underset{}{\rightleftharpoons}} CF_3CH_2F + HCl$$

This route may have been particularly attractive to ICI because they have practiced the first step for many years. In the ICI process to make the anaesthetic Halothane ($CF_3CHBrCl$), CF_3CH_2Cl, produced as above, is brominated photochemically to replace one hydrogen with bromine [36].

In the first step of the ICI process for CF_3CH_2F, trichloroethylene (Section 12.1) is treated with liquid HF in the presence of an antimony catalyst to give CF_3CH_2Cl. This intermediate, in turn, is treated with gaseous HF over a solid chromium oxide catalyst. The liquid-phase fluorination involves reaction of trichloroethylene with anhydrous HF at about 100°C and autogeneous pressure in the presence of $SbCl_{5-x}F_x$. The pressure is maintained at about 15 atmospheres by continuous venting of HCl, just as in the fluorination of CCl_4. Distillation of excess HF and products away from the catalyst gives CF_3CH_2Cl in 60-70% yield [37]. This product may also be obtained by fluorination of Cl_3CCH_2Cl under similar conditions [38].

In contrast to the CCl_4 fluorination, a significant side reaction is the formation of an olefin, $F_2C=CCl_2$, which must be removed by distillation. Also in contrast to the CCl_4 reaction, the Sb(V) catalyst deactivates fairly rapidly by reduction to Sb(III) and by the formation of tars. These catalyst deactivation processes may be reduced by the addition of a metal halide such as $HgCl_2$ [39]. Another effective approach to catalyst stabilization is to add Cl_2 to the reactants to keep the antimony in the +5 oxidation state. This play has the disadvantage of reducing the yield by chlorination of the desired CF_3CH_2Cl to form CF_3CHCl_2 and CF_3CCl_3.

Many of the deficiencies of the $SbCl_5$ catalyst are overcome by use of TaF_5, which is more resistant to reduction. For example, the reaction of trichloroethylene with liquid HF at 102°C gives about a 95% yield of CF_3CH_2Cl along with some C_4-oligomers [40]. The tantalum may also be supplied to the reactor as $TaCl_5$ or $TaBr_5$ [41]. Presumably the catalyst species in the presence of large amounts of HF and HCl in the reaction solution is a mixed halide $TaCl_{5-x}F_x$.

The fluorination of a chloroolefin by HF appears to involve an initial HF addition to the C=C bond, followed by a series of Cl/F exchanges [34]:

$$Cl_2C = CHCl \xrightarrow{\text{HF}} FCl_2C\text{-}CH_2Cl \xrightarrow[\text{-HCl}]{\text{HF}} \xrightarrow[\text{-HCl}]{\text{HF}} F_3C\text{-}CH_2Cl$$

The selectivity for HF addition is such that a maximum number of halogens are attached to one carbon. This trihalomethyl group is more susceptible to Cl/F exchange than is the CH_2Cl group.

2,2-Dichloro-1,1,1-trifluoroethane

The title compound (CF_3CHCl_2) was considered a prime candidate to replace $CFCl_3$ as a "blowing agent" for polyurethane foams [42] until it was discovered to produce benign tumors in rats on prolonged exposure. It remains a useful refrigerant for commercial coolers and is a valuable intermediate for the manufacture of CF_3CH_2F via the sequence

$$CF_3CHCl_2 \xrightarrow[\text{-HCl}]{\text{HF(g)}} CF_3CHClF \xrightarrow[\text{-HCl}]{\text{H}_2} CF_3CH_2F$$

The title compound can be made by fluorination of tetrachloroethylene ("perclene") with either liquid or gaseous HF [34]. In the liquid-phase process, perclene is treated with HF in the presence of an antimony [43] or tantalum halide catalyst. With TaF_5 as the catalyst, perclene is heated at 150°C for 2 hours with about a twenty-fold excess of HF. The yield of the desired CF_3CHCl_2 is 97% along with small amounts of under-fluorinated products, which may be recycled to the reactor [40].

Attempts to discern the detailed reaction mechanism in the TaX_5 catalyst system have been unrewarding [32]. In the CF_3CHCl_2 synthesis just as in the CF_3CH_2Cl preparation described earlier, it appears that the reaction begins with HF addition to the olefin, followed by a series of Cl/F exchanges:

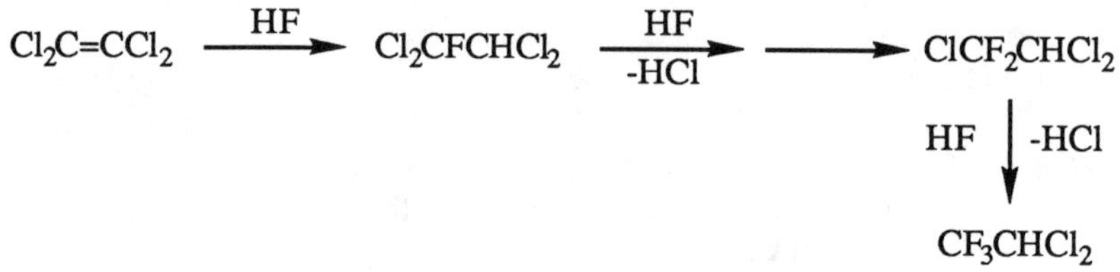

$$Cl_2C=CCl_2 \xrightarrow{\text{HF}} Cl_2CFCHCl_2 \xrightarrow[\text{-HCl}]{\text{HF}} \longrightarrow ClCF_2CHCl_2$$

$$HF \downarrow \text{-HCl}$$

$$CF_3CHCl_2$$

1,1-Difluoroethane

The title compound has been proposed as a replacement for CF_2Cl_2 as a refrigerant and blowing agent. It suffers the disadvantage of being flammable, but should be cheaper and more readily accessible than CF_3CH_2F, the currently preferred replacement. It may find extensive use as a component of a nonflammable blend of materials that can be used in existing refrigeration equipment.

1,1-Difluoroethane is accessible from a variety of starting materials by HF addition or Cl/F exchange reactions [34]:

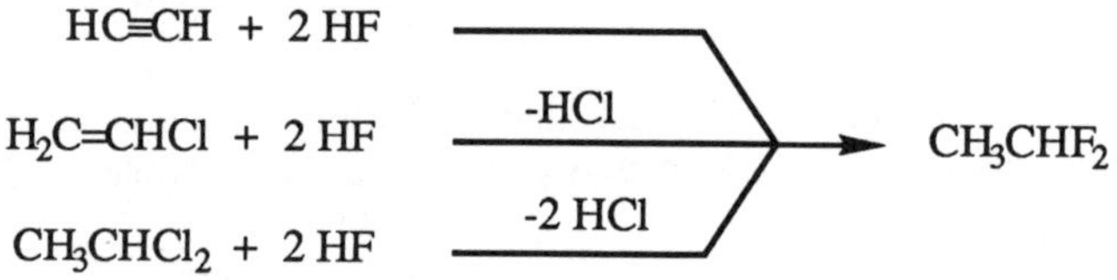

$$HC\equiv CH + 2\ HF$$
$$H_2C=CHCl + 2\ HF \xrightarrow{\text{-HCl}}$$
$$CH_3CHCl_2 + 2\ HF \xrightarrow{\text{-2 HCl}} \longrightarrow CH_3CHF_2$$

The addition of two moles of HF to acetylene has the advantage that it does not generate HCl as a byproduct, but is limited by the industrial availability of acetylene. The addition is catalyzed by strong acids such as BF_3 [44] or FSO_3H [45]. With the former catalyst, a chilled autoclave is charged with liquid HF containing 30% BF_3 and is maintained at 0°C while acetylene is added until absorption ceases. The yield is essentially quantitative if the temperature is kept at 0°C or lower to prevent polymerization of the acetylene or the presumed intermediate, vinyl fluoride.

GENERAL REFERENCES

J. C. Tatlow, R. E. Banks, and B. E. Smart, Eds., *Organofluorine Chemistry: Principles and Commercial Applications*, Plenum Press, New York, 1993.

R. E. Banks, Ed., *Preparation, Properties and Industrial Applications of Organofluorine Compounds*, Ellis Horwood Ltd., London, 1982.

R. E. Banks, Ed., *Organofluorine Chemicals and Their Industrial Applications,* Ellis Horwood Ltd., London, 1979.

SPECIFIC REFERENCES

1. "Chlorinated Hydrocarbons," in *Ullmann's Encyclopedia of Industrial Chemistry*, Vol. A6, Verlag Chemie, 1985, pp. 260-306.
2. J. S. Naworski and E. S. Velez, "Oxychlorination of Ethylene," in B. E. Leach, Ed., *Applied Industrial Catalysis*, Vol. 1, Academic Press, New York, 1983, pp. 239-73.
3. E. Cavaterra, *Hydrocarbon Proc.*, 63-7 (Dec. 1988).
4. J. D. Scott, U.S. Patent 5,011,808 (1991).
5. L. F. Albright, *Chem. Eng.*, **74**(8), 219-26 (1967).
6. J. van Dalfsen and J. P. Wibaut, *Rec. Trav. Chim.*, **51**, 636-40 (1932).
7. British Patent 1,106,533 (1968); R. Stephan and H. Richtzenhain, U.S. Patent 3,707,574 (1972).
8. H. Alper *et al.*, *Organometallics*, **10**, 1665-71 (1991).
9. G. T. Martirosyan and A. Ts. Malkhasyan, *Zh. Vses. Khim, O-va. im. D. I. Mendeleeva*, **30**, 263-7 (1985); *Chem. Abstr.* **104**, 6948 (1986).
10. C. E. Hollis, "Chloroprene and Polychloroprene Rubbers," *Chem. Ind.*, 1030-41 (1969).
11. V. D. Luedecke, "Adiponitrile," in J. J. McKetta and W. A. Cunningham, Eds., *Encyclopedia of Chemical Processing and Design*, Vol. 2, Marcel Dekker, 1977, pp. 146-162.
12. A. T. Harris, European Patent Application EP429,967 (1991); *Chem. Abstr.* **115**, 70905 (1991).
13. L. J. Maurin, U.S. Patent 4,418,232 (1983).
14. L. A. Smith, U.S. Patent 4,089,751 (1978).
15. F. J. Bellringer and C. E. Hollis, *Hydrocarbon Proc.*, **47** (11) 127-9 (1968).
16. D. D. Wild, U.S. Patent 3,515,760 (1970).
17. B. T. Nakata and E. D. Wilhoit, Ger. Offenleg. 2,248,668 (1973).
18. G. Henrici-Olive and S. Olive, *J. Organomet. Chem.*, **29**, 307-11 (1971).
19. L. F. Hatsch and R. R. Estes, *J. Am. Chem Soc.*, **67**, 1730-3 (1945).
20. J. Levisalles, M. Rudler-Chauvin, and H. Rudler, *J. Organomet. Chem.*, **136**, 103-10 (1977).
21. I. D. Webb and G. Tabet, U.S. Patent 2,477,672 (1949).
22. G. Siegemund, W. Schwertfeger, A. E. Feiring, B. E. Smart, F. Behr, H. Vogel, and B. C. McKusick, "Organic Fluorine Compounds," in *Ullmann's Encyclopedia of Industrial Chemistry*, Vol. A11, Verlag Chemie, 1988, pp. 349-392.
23. A. K. Barbour, L. J. Belf, and M. W. Buxton, "The Preparation of Organic Fluorine Compounds by Halogen Exchange," in M. J. Stacey, J. C. Tatlow, and A. G. Sharpe, Eds., *Advances in Fluorine Chemistry*, Vol. 3, Butterworths, London, 1963, pp. 181-270.
24. J. M. Hamilton, "The Organic Fluorochemicals Industry" in M. J. Stacey, J. C. Tatlow and A. G. Sharpe, Eds., *Advances in Fluorine Chemistry*, Vol. 3, Butterworths, 1963, pp. 181-270.
25. L. C. Holt and E. L. Mattison, U.S. Patent 2,005,713 (1935).
26. E. Santacesaria, M. Di Serio, G. Basile, and S. Carra, *J. Fluorine Chem.*, **44**, 87-111 (1989).
27. F. Swartz, *Bull. Acad. Roy., Belg.*, **24**, 309 (1892).
28. T. Midgely and A. L. Henne, *Ind. Eng. Chem.*, **22**, 542-5 (1930).
29. H. W. Daudt and M. A. Yonker, U.S. Patents 2,005,705; 2,005,708; 2,005,710 (1935).

30. M. Blanchard and S. Brunet, *J. Mol. Catal.*, **62**, L33-7 (1990).

31. G. A. Olah, *Friedel-Crafts Chemistry*, Wiley, New York, 1973.

32. A. E. Feiring, *J. Fluorine Chem.*, **13**, 7-18 (1979).

33. L. E. Manzer, *Science*, **249**, 31-5 (1990).

34. V. N. M. Rao, "Alternatives to Chlorofluorocarbons," in J. C. Tatlow, R. E. Banks, and B. E. Smart, Eds., *Organofluorine Chemistry: Principles and Commercial Applications*, Plenum Press, New York, 1993.

35. E. Chynowyth, *European Chem. News*, **8** (17 Apr. 1991); *ibid.*, 21 (15 July 1991).

36. W. G. M. Jones in R. E. Banks, Ed., *Preparation, Properties and Industrial Applications of Organofluorine Compounds*, Ellis Horwood Ltd., London, 1982, p. 160.

37. R. L. McGinty, U.S. Patent 3,003,003 (1961).

38. A. F. Benning, U.S. Patent 2,230,925 (1941).

39. K-H. Mitschke and H. Niederprum, British Patent 1,585,938 (1981).

40. W. H. Gumprecht, W. G. Schindel, and V. M. Felix, U.S. Patent 4,967,024 (1990).

41. V. N. M. Rao, European Patent Application 0349,190 (1990).

42. G. Woods, *The ICI Polyurethane Book*, 2nd ed., Wiley, New York, 1990.

43. G. Fernschield, W. Rudolph, and C. Brosch, Ger. Offen. 4,005,944 (1991).

44. R. E. Burk, D. D. Coffman, and G. H. Kalb, U.S. Patent 2,425,991 (1947).

45. J. D. Calfee and F. H. Bratton, U.S. Patent 2,462,359 (1949).

APPENDIX: PATENTS, SCIENTIFIC LITERATURE, AND ON-LINE SEARCHING

In this book, many of the references are to patents, often U.S. patents. Patents can give industrial scientists and engineers who are familiar with the patent system considerable insight into what their competition is doing. Thereby, they constitute a very important form of "competitive intelligence." There are several valuable books on U.S. and non-U.S. patents. In the very brief space of this appendix, we hope to provide some insight into patents and literature searching

Searches of the *Chemical Abstracts* data bases CA and CASREACT through STN® International are powerful because they cover open literature as well as patent references in a massive data base. STN-Express® provides easy access to these data bases for the novice. It also provides useful drawing tools for structural searches for the more advanced user. The current inability to modify searches in progress, to tailor searches to a single company, to use the powerful EXPAND command, or to use NOT in addition to AND and OR when designing searches are all serious limitations which will undoubtedly be corrected in future versions. Nonetheless, it is a powerful tool as searchers develop their skills. A useful guide to *Chemical Abstracts* and hands-on on-line searching is provided in a book by Schulz [1].

Patent information is also available through STN International® from the IFIPAT, IFICDB, and INPADOC data bases. Derwent is also a valuable source of patent information. These sources allow one to locate primary citations as well as equivalent patents in a variety of languages. Additional information available from these sources include first claims of the patents. Because some non-U.S. countries restrict claims more closely to the conditions actually demonstrated, the first claims in their patents are focused on the actual desired conditions. This makes the information more useful for the practicing scientist. Nonetheless, there is no substitute for having an actual copy of the patent to be able to read the Examples.

Understanding the content of patents requires a certain perspective. Having read some Du Pont patents on hydrocyanation, Roy Jackson once announced to an audience at a national meeting of the American Chemical Society that he had come to the conclusion that "patents are not written for the enlightenment of the academic community." He was correct. Patents are legal documents, written for the legal community. By law, they must contain a certain amount of scientific or technical information - they must "teach" - but they are written for the courts, not scientists. In fact, patents are legal contracts between the government as the grantor of monopolistic rights to an invention and the inventor as grantee. The period of this contract is 17 years.

U.S. PATENTS

In addition to the "open literature" of journal references, we have often cited U.S. patents in this book. This is a result of our primary interest in chemistry in the United States and our grasp of the English language. Nonetheless, in keeping with the global nature of the chemical industry, we have cited a great number of non-U.S. patents also. There are several guides to U.S. patents, but one is particularly useful for chemical patents [2].

United States patents include several common elements:
1. The "Field of Invention" or subject of the patent.
2. A discussion of "Prior Art" or background information. This bears little resemblance to the references of a scientific paper. Instead, it is a description of previous attempts to solve the same problem.
3. The "Object" or purpose of the invention.
4. The "Summary" of the invention or a general technical description.
5. An elaboration of the invention, describing all technical aspects.
6. The "Utility" of the invention. It is not sufficient to invent something new; it must also be useful.
7. Detailed descriptions of actual "Examples" of the invention, usually including the first if somewhat imperfect rendering of the invention as well as the best examples. The best practice must be included.
8. The "Claims," which are a legal description of the exclusive rights granted to the inventor or his assignee by the government.

For a scientist, the most useful portions of a patent are the Examples, because they clearly state what has been done, giving enough detail that "one skilled in the art" is able to reproduce the experiments. The examples will include the best embodiment of the invention at the time of filing. Claims also provide useful information, but in U.S. patents, the claims can extend well beyond the experimental work to include reasonable guesses of what else might work in a particular system. Thus, the initial claims include everything that might reasonably work, but subsequent claims will focus more carefully on the preferred embodiment of the invention.

The patents also include other useful information. If the work has been done for a corporation, they will be mentioned as the assignee of the patent rights. Dates, earlier cases, and divisional filings provide a useful trail when searching patents. Copies of U.S. patents may be obtained from the U.S. Patent Office for a fee of $5.00. They are also available in a number of libraries around the United States.

NON-U.S. PATENTS

United States patents are published only after they have been granted; the average is two years between filing and publication of granted patents. In a number of other countries, patents are opened to inspection before they are granted, often within six months. As a result, most important U.S. patents appear first as non-U.S. filings. Chemical Abstracts Service abstracts patents only at their first appearance, and as a result, there are few citations of U.S. patents in *Chemical Abstracts*.

Non-U.S. patents can come in a bewildering array of forms with a bewildering array of numbers. The British Library has published a very useful guide to understanding what one is reading when one is reading non-U.S. patents [3]. Because German and Japanese filings and patents are referenced in this book, a few comments will serve to illustrate differences between them and U.S. patents

Japanese patent numbers are given a prefix indicating the year. Japanese years are based on the number of years since the current emperor came to the throne, that is, in the year of the Emperor. As a result, Hirohito's death in 1989 started the prefix over one week into the year. The running numbers, however, were continued sequentially. Japanese patents come as filed applications, unexamined applications or "kokai" (labeled "A") and examined applications (labeled "B2"). Each of these types is assigned its own sequential number, so the same patent numbers can be used up to three times in chemical patents. There is a similar series of three for mechanical patents, so numbers can be used up to six times, but the mechanical patents are rarely cited in CA. *Chemical Abstracts* usually cites the unexamined "kokai" applications. Derwent and INPADOC cover both unexamined and examined patents.

German patent applications are assigned sequential numbers. For patents from 1968 on, the year of filing can be determined by adding 50 to the first two numbers, that is, patents filed in 1991 begin with the digits 4,1--,---. Three different stages of patents are encountered. The first is "Offenlegungsschrift" and is the one most frequently encountered in *CA*. "Auslegeschrift" were discontinued in 1981. "Patentschrift" is the granted patent and may be the only published form if processing is exceptionally fast. The patent numbers are accompanied by numbers such as **A1** or **C3**. These numbers indicate the stage at which the patent stands (A is initial application and C is in the granting stage; 1 is the first publication and 3 is the third).

Since 1978, the European Patent Convention allows single application for up to 13 designated European countries. An "A" specification is published 18 months after the priority date. Because application may be delayed for up to a year after application in countries like the United States, this may mean that the application is published within six months. "A1" is published with a search report and "A2" is published without. A revision may be published as "A3." A "B1" publication is a granted patent and "B2" is granted patent after any amendments resulting from opposition.

References

1. H. Schulz, *From CA to CAS ONLINE*, VCH Verlagsgesellschaft, Weinheim, Germany, 1988.
2. J. T. Maynard and H. Peters, *Understanding Chemical Patents - A Guide for the Inventor*, 2nd Edition American Chemical Society, Washington, D. C., 1991.
3. S. van Dulken, Ed., *Introduction to Patent Information*, for Science Reference and Information Service, British Library, by Gresham Press, Old Woking, Surrey, England, 1990.

INDEX

315